Protection of Built Environment Against Earthquakes

Matjaž Dolšek
Editor

Protection of Built Environment Against Earthquakes

Springer

Editor
Matjaž Dolšek
Faculty of Civil and Geodetic Engineering
University of Ljubljana
Jamova 2, Ljubljana SI 1000
Slovenia
mdolsek@ikpir.fgg.uni-lj.si
mdolsek@gmail.com

ISBN 978-94-007-1447-2 e-ISBN 978-94-007-1448-9
DOI 10.1007/978-94-007-1448-9
Springer Dordrecht Heidelberg London New York

Library of Congress Control Number: 2011930256

Cover illustration: Figure 12.10 from this book

Printed on acid-free paper

Springer is part of Springer Science+Business Media (www.springer.com)

Preface

High observed losses throughout the world indicate that seismic risk is beyond a tolerable level. In order to make a contribution towards the finding of a solution to this problem, the Institute of Structural Engineering, Earthquake Engineering and Construction IT at the Faculty of Civil and Geodetic Engineering of the University of Ljubljana invited several researchers from Europe and other parts of the world to participate at an international workshop entitled "Protection of the built environment against earthquakes". This workshop was held in Ljubljana at the Faculty of Civil and Geodetic Engineering, on August 27 and 28, 2010 (http://ice4risk.slo-projekt.info/workshop/).

Based on the success of this event, the authors decided to prepare a book which would address the seismic performance and risk assessment of structures. The first part of this peer-reviewed book contains four chapters, which are concerned with a global earthquake model, seismic hazard and ground motions, whereas the second, more extensive part contains eleven chapters. The book starts with an introduction to the five global risk components of the Global Earthquake Model initiative. The use of a generalised conditional intensity measure approach, as a basis for the holistic selection of ground motions, is then discussed, followed by the disaggregation of seismic hazard, which can be a useful tool for the definition of design earthquake scenarios to be used in engineering practice. The first part of the book concludes with a chapter which discusses how the orientation-dependence of earthquake ground motions could be considered within code specifications for seismic demand. The authors of this chapter were invited to the project after the workshop.

The second part of the book presents recent advances in methods and tools for the seismic performance and risk assessment of various types of structures, with an emphasis on probabilistic approaches. Two chapters focus on comparative studies, i.e. evaluation of the available choices for judging the seismic performance of structures, and of several simplified nonlinear methods for the seismic performance assessment of buildings, whereas the other chapters are arranged according to the type of structural system involved. Thus this part of the book begins with a study which utilizes the FEMA P695 methodology for the evaluation of design modification factors for steel moment resisting frames, followed by an introduction to the tools and strategies which are available for the seismic performance assessment of masonry buildings.

The next four chapters deal with reinforced concrete buildings. Two of the presented studies are concerned with optimization methods. The first one presents a method for the designing of reinforced concrete buildings while meeting different probability-based performance requirements, whereas the second involves the evaluation of design procedures for frames with regard to life-cycle cost analysis. The problem of seismic risk, with the consideration of structural ageing, is addressed by using a probabilistic framework and a pushover-based nonlinear method. In the last of these four chapters, dealing with reinforced concrete buildings, a toolbox for the seismic performance assessment of such buildings is presented, together with a web application for the prediction of approximate IDA curves.

The final three chapters of the second part of the book are concerned with advanced seismic performance and risk assessment of engineering structures, starting with an overview of recent progress in the seismic analysis and design of reinforced concrete bridges in Slovenia. This is followed by a multi-platform simulation concept, which can be used for the seismic performance assessment of bridges. The last chapter focuses on the development and validation of a probabilistic methodology for evaluating the seismic risk of concrete gravity dams.

This book provides an overview of current knowledge and state-of-the-art developments of topics related to the seismic performance and risk assessment of different types of structures and building stock. The importance of informing the public about seismic risk is recognized, which could result in increased awareness of this risk and its possible reduction. Structural engineers need not only advanced knowledge regarding the assessment of structural performance, but also an ever increasing amount of information related to seismic hazard and ground motions, since the accurate assessment of seismic risk involves probabilistic methods and nonlinear methods of analysis. It is hoped that this book as a whole will provide insight into advanced methods and tools which can be use to achieve well-informed decision-making, which is the key element for the protection of the built environment against earthquakes.

This scientific monograph is primarily intended for researchers and postgraduate students working in the field of earthquake engineering, as well as for structural engineers who are involved in the advanced seismic design and assessment of structures. On the other hand, the book may also be attractive to the general public, since some parts of individual chapters are devoted to a general description of the problems involved.

I would like to express my deep gratitude to all the invited researchers, who responded promptly, showing their willingness to participate in the workshop, and especially to Professor Helmut Krawinkler, who presented a keynote lecture. Many thanks also to all the authors who wrote individual chapters of this book. The invaluable help of post-doctorate and Ph.D. students in the organising of the workshop, together with the support of this Institute's Professors, is hereby greatly appreciated, as well as the financial support of the Slovenian Research Agency, which was provided within the framework of the project "High-throughput computing environment for seismic risk assessment" (http://ice4risk.slo-projekt.info/) (J2-0845-0792-08). Finally, I would like to express my gratitude to the Springer Publishing Group, who agreed to publish this book, and particularly to the latter's staff for all their kind help and cooperation in this project.

Ljubljana, February 2011 Matjaž Dolšek

Contents

Contributors

Mark Aschheim
Department of Civil Engineering, Santa Clara University,
Santa Clara, CA, USA
maschheim@scu.edu

Brendon A. Bradley
Department of Civil and Natural Resources Engineering,
University of Canterbury, Christchurch, New Zealand

Chuo University, Institute of Science and Engineering, Tokyo, Japan
brendon.bradley@canterbury.ac.nz

Marko Brozovič
Faculty of Civil and Geodetic Engineering, University of Ljubljana,
Jamova 2, Ljubljana SI 1000, Slovenia
Mbrozovic@ikpir.fgg.uni-lj.si

Carlo Callari
Facolta di Ingegneria, University of Molise, Termoli (CB), Italy
carlo.callari@unimol.it

Daniel Celarec
Faculty of Civil and Geodetic Engineering, University of Ljubljana,
Jamova 2, Ljubljana SI 1000, Slovenia
dcelarec@fgg.uni-lj.si

Helen Crowley
GEM Foundation, Via Ferrata 1, Pavia, Italy
helen.crowley@globalquakemodel.org

Matevž Dolenc
Faculty of Civil and Geodetic Engineering, University of Ljubljana,
Jamova 2, Ljubljana SI 1000, Slovenia
matevz.dolenc@itc.fgg.uni-lj.si

Matjaž Dolšek
Faculty of Civil and Geodetic Engineering, University of Ljubljana,
Jamova 2, Ljubljana SI 1000, Slovenia
mdolsek@ikpir.fgg.uni-lj.si; mdolsek@gmail.com

Matej Fischinger
Faculty of Civil and Geodetic Engineering, University of Ljubljana,
Jamova 2, Ljubljana SI 1000, Slovenia
mfischin@ikpir.fgg.uni-lj.si

Michalis Fragiadakis
Department of Civil and Environmental Engineering,
University of Cyprus, Nicosia, Cyprus
mfrag@mail.ntua.gr

Paolo Franchin
Department of Structural Engineering & Geotechnics,
University of Rome "La Sapienza", Via Gramisci, 53-00197 Rome, Italy
paolo.franchin@uniroma1.it

Damian N. Grant
Department of Civil, Environmental and Geomatic Engineering,
University College London, Gower Street, WC1E 6BT London, UK
d.grant@ucl.ac.uk

Paul D. Greening
Department of Civil, Environmental and Geomatic Engineering,
University College London, Gower Street, WC1E 6BT London, UK
paul.greening@ucl.ac.uk

Iunio Iervolino
Dipartimento di Ingegneria Strutturale, Università degli Studi
di Napoli Federico II, Via Claudio 21, 80125 Naples, Italy
iunio.iervolino@unina.it

Tatjana Isaković
Faculty of Civil and Geodetic Engineering, University of Ljubljana,
Jamova 2, Ljubljana SI 1000, Slovenia
tisak@ikpir.fgg.uni-lj.si

Robert Klinc
Faculty of Civil and Geodetic Engineering, University of Ljubljana,
Jamova 2, Ljubljana SI 1000, Slovenia
robert.klinc@itc.fgg.uni-lj.si

Helmut Krawinkler
Department of Civil and Environmental Engineering,
Stanford University, Stanford, CA 94305, USA
krawinkler@stanford.edu

Nikos D. Lagaros
Institute of Structural Analysis & Antiseismic Research,
National Technical University of Athens, 9, Iroon Polytechniou Street,
Zografou Campus, Athens 15780, Greece
nlagaros@central.ntua.gr

Dimitrios G. Lignos
Department of Civil Engineering and Applied Mechanics,
McGill University, Montreal, QC H3A 2K6, Canada
dimitrios.lignos@mcgill.ca

Alessio Lupoi
Department of Structural Engineering & Geotechnics,
University of Rome "La Sapienza", Via Gramisci, 53-00197 Rome, Italy
alessio.lupoi@uniroma1.it

Chara Ch. Mitropoulou
Institute of Structural Analysis & Antiseismic Research,
National Technical University of Athens, 9, Iroon Polytechniou Street,
Zografou Campus, Athens 15780, Greece
chmitrop@central.ntua.gr

Diego Padilla
Department of Civil, Environmental and Geomatic Engineering,
University College London, Gower Street, WC1E 6BT London, UK
padilla-diego@hotmail.com

Manolis Papadrakakis
Institute of Structural Analysis & Antiseismic Research,
National Technical University of Athens, 9, Iroon Polytechniou Street,
Zografou Campus, Athens 15780, Greece
mpapadra@central.ntua.gr

Andrea Penna
European Centre for Training and Research
in Earthquake Engineering, Via Ferrata 1, Pavia, Italy
andrea.penna@eucentre.it

Iztok Peruš
Faculty of Civil and Geodetic Engineering, University of Ljubljana,
Jamova 2, Ljubljana SI 1000, Slovenia
iperus@ikpir.fgg.uni-lj.si

Rui Pinho
GEM Foundation, Via Ferrata 1, Pavia, Italy
rui.pinho@globalquakemodel.org

Paolo E. Pinto
Department of Structural Engineering & Geotechnics,
University of Rome “La Sapienza”, Via Gramisci, 53-00197 Rome, Italy
pinto@uniroma1.it

Anastasios G. Sextos
Department of Civil Engineering, Division of Structural Engineering,
Aristotle University of Thessaloniki, Thessaloniki, Greece
asextos@civil.auth.gr

Dimitrios Vamvatsikos
Metal Structures Laboratory, School of Civil Engineering,
National Technical University of Athens, Athens, Greece
divamva@mail.ntua.gr

Farzin Zareian
Department of Civil and Environmental Engineering,
University of California, Irvine, CA 92697, USA
zareian@uci.edu

Part I
Global Earthquake Model, Seismic Hazard and Ground Motions

Chapter 1
Global Earthquake Model: Community-Based Seismic Risk Assessment

Helen Crowley and Rui Pinho

Abstract The Global Earthquake Model (GEM) initiative aims to develop a global model of earthquake risk as an open source, community-driven project. In order to begin this in a structured way, a number of Global Components that cover the scientific modules of the model have been defined, and Requests for Proposals have been released, requesting international consortia to bid to lead these projects. Within the risk domain, 5 Global Components have been identified, with the following objectives:

- GEM Ontology and Taxonomy: to define the framework for calculating seismic risk and classify the components therein.
- Global Earthquake Consequences Database: to collect post-earthquake data such as loss of life, injuries, damage and economic loss into a common web-based repository.
- Global Exposure Database: to construct a global building and population inventory
- Global Vulnerability Estimation Methods: to define levels of damage and loss as a function of ground motion intensity, for a global taxonomy of buildings.
- Inventory Data Capture Tools: to support the population of the exposure and consequences databases through innovative open-source tools.

The consortia leading these global components will be required to define standards and best practice related to the methodologies used in seismic risk assessment and in particular the collection and storage of data needed therein. This chapter introduces these risk global components and describes the open source software and development platform that is being developed to provide access to and community interaction with the proposed data and tools.

H. Crowley (✉) • R. Pinho
GEM Foundation, Via Ferrata 1, Pavia, Italy
e-mail: helen.crowley@globalquakemodel.org; rui.pinho@globalquakemodel.org

M. Dolšek (ed.), *Protection of Built Environment Against Earthquakes*,
DOI 10.1007/978-94-007-1448-9_1, © Springer Science+Business Media B.V. 2011

Keywords Global • Seismic risk • Vulnerability • Exposure • Community • Open source

1.1 The GEM Initiative

Almost half a million people died in the last decade due to earthquakes. Most of these casualties were felt in the developing world, where risk is increasing due to rapid population growth and urbanisation. In particular many of the world's megacities of ten million inhabitants and more, such as Delhi, Bogota, Jakarta, and Lima, are situated in highly seismic active areas. A significant proportion of the world's population is therefore at risk from earthquakes. The 2010 Haiti and Chile earthquakes painfully reminded the world of the destructive impact of seismic events and the importance of reliable earthquake risk information. However, in many earthquake-prone regions no risk models exist, and even where models do exist, they are often inaccessible due to their proprietary nature or complex user-interface.

State-of-the-art information on earthquake risk (including socio-economic impact) covering all areas of the world is a necessary first step towards risk awareness and the undertaking of mitigating action. Such information is therefore a critical puzzle-piece for minimising loss of life, property damage and social and economic disruption due to earthquakes, by leading to better buildings codes and construction, land use planning for sustainable development, improved emergency response, protection of critical infrastructures and greater access to insurance.

There is hence a need for such earthquake risk information to become accessible to a wide spectrum of end-users and beneficiaries. This need has been underlined by a call from the Organisation for Economic Cooperation and Development's (OECD) Global Science Forum for the development of open-source risk assessment tools, and has been confirmed by a variety of institutions and organisations, the scientific community, and public opinion. In response to the needs outlined above, the GEM initiative has been launched to establish independent, uniform standards to calculate and communicate earthquake risk worldwide, based on a common framework: a Global Earthquake Model.

1.1.1 Global Earthquake Risk Model

By functioning as a community effort, GEM aims to build a state-of-the-art, widely accepted open source model for the assessment of seismic risk on a global scale and develop a supporting IT infrastructure by 2013. In order to serve GEM's stakeholders best, the model needs to transparent, modular, flexible and dynamic.

Users across the globe (which are envisaged to include engineers, researchers, risk managers, urban planners, insurers/reinsurers, civil protection departments, amongst others) will be able to access the model and accompanying tools through GEM's risk assessment platform (OpenGEM), allowing them not only to perform hazard, risk and socio-economic impact analyses, but also to collaborate and exchange data, results and opinions amongst each other. It is envisaged that the model will however be continuously evolving after 2013, and hence it will be built in such a way that it can capture the world's best understanding of data and likely behaviour of the earth and the built environment at any given moment in the future. In the years after 2013, GEM's growing user-community will be able to add data, exchange knowledge on methods for data analysis and other issues, and will in that way contribute to the continuous improvement of the model.

The ultimate aim of GEM will be for the Global Earthquake Risk Model to cover the entire globe as uniformly as possible, and to use models that incorporate the latest logic-tree approaches and that permit parameter changes so that seismic hazard and risk can be explored under different assumptions. In 5 years time however, it will not be possible to incorporate all the data effectively available, nor cover all the countries in the world in a uniform way. Also, it is unlikely that secondary hazards such as tsunamis or landslides will be covered within the model. During its first Working Programme (2009–2013) GEM will hence create tools and standardised methods for obtaining and analysing data, and GEM collaborators/affiliated researchers will use these tools to begin the process of assembling the needed (global) datasets, hereby producing data coverage for the world that is more uniform and complete than before.

1.1.2 Scientific Modules

Three scientific modules form the core of GEM's model; Seismic Hazard, Seismic Risk and Socio-Economic Impact. Within GEM, seismic risk is defined as a product of seismic hazard (the probability of levels of ground shaking, resulting from earthquakes, within a given time span), seismic vulnerability (the probability of loss given a level of ground shaking), and exposure (the elements at risk, mainly buildings, critical infrastructure and humans). Earthquakes however have an impact that goes beyond physical damage or casualties. Earthquakes can severely damage the economy and influence society and social well-being. Therefore, GEM's Global Earthquake Risk Model will include methods (models, indicators, tools) for analysis and evaluation of the impacts of earthquakes on the short, medium and long term, on local and global scales. Insight into earthquake effects over time will support decisions on short-term needs after an event (relief, shelter), medium-term needs (recovery and reconstruction) and long term needs related to policies and activities aimed at risk mitigation (Fig. 1.1).

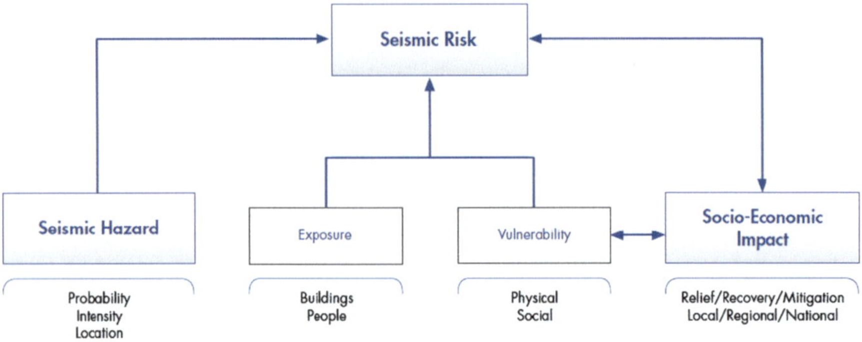

Fig. 1.1 Scientific modules of GEM

1.1.3 Collaborative Development

GEM believes that a global model can only be built in cooperation with the community and be authoritative because the community thus believes in it. Development of the model is therefore for a great part carried out by the community, both on global and regional scales. Hundreds of institutions and professionals in the relevant disciplines carry out research and develop methods, tools and discuss on standards. Developments on the forefronts of scientific and engineering knowledge as well as IT processes and infrastructure are being integrated into the model, also thanks to the strong community component.

The global earthquake model is being constructed by means of various 'building blocks'. The first building block was the GEM1 pilot project, which ended in March 2010 and delivered a proof-of-concept in terms of preliminary global hazard and risk calculations and an initial model-building infrastructure. Currently, development of the global earthquake risk model is based upon the efforts of the Global Components, Regional Programmes and the Model Facility (see Fig. 1.2).

Within GEM's Model Facility, the IT architecture is developed which will enable global risk calculations and communication and dissemination of output through user-interfaces. Global Components constitute the core of the Global Earthquake Model, and will comprise global datasets, methods, models, standards and tools related to seismic hazard, risk and socio-economic impact. International consortia, involving reputable institutions from around the globe and the best international and local experts in their field, are working on the development of the global components, in interaction where possible with the community. The selected consortia responded to open calls for proposals of which, for GEM's Risk and Socio-Economic Impact module, drafts were published online before the call to allow the public to provide feedback and improve them where possible.

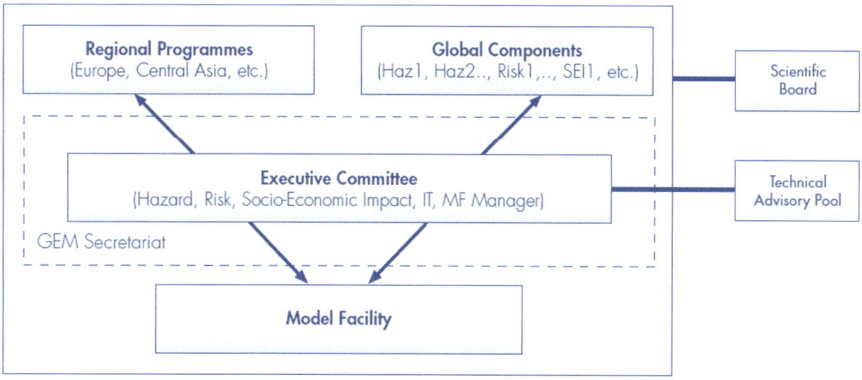

Fig. 1.2 GEM model infrastructure

Consortia are selected only after careful peer-review of their project proposal by a number of international experts, as well as GEM's Scientific Board. Consortia receive funding from GEM to carry out their activities, but often contribute in-kind themselves as well; the value of what they produce will extend beyond GEM as they will benefit the community and scientific development at large. Global Component consortia need the Regional Programmes to provide them with feedback on the standards and classifications for data collection they developed, considering applicability in each given region. GEM Regional Programmes are independently-run regional projects that are carried out under the GEM umbrella; in conformance with GEM standards and goals. Some are set up as dedicated bottom-up projects; in other occasions collaboration is sought with ongoing projects. GEM Regional Programmes involve local institutions and experts from as many of the countries of the region as possible. Besides providing feedback on the global components, they deliver essential contributions in the form of more detailed local data and will serve as a starting point for workshops and technology transfer in the region.

The Global Components will populate global harmonised databases to the extent possible and develop tools that will allow continuous updating and improvement of the databases, methods and standards, as the common understanding on earthquake risk evolves over time. The Model Facility will then bring together the final scientific/technical outcome of the consensus reached within and between the Global Components consortia and the Regional Programmes.

All contributions, including models, tools, and data, will be rigorously vetted to ensure uniformity and conformance with the highest scientific standards, and before releasing the "GEM stamped" model to the community, it will be extensively tested and evaluated.

1.2 Risk Global Components

The Requests for Proposals for the following 5 Risk Global Components were drafted by a focus group and subsequently modified based on the feedback from the Risk community through the online commenting system on the GEM website:

- GEM Ontology and Taxonomy
- Global Exposure Database
- Global Vulnerability Estimation Methods
- Global Earthquake Consequences Database
- Inventory Data Capture Tools

A key Global Component is the GEM Ontology and Taxonomy, whose consortium will be required to collaborate intensively with those of the 4 other risk global components (in addition to the global components in Hazard and Socio-Economic Impact) in order to define the framework that will guide GEM's development and the classification and order of the parameters involved in risk assessment. An illustration of how GEM envisages that the Risk Global Components will fit together is shown in Fig. 1.3, though it is noted that there will obviously be many more interactions and collaborations between the consortia than those shown in the diagram.

It is noted that during this first 5 year phase of GEM, the emphasis will be placed on buildings (residential and commercial) due to budget and time constraints, whilst future extensions are expected to consider infrastructure/lifelines.

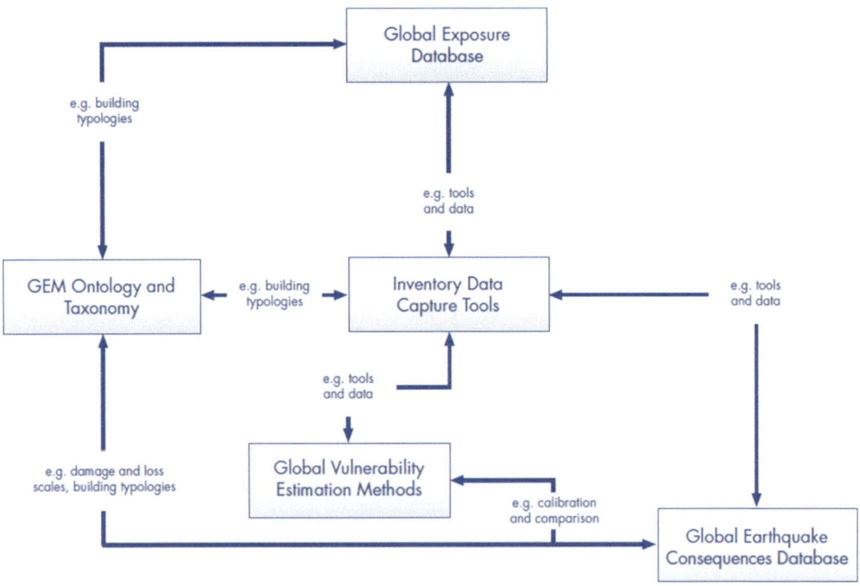

Fig. 1.3 Integration of risk global components

1.2.1 Ontology and Taxonomy

The development of the Global Earthquake Model requires a solid methodological foundation and terminology in order to achieve a shared understanding across the many fields and endeavours that it will address. The methodological foundation and terminology are termed an Ontology and Taxonomy (respectively, O&T), where Ontology refers to the entire framework that will guide GEM's development – the set of concepts, and the relationship between those concepts that will allow determination and communication of earthquake risk, and Taxonomy is a part of the ontology, and refers to the classification of things in an ordered system. The GEM taxonomy should address hazard, asset, risk and consequence related 'things'. The taxonomy developed during the first 5 years of GEM will be restricted to buildings, as mentioned previously (www.globalquakemodel.org/risk-global-components/ontology-taxonomy).

The project is of 2 years duration with an accelerated first phase during the first 6 months, in order to meet the needs of the other global components (Fig. 1.4).

The project consists of five major tasks:

Task 1: Development Plan. The project is managed by a core management group. Two teams comprise the core expert groups – an Ontology team, and a Taxonomy team. The Ontology team is somewhat larger than the Taxonomy team, since the Taxonomy team will draw on a broader group through the World Housing Encyclopedia. Both teams work closely with and are advised by GEM's Executive Committee and Scientific Board.

Task 2: Ontology. The Ontology is being developed in an open framework, consisting of (a) review of relevant literature; (b) commissioning of selected white papers; and (c) convening of two workshops that will take the white papers as a starting point, and seek development of a preliminary and then accepted ontology. These activities will be performed in an open manner, via a wiki with wide access to the earthquake risk community.

Task 3: Taxonomy. The Taxonomy is being developed in a similar manner to the Ontology, although the key to the Taxonomy development will be the close integration of this task with the World Housing Encyclopedia (WHE (www.world-housing.net)). The plan is to involve many competent professionals around the world to identify the specific and detailed nuances of the building stock in their countries.

Task 4: Evaluation and Testing (E&T). Following an initial 6 month intensive O&T development phase, the project will undertake Evaluation and Testing (E&T) of the O&T. The purpose of the E&T is demonstration and confirmation of the appropriateness, usefulness, accuracy and comprehensiveness of the O&T.

Task 5: Information Programme. Following its development, E&T and general acceptance by GEM, the project in its 1st and 2nd years will undertake an Information Programme with the goals of reaching out to users of varying background, from technically sophisticated to novices.

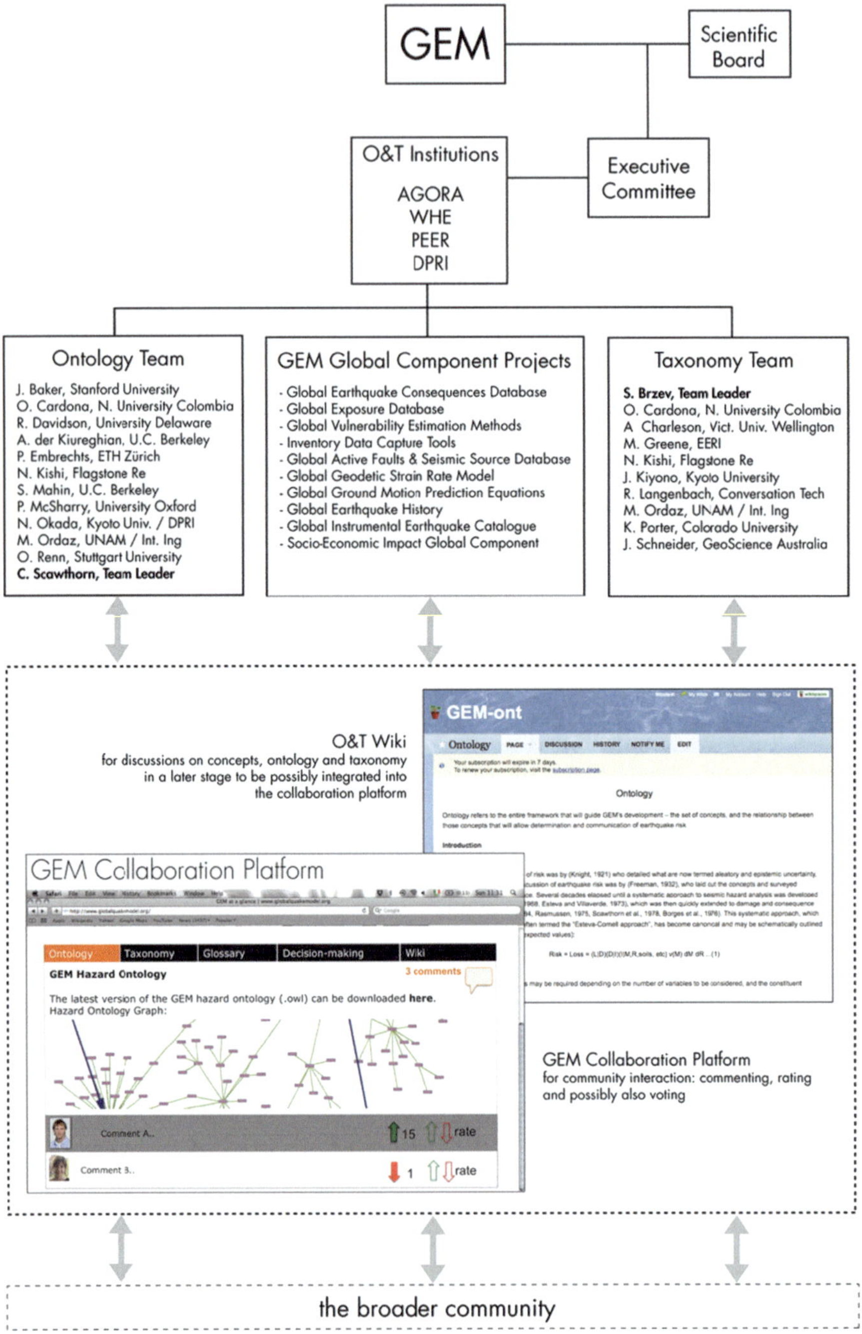

Fig. 1.4 Illustration of Ontology and Taxonomy global component

1.2.2 Global Exposure Database

This project aims to create the first open database of global building stock and population distribution containing the spatial, structural, and occupancy-related information necessary for damage, loss and human casualty (estimation) models to be deployed in GEM. The project goes by the name of GED4GEM (www.globalquakemodel.org/risk-global-components/exposure-database).

The overall structure of the project is aimed at managing three main tasks:

- Collection, analysis and homogenisation of global databases that may be available and useful for the global exposure database;
- Definition and implementation of a global exposure database based on the output of Task 1;
- Definition of "best practices" aimed at populating missing layers and/or information in specific geographical areas.

The consortium will start by building on existing databases (e.g. UN, regional and other public organisations, governmental building census data, national statistics) and published literature. It will then aim to collect population and building stock data for all countries of the world. The database should be able to incorporate data assembled using the Inventory Data Capture Tools. The consortium will furthermore devise and document a systematic and flexible approach for global application. Finally, an open data development environment is to be created for future modification and improvement of the database (Fig. 1.5).

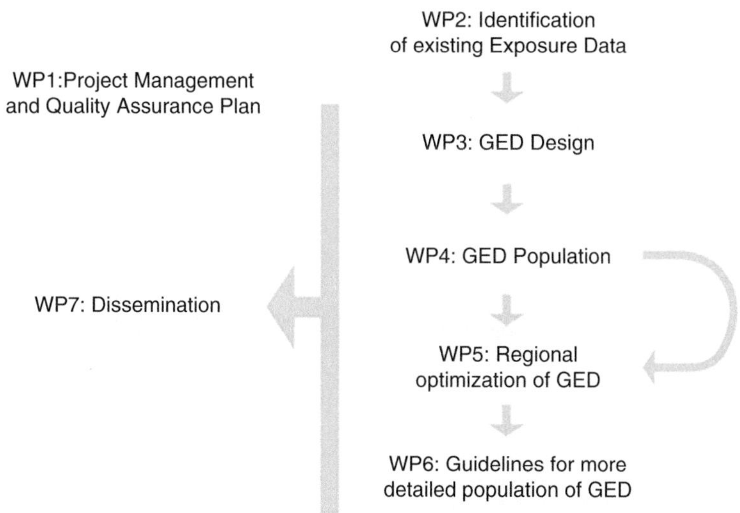

Fig. 1.5 Global Exposure Database work packages and integration

The schedule of the project has two distinct phases: the identification and collection of existing databases, mainly coming from in situ surveys, are expected to be completed in the first 9 months of the 3 years project. The months from 6 to 12 will be used to define, in accordance with the "Ontology and Taxonomy" and the "Global Vulnerability Estimation Methods" consortia, the structure of the exposure database. The database will include as much as possible existing data sets at the building level, but also all other data sets useful to infer, through mapping schemes, information about the buildings, such as population and boundary datasets, national and sub-national socioeconomic data, e.g., on land use, age structure, economic activities, poverty distribution, educational level, and health status, that may be correlated with or important to hazard exposure, as well as physical and social vulnerability. The main core of the project, that is the population of the exposure database and the development of the procedures for its management, updating and improvement will cover the rest of the time span of the project.

1.2.3 Global Vulnerability Estimation Methods

GEM is sponsoring the largest-ever public effort to understand the seismic vulnerability of buildings around the world. The project has 3 years duration and involves 9 partners worldwide. The project focuses on relationships between earthquake shaking intensity and building damage or loss, relationships often called seismic vulnerability functions. The project has two central objectives: to develop procedures for deriving vulnerability functions, and to actually implement those procedures and produce seismic vulnerability functions for a wide variety of building types. The project will not produce seismic vulnerability functions for every building type everywhere in the world, but it will most likely provide a major advance, both in terms of a library of open source seismic vulnerability functions and standardised procedures for adding to that library (www.globalquakemodel.org/risk-global-components/vulnerability-estimation). The project will address four distinct approaches to creating seismic vulnerability functions:

Empirical. This approach uses regression analysis to derive seismic vulnerability functions from past observations of earthquake loss experienced by real buildings of a particular type. This is the gold standard of seismic vulnerability functions, in that empirical relationships have the highest degree of credibility. The project will create guidelines for performing and documenting this type of analysis, and then derive as many empirical seismic vulnerability functions as practical.

Analytical. This approach uses first principles of structural engineering to relate damage and loss to shaking intensity. Where empirical data are insufficient, analytical techniques can produce valuable insight into the seismic vulnerability of buildings. The project will create guidelines for deriving analytical seismic vulnerability functions, and then create as many as possible, in an attempt to fill many of the gaps left by empirical effort.

Expert opinion. In this approach, the analyst elicits the judgment of experts familiar with the building type of interest to produce a seismic vulnerability function. It is an extremely efficient approach to estimating seismic vulnerability, and can be valuable in the absence of empirical loss data and where insufficient resources are available for the analytical approach. The project will develop new procedures for developing and documenting expert-opinion-based seismic vulnerability functions, and then implement those procedures for many of the gaps left by both the empirical and analytical efforts.

Empirical-national. In this approach, seismic vulnerability functions are developed for entire countries, or large sub- and supranational regions, without regard to building type, to best fit past loss data. The project will borrow empirical-national seismic vulnerability functions from the USGS' Prompt Assessment of Global Earthquakes for Response (PAGER) project to fill gaps left by empirical, analytical, and expert-opinion efforts.

Other elements of the project include the following efforts: to develop procedures for selecting among competing seismic vulnerability functions; to relate building damage to human casualties; to deal rigorously with uncertainty; to update seismic vulnerability functions as new data become available; and to ensure that the results of the project serve the needs of loss-estimation practitioners. The 3 year project will focus its effort in year 1 on developing guidelines; year 2 will be largely devoted to implementing those vulnerability guidelines and creating global seismic vulnerability functions; and the year-3 effort will focus on peer review and efforts to disseminate the products of the research to regional efforts around the world (Fig. 1.6).

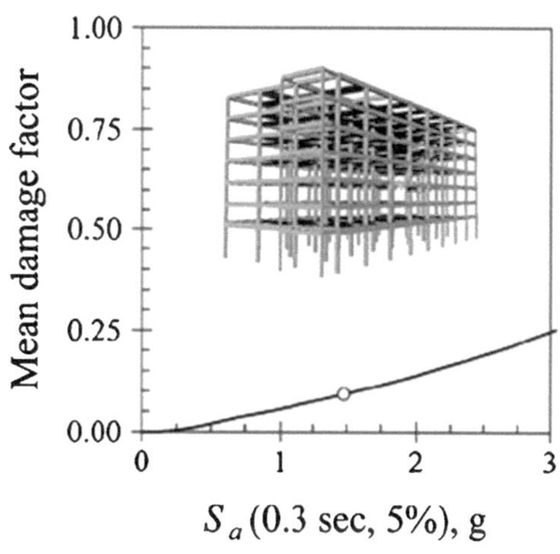

Fig. 1.6 Example vulnerability function

1.2.4 Global Earthquake Consequences Database

An international consortium is carrying out the GEMECD, or GEM Earthquake Consequences Database project, with as central product a GIS relational database. This database will serve to inform users on consequences from past events, as a benchmarking tool for analytical loss models and to support the development of tools to create vulnerability data appropriate to specific countries, structures, or building classes. Preparation of an interface enabling impact damage from future earthquakes to be captured and uploaded to the database is also part of the project (www.globalquakemodel.org/risk-global-components/consequence-database).

The project has 3 years duration and involves 10 partners worldwide.

The legacy of the project will be a web accessible database for the ongoing service of GEM users. For current events, GEMECD will serve as a clearing-house of information, posted by users based on the standards and protocols set in the GEMECD documentation. In the long term, GEMECD will be a repository of the most relevant and validated data on consequences of the significant events of the last 40 years around the world.

The project can be described as consisting of three components, namely (I) Database Design, (II) Consequences Data Assembly and (III) Database Use.

The main features of the project are as follows:

- In GEMECD, there will be a separation of presentation layer and storage/analysis layer and interfacing between the two using Web Service Architecture, adopting OGC (Open Geospatial Consortium) and OpenGIS Simple Features for manipulation of geospatial data.
- Working closely with the ontology and taxonomy (O&T), vulnerability and inventory capture consortia, the building typologies and grading of damage levels will be harmonised.
- Damage assessments carried out by remote sensing techniques will be critiqued and incorporated.
- Ground motion parameters from the USGS ShakeMaps (in addition to intensity adding PGA, PGV, and spectral acceleration) will be included.
- Qualitative observations of secondary hazards, infrastructure damage and social disruption from published sources will be embedded.
- CRED and MunichRe will ensure that the definitions and quality of social and economic data are consistent with user needs.

1.2.5 Inventory Data Capture Tools

The Inventory Data Capture Tools (IDCT) project addresses the inventory and damage data development needs of the GEM user community, in developing input to the Global Exposure and Global Earthquake Consequences databases that

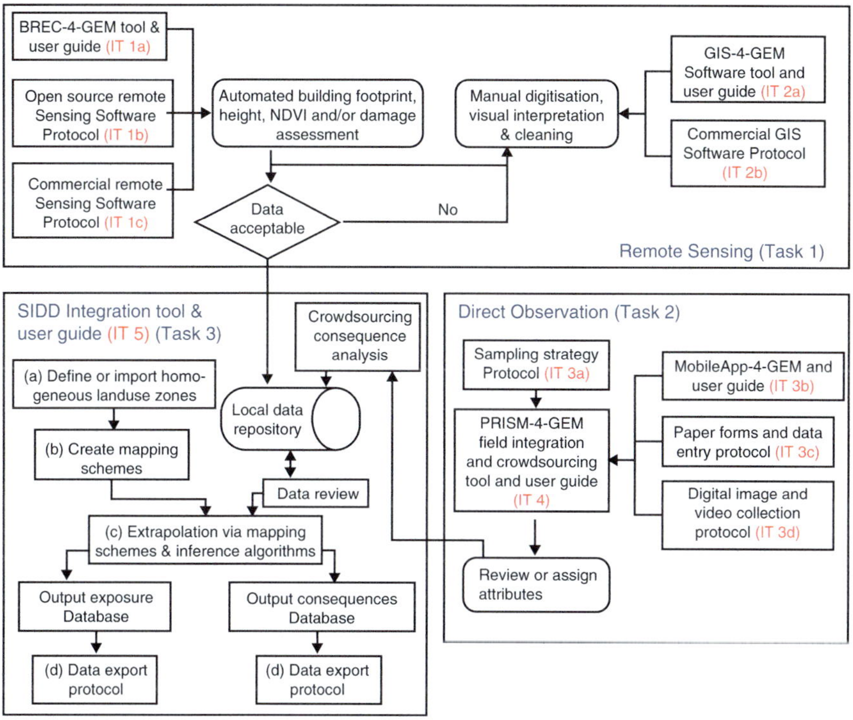

Fig. 1.7 Workflow of inventory data capture tools

underpin GEM's seismic risk module (see Fig. 1.7) (www.globalquakemodel.org/risk-global-components/inventory-capture).

The project aims to provide a fully operational, flexible and integrated suite of tools, protocols and guidelines that are scientifically-founded, yet straightforward to use. Remote sensing is at the heart of the project's flexible and integrated approach, which is scalable as to allow for data production that is tailored to GEM's end users and beneficiaries. The project capitalises on the state-of-the-art in remote sensing, GIS, inference and extrapolation methodologies and data integration techniques. The project has a duration of 30 months.

The project plan is comprised of four work packages:

WP0 Set-up and Management (SUM). Coordination with the other risk global component projects, project management planning, and quality assurance.

WP1 Inventory Toolkit (IT) Development. Building on existing software solutions developed by team members, WP1 will result in four GEM-specific freely-available, open source tools: (a) BREC-4-GEM to automatically extract key building inventory parameters including footprints, height, shape and damage occurrence from high-resolution optical and LiDAR imagery; (b) MobileApp-4-GEM mobile in-field data collection system; (c) PRISM-4-GEM field data fusion tool; and the (d) SIDD Spatial

Inventory and Damage Data integration system, which through data fusion, mapping scheme-based extrapolation, and logical inference completes the dataset for any given building, and facilitates upload to the consequences and global exposure databases. With flexibility in mind, the project will also produce a library of protocol documents such as GoogleEarth-4-GEM, providing step-by-step instructions for imagery pre-processing and data development in both freely available and commercial software including Google Earth, ArcGIS, Feature Analyst, and ENVI.

WP2 Pilot Studies (PS) and Validation. Through four pilot studies, the tools and protocols will be developed, beta tested by project team members and then demonstrated operationally by selected members of the advisory committee. Tool development will focus on Padang in Indonesia (inventory and damage), beta testing on Istanbul (inventory) and Haiti (damage), and the demonstration case will be determined through consultation with GEM users; options include Mexico City, Manila, Cologne, Lima, Cairo.

WP3 Training and Application Guidelines (TAG). Training and application guidelines will form a two-volume series. Volume I – Application Guidelines – is an easy-to-use manual that introduces the IDCT framework, and provides decision support for potential users, with summaries of expected costs, technical and expertise requirements, expected outcomes, and a discussion of benefits and limitations. It also includes a demonstration of the tools in action, based on pilot study results. Volume II – Training Guide and Programme – will guide users through tools and protocol implementation with step-by-step instructions and sample results, culminating in the upload of data into the GEM global exposure and/or consequences databases.

The international project team brings together domain experts from leading universities and research organisations, whom are at the cutting edge of both inventory and damage data development from both research and operational standpoints (much of the state-of-the art literature has been written by, and the building blocks for these tools have largely been created by, team members). This scientific prowess is combined with the operational perspective of the commercial sector to ensure that the IDCT achieve both the scientific and operational standards demanded by GEM. An advisory committee composed of academic and industry leaders will ensure quality assurance. Furthermore Google are official collaborators on this project.

1.3 OpenGEM and OpenQuake

GEM has set up a Model Facility (MF) whose mandate is to undertake the development of the OpenGEM platform (and its underlying OpenQuake software) that will integrate hazard, risk, and socio-economic impact assessment tools and data and provide these to the community, and to enable and support all modelling developments related to the mission of GEM.

A first sandbox version (v0.2) of the software that will power OpenGEM – OpenQuake – was released in January 2011 through an open source development platform (GitHub), allowing for further development of the engine (and in a later stage other software features) by a community of developers and experts (http://openquake.org). In order to achieve that goal, the IT development follows the Agile framework. This approach caters for readily tuning of requirements and priorities during code development. A number of online tools are being used to support the collaborative but distributed effort, such as Pivotal Tracker, GitHub, Etherpad and IRC channels. From 2011 onwards, there will be time-based releases of alpha and beta-versions of OpenQuake (based on test-data).

For what concerns risk calculations, OpenQuake will allow users to obtain the following results, through a number of different methodologies:

- Loss exceedance curves (loss versus probability of exceedance in a given time span, both for single assets and aggregated losses for multiple assets);
- Conditional loss maps describing the geographical distribution of values of loss with a fixed probability of exceedance in a given time span;
- Mean loss maps describing the geographic distribution of mean loss within a given time span;
- Damage maps describing the geographical distribution of mean number/area/percentage of damage at a given limit state for a given deterministic event;
- Loss statistics per event or across all events (mean loss, standard deviation of loss etc.)

Open source development offers GEM many opportunities. One of them is collaborative development with (external) organisations and individuals on one or more of the modules that are part of OpenQuake and have a wider application of use. GEM will for example collaborate with a number of partners on the open source project RiskLib, which is a basic engine for natural hazards risk modelling, and on the joint development of risk-related 'apps' that can be built on top of OpenQuake's API, such as tools that support risk mitigation/reduction. OpenQuake will hence make use of parts of RiskLib, and it is envisaged that much more of such collaborations will take place, directly within the OpenQuake (open source) community, or through other projects. Development of the open source community is an important task that the MF collaborators will dedicate considerable time to, from January 2011 onwards. Future activities include development of web interface technologies and making the engine suitable for a cloud-computing infrastructure.

GEM will build the OpenGEM platform in such a manner that data is not automatically open. Data is usually subjected to licenses and there are different ways in which organisations may allow GEM to use it: (1.) full use of data without restrictions; users can download and use datasets; (2.) providing it as input data to the engine; users will only be able to produce output, but not see the input data; (3.) running calculations through the GEM engine, using own data that will not be submitted to GEM's Model Facility.

Development of the web-based risk assessment platform OpenGEM will occur in phases. A demo OpenGEM platform will be released in the Spring of 2011, to

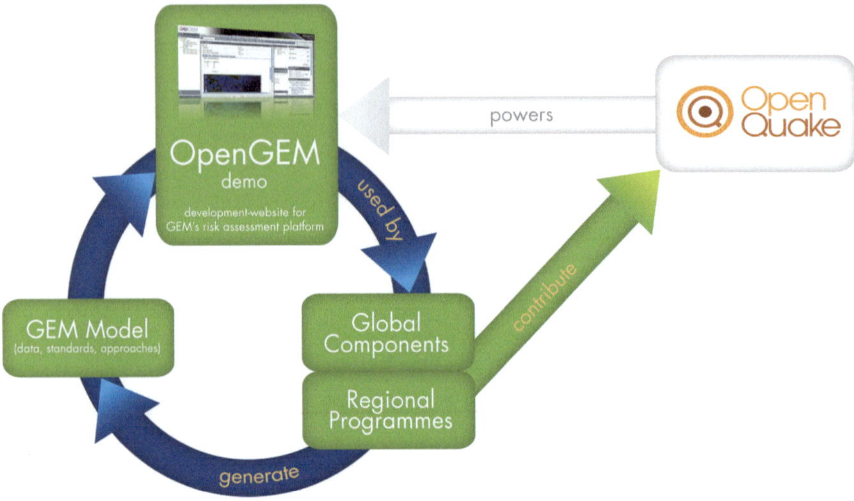

Fig. 1.8 Interaction between OpenQuake, OpenGEM, the Regional Programmes and Global Components for the development of the GEM model

provide the Global Components and Regional Programmes with the tools they need to work on GEM's model (data, approaches, etc.). Support of the Global Components in the construction of tools for development and updating of databases is hence also part of MF activities (e.g., development of an interface for scientists to submit active faults to GEM's database). The demo will hence gradually include more and more features, and GEM's partners will be able to start testing the platform (Fig. 1.8).

1.4 Closure

The Global Earthquake Model strives to be a community-driven initiative, and provides a number of mechanisms for those interested in seismic risk to become a part of the development process. This chapter has focused on the global components (in particular those related to Risk which began at the end of 2010), which involve a large number of international organisations and individuals with expertise in activities related to hazard, risk and socio-economic impact. The global components have kick-started the scientific activities, but GEM will rely on a large network of parties to develop its model, e.g. the members of the Governing Board (which includes both public and private participants), the Scientific Board (with experts in hazard, risk, socio-economic impact and IT), the Technical Advisory Pool (http://www.globalquakemodel.org/organisation), those involved in Regional Programmes, the users of the Forum and Wiki, those providing comments on the website, software developers from the open source community, as well as governments, organisations or even individuals providing data to the OpenGEM platform.

References

http://openquake.org
www.globalquakemodel.org/organisation
www.globalquakemodel.org/risk-global-components/consequence-database
www.globalquakemodel.org/risk-global-components/exposure-database
www.globalquakemodel.org/risk-global-components/inventory-capture
www.globalquakemodel.org/risk-global-components/ontology-taxonomy
www.globalquakemodel.org/risk-global-components/vulnerability-estimation
www.world-housing.net

Chapter 2
Frameworks for the Consideration of Ground Motion and Seismic Response Uncertainties in Seismic Performance Assessment

Brendon A. Bradley

Abstract Quantification of the seismic performance of engineered structures is complicated by the significant uncertainties in the future strong ground motions such structures will be subject to and engineering models of the consequent seismic response. As a result, rigorous seismic performance quantification must embody a probabilistic approach in which a key aspect is uncertainty characterization. This chapter discusses two recently proposed frameworks which provide a rigorous and holistic characterization of uncertainties in ground motion selection and seismic response analysis.

Keywords Seismic performance • Uncertainties • Ground motion selection • Seismic response • Generalised conditional intensity measure (GCIM) approach • Seismometer arrays

2.1 Introduction

The continuing evolution toward the seismic design of engineered facilities based on their expected seismic performance places increasing emphasis on the use of computational models to predict the seismic response of such facilities. Despite our best efforts in the design and assessment of facilities to reduce their vulnerability to earthquake-induced hazards, the occurrence of every large earthquake seems to provide new evidence of the complex phenomenon producing strong ground motions at the earth's surface, and weaknesses in these contemporary seismic

B.A. Bradley (✉)
Department of Civil and Natural Resources Engineering,
University of Canterbury, Christchurch, New Zealand

Chuo University, Institute of Science and Engineering, Tokyo, Japan
e-mail: brendon.bradley@canterbury.ac.nz

M. Dolšek (ed.), *Protection of Built Environment Against Earthquakes*,
DOI 10.1007/978-94-007-1448-9_2, © Springer Science+Business Media B.V. 2011

design and/or assessment methods (Japanese Geotechnical 1998; Gates and Morden 1995; Hanks and Brady 1991). Our inability to precisely predict the nature of future ground motions at a site and the seismic response of co-located facilities results in the need for rigorous seismic performance quantification to embody a probabilistic approach. In such approaches a key aspect is uncertainty characterization in the input ground motions and the resulting seismic response.

Herein, two recently proposed frameworks which provide a rigorous and holistic characterization of uncertainties in ground motion selection and seismic response analysis are discussed. The first framework utilizes a generalised conditional intensity measure (GCIM) approach as the basis for the holistic selection of ground motions for any form of seismic response analysis. The essence of the method is the construction of the conditional multivariate distribution of any set of ground motion intensity measures using probabilistic seismic hazard analysis (PSHA) results. The approach therefore allows any number of ground motion intensity measures identified as important in a particular seismic response problem to be considered. Ground motions are then selected, modified, and/or simulated based on the statistical comparison, for each intensity measure, of the empirical distribution of the ground motion suite with the 'target' GCIM distribution. The second framework makes use of observations from seismometer arrays for the validation of computational seismic response models. The framework explicitly accounts for the epistemic uncertainty related to the unknown characteristics of the 'site' (i.e. the problem under consideration) and constitutive model parameters. Multiple prediction-observation pairs are used to improve the statistical significance of inferences regarding the accuracy and precision of the computational seismic response methodology and constitutive model. Among other things, the benefit of such a formal validation framework includes an improved understanding of the uncertainties in computational model assumptions, constitutive models and their parameters.

2.2 The Generalised Conditional Intensity Measure (GCIM) Approach

2.2.1 Limitations of Conventional Ground Motion Selection Approaches

Current ground motion selection methodologies are burdened by the limited amount of information on the severity of a ground motion that is conveyed (often using a single ground motion intensity measure (IM)). For example, ground motion selection methods utilized in both research and engineering practice commonly select motions on the basis of their 'match' to some target response spectrum. It is well recognized however that the severity of a ground motion, in general, depends on its intensity, frequency content and duration. Spectral accelerations, by definition, represent the peak response of a single-degree-of-freedom oscillator of a specific period, and therefore do not explicitly account for many other important features, such as the duration, and energy (both total and its temporal accumulation) of ground shaking, among others.

The neglection of ground motion characteristics other than those reflected in a ground motions response spectrum has been somewhat justified out of both the convenience of the response spectrum as a tool for (predominately structural earthquake) engineers to understand ground motion's, and that several studies have concluded that, regarding the seismic response of structures, other features of a ground motion are of secondary concern to that of the ground motion's response spectrum. Examples of such studies include (Shome et al. 1998; Baker 2005; Luco and Bazzurro 2007). Baker (2005) found that when predicting the peak interstorey drift over all floors for ductile structures, scaling ground motions to the conditional mean spectrum produced seismic response estimates with lower uncertainty and less bias than: (i) randomly selecting records, (ii) selecting records based solely on magnitude and distance, or (iii) selecting records based on their epsilon value.

Despite the findings of the aforementioned studies, they must be kept in context with what was being analysed and what was being measured. In all these studies either single-degree-of-freedom or multi-degree-of-freedom representations of multi-storey structures were analysed, and the response parameter of interest was the peak interstorey drift ratio over all floors. The obvious question is therefore: does the relative unimportance of aspects other than the response spectrum of a ground motion change when measuring an arbitrary response measure for any arbitrary system? It is argued that the answer to such a question is most certainly yes, reasons for which are elaborated upon in Bradley (2010).

Clearly, existing literature on the dependence of the results of seismic response analyses on particular ground motion intensity measures all have limitations which may or may not be applicable for the problem under an analyst's consideration. It is therefore desirable to have a holistic method for the selection of ground motions for any seismic response problem. To cater for the inevitably-wide variety of complexity in engineering design (and consequently in ground motion selection and seismic response analyses) such a method should allow for simplifications in the selection of ground motions, but analysts should have an explicit appreciation for the simplifications they make and be able to determine (in a simple manner) if such simplifications significantly affect the seismic response analysis results for the problem at hand.

2.2.2 *Theory of the GCIM Approach*

Bradley (2010) proposed that for a given earthquake scenario (i.e. given a causal earthquake magnitude, source-to-site distance, local site properties etc.) any arbitrary vector of ground motion intensity measures, ***IM***, has a multivariate lognormal distribution (i.e. $\boldsymbol{IM}|Rup$ has a multivariate lognormal distribution, where "$|Rup$" indicates conditioning on a specific earthquake rupture scenario). ***IM*** may include any scalar intensity measures of a ground motion, e.g. spectral acceleration of various vibration periods, Sa_i, spectrum intensity, SI, acceleration spectrum intensity, ASI, Arias intensity, I_a, significant duration, SD, and vertical spectral acceleration at any vibration period, $Sa_{V,i}$, among others. Based on the properties of a multivariate lognormal distribution it then follows that the conditional distribution of $\boldsymbol{IM}|Rup$ given the occurrence of $IM_j = im_j$ is also a multivariate lognormal distribution. The distribution of each IM_i in $\boldsymbol{IM}$ conditioned on $IM_j = im_j$ can therefore be obtained from:

$$f_{IM_i|IM_j}(im_i|im_j) = \sum_{k=1}^{N_{Rup}} f_{IM_i|Rup,IM_j}(im_i|rup_k, im_j)\mathrm{P}_{Rup|IM_j}(rup_k|im_j) \tag{2.1}$$

where $f_{IM_i|IM_j}(im_i|im_j)$ is the probability density function (pdf) of IM_i given $IM_j = im_j$; $f_{IM_i|Rup,IM_j}(im_i|rup_k, im_j)$ is the pdf of IM_i given $Rup = rup_k$ and $IM_j = im_j$; $\mathrm{P}_{Rup|IM_j}(rup_k|im_j)$ is the probability that $Rup = rup_k$ caused the observed ground motion with $IM_j = im_j$ (i.e. that obtained from seismic hazard disaggregation); and N_{Rup} is the number of ruptures. From the assumption that $\boldsymbol{IM}|Rup$ has a multivariate lognormal distribution, it follows that $\boldsymbol{IM}|Rup,IM_j$ also has a multivariate lognormal distribution, and in particular, that for each IM_i in $\boldsymbol{IM}$, $IM_i|Rup,IM_j$ has a univariate conditional lognormal distribution which can be expressed as:

$$f_{IM_i|Rup,IM_j}(im_i|rup_k, im_j) \sim LN\begin{pmatrix} \mu_{lnIM_i|Rup,IM_j}(rup_k|im_j), \\ \sigma^2_{lnIM_i|Rup,IM_j}(rup_k|im_j) \end{pmatrix} \tag{2.2}$$

where $f_X(x) \sim LN(\mu_{lnx}, \sigma^2_{lnx})$ is short-hand notation for X having a lognormal distribution with mean μ_{lnx} and variance σ^2_{lnx}. The conditional mean and standard deviation (square root of the variance) in Eq. 2.2 can then be obtained from Eqs. 2.3 and 2.4, respectively:

$$\begin{aligned} \mu_{lnIM_i|Rup,IM_j}(rup_k|im_j) &= \mu_{lnIM_i|Rup}(rup_k) \\ &+ \sigma_{lnIM_i|Rup}(rup_k)\rho_{lnIM_i|Rup,lnIM_j|Rup}\varepsilon_{lnIM_j} \end{aligned} \tag{2.3}$$

$$\sigma_{lnIM_i|Rup,IM_j}(rup_k|im_j) = \sigma_{lnIM_i|Rup}(rup_k)\sqrt{1 - \rho^2_{lnIM_i|Rup,lnIM_j|Rup}} \tag{2.4}$$

where $\mu_{lnIM_i|Rup}(rup_k)$ and $\sigma_{lnIM_i|Rup}(rup_k)$ are the mean and standard deviation of $lnIM_i/Rup = rup_k$ (i.e. from a ground motion prediction equation); $\rho_{lnIM_i|Rup,lnIM_j|Rup}$ is the correlation between $lnIM_i$ and $lnIM_j$ given $Rup = rup_k$; and the parameter "epsilon", ε_{lnIM_j}, in Eq. 2.3 is given by:

$$\varepsilon_{lnIM_j} = \frac{lnIM_j - \mu_{lnIM_j|Rup}(rup_k)}{\sigma_{lnIM_j|Rup}(rup_k)} \tag{2.5}$$

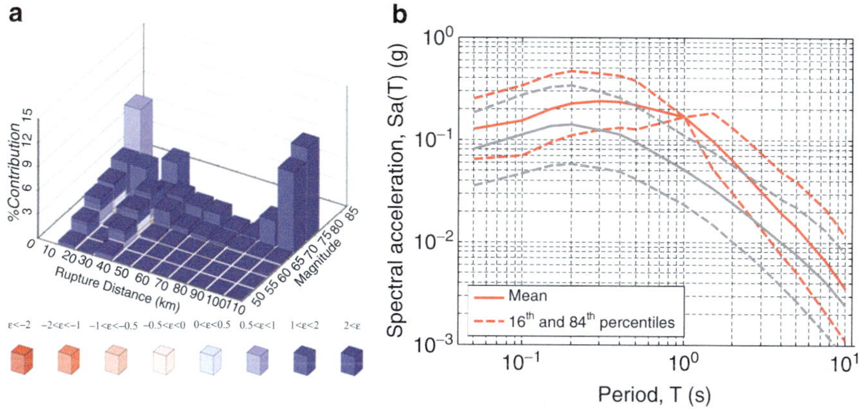

Fig. 2.1 (**a**) Magnitude-distance-epsilon disaggregation of PSHA for Christchurch, New Zealand for $Sa(1.0) = 0.165$ g which has a 2% in 50 year probability of exceedance; (**b**) mean, 16th and 84th percentiles of the conditional distribution of Sa given Sa(1.0) = 0.165 g (*red*) and the 'unconditional' distribution of Sa (*grey*)

Figure 2.1a illustrates the seismic hazard disaggregation for a rock site ($Vs_{30} = 760$ m/s) in Christchurch, New Zealand for $Sa(1.0) = 0.165$ g, which has an annual exceedance rate of 4.04×10^{-4} (i.e. a 2% exceedance probability in 50 years). It can be seen that the seismic hazard is contributed to by a range of different potential casual earthquake ruptures.

Figure 2.1b illustrates the mean (of *lnSa*), 16th and 84th percentiles of the conditional response spectrum obtained based on Eqs. 2.1–2.5 for the site in question. Also shown is the 'unconditional' distribution of spectral accelerations, which has been computed by replacing $f_{IM_i|Rup,IM_j}$ in Eq. 2.1 with $f_{IM_i|Rup}$. That is, the 'unconditional' distribution of IM_i is obtained by neglecting the correlation between IM_i and IM_j. It can be seen that the conditional mean spectrum is largest relative to the 'unconditional' mean spectrum at $T = 1.0$s (as a result of most potential ruptures have an associated ε value greater than zero (Fig. 2.1a)) and tends toward the unconditional spectrum as the period of vibration tends away from $T = 1.0$s. The uncertainty in the response spectrum distribution also increases as the period of vibration tends away from $T = 1.0$s. Both the above two observations are the result of the correlation of spectral accelerations generally decreasing as the difference between the periods of interest increases.

Figure 2.2 illustrates the conditional distributions of four other ground motion intensity measures, *SI*, *ASI*, I_a, and *SD*, which can be computed using Eqs. 2.1–2.5 for the site in question. The ground motion prediction equations of Bradley et al. (2009), Bradley (2009), Travasarou et al. (2003), and Abrahamson and Silva (1996), were used for computing *SI*, *ASI*, I_a, and *SD*, respectively, while the correlation equations in Bradley (2009), and Baker (2007) were also adopted. Inspection of Fig. 2.2 reveals that for some IM_i, the conditional and 'unconditional' distributions are relatively similar, yet significantly different for others. For example, the median

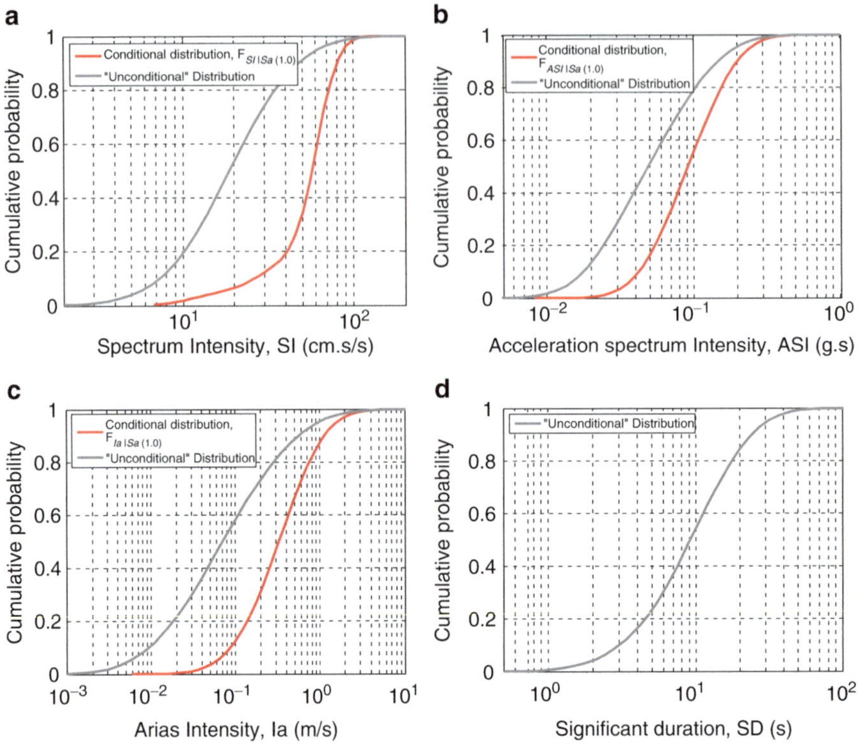

Fig. 2.2 Conditional and 'unconditional' distributions of various IMs obtained from the conditional IM approach for Sa(1.0) = 0.165 g at a site in Christchurch, NZ: (**a**) Spectrum Intensity, SI; (**b**) Acceleration Spectrum Intensity, ASI; (**c**) Arias Intensity, Ia; and (**d**) Significant duration, SD

of the unconditional distribution of *SI* is approximately 20 cm.s/s, while the conditional median is approximately 60 cm.s/s. This difference, or lack thereof, between the conditional and 'unconditional' distributions is obviously a function of the correlation between IM_i and IM_j.

2.2.3 *Ground Motion Selection Using the GCIM Distributions*

The benefit of the GCIM approach is that $\boldsymbol{IM}|IM_j$ provides the exact distribution (for the given inputs in a PSHA) of intensity measures, from potential ground motions with $IM_j = im_j$, which may be observed at the site. The GCIM distributions are therefore the 'target' which should be used in selecting a suite of ground motions for seismic response analysis. The aim of this section is to briefly discuss a holistic method by which ground motions can be selected to match $\boldsymbol{IM}|IM_j$ for any seismic response analysis problem. More elaborate discussion is given in Bradley (2010).

Here, it is desired to perform ground motion selection which is completely consistent with the results of a PSHA. Therefore when selecting ground motions for seismic response analyses based on $\boldsymbol{IM}|IM_j = im_j$, the first step is that all potential ground motions must be scaled to have $IM_j = im_j$. Although this requirement may seem restrictive, it is also convenient in that it uniquely specifies the scaling of the (either as-recorded, modified or simulated) ground motion, and therefore the only task left to do is select a suite of such ground motions which are representative of $\boldsymbol{IM}|IM_j = im_j$.

A primary relaxation made here is that ground motions are selected to match all of the univariate distributions of $\boldsymbol{IM}|IM_j$ (i.e. $IM_i|IM_j$ for all i), but not the complete multivariate distribution, $\boldsymbol{IM}|IM_j$, itself. This relaxation is considered pragmatic because of the significant reduction in complexity that it entails and is unlikely to affect the final ground motions selected (Bradley 2010). Because ground motion selection is desired for a finite number of N_{gm} ground motions, then a comparison of the appropriateness of the N_{gm} ground motions as representative of $IM_i|IM_j$ (for all i) must be done so using statistical goodness-of-fit tests. For continuous and discrete IM_i's of interest the Kolmogorov-Smirnov and Chi-Square goodness-of-fit tests can be used to examine the adequacy of a particular suite of ground motions with respect to the GCIM distribution of a single IM_i (Bradley 2010).

In order to illustrate the aforementioned concepts two suites of ground motions were selected for the site in Christchurch previously discussed. In order to elucidate the features of the proposed procedure the two ground motion suites (Bradley 2010) were selected based on different aspects of the seismic hazard disaggregation at the site (Fig. 2.1a). Suite 1 was selected on the basis of ground motions having causal earthquake magnitudes less than 6.0 and source-to-site distances less than 20 km, while Suite 2 was selected on the basis of ground motions having causal earthquake magnitudes greater than 7.0 and source-to-site distances greater than 50 km. Hence comparison of the basis for selecting these suites with the disaggregation given in Fig. 2.1a, indicates that, when scaled to have Sa(1.0) = 0.165 g, the two suites may not provide an appropriate representation of all the ground motions expected at the site. This is however a weak statement because of the fact that many features, in addition to magnitude and source-to-site distance, affect ground motion characteristics. For example, based on the previously mentioned studies of Shome et al. (1998), Baker (2005), and Luco and Bazzurro (2007), magnitude and source-to-site distance are of secondary importance to the spectral shape of the ground motion.

Figure 2.3a illustrates the un-scaled response spectra of the two ground motion suites compared to the mean (of *lnSa*), 16th and 84th percentiles of the conditional response spectrum obtained from the GCIM approach. Figure 2.3b shows the corresponding response spectra of the two ground motion suites once they have been scaled in amplitude to have Sa(1.0) = 0.165 g. It can be seen from Fig. 2.3b, that the Suite 1 ground motions (which have relatively low causal magnitudes) have notably larger response spectral ordinates at periods less than 1s, and weaker spectral ordinates at periods greater than 1s, relative to the conditional response spectrum. Conversely, the Suite 2 ground motions have slightly lower response spectra for periods less than 1s, but notably larger response spectra for periods

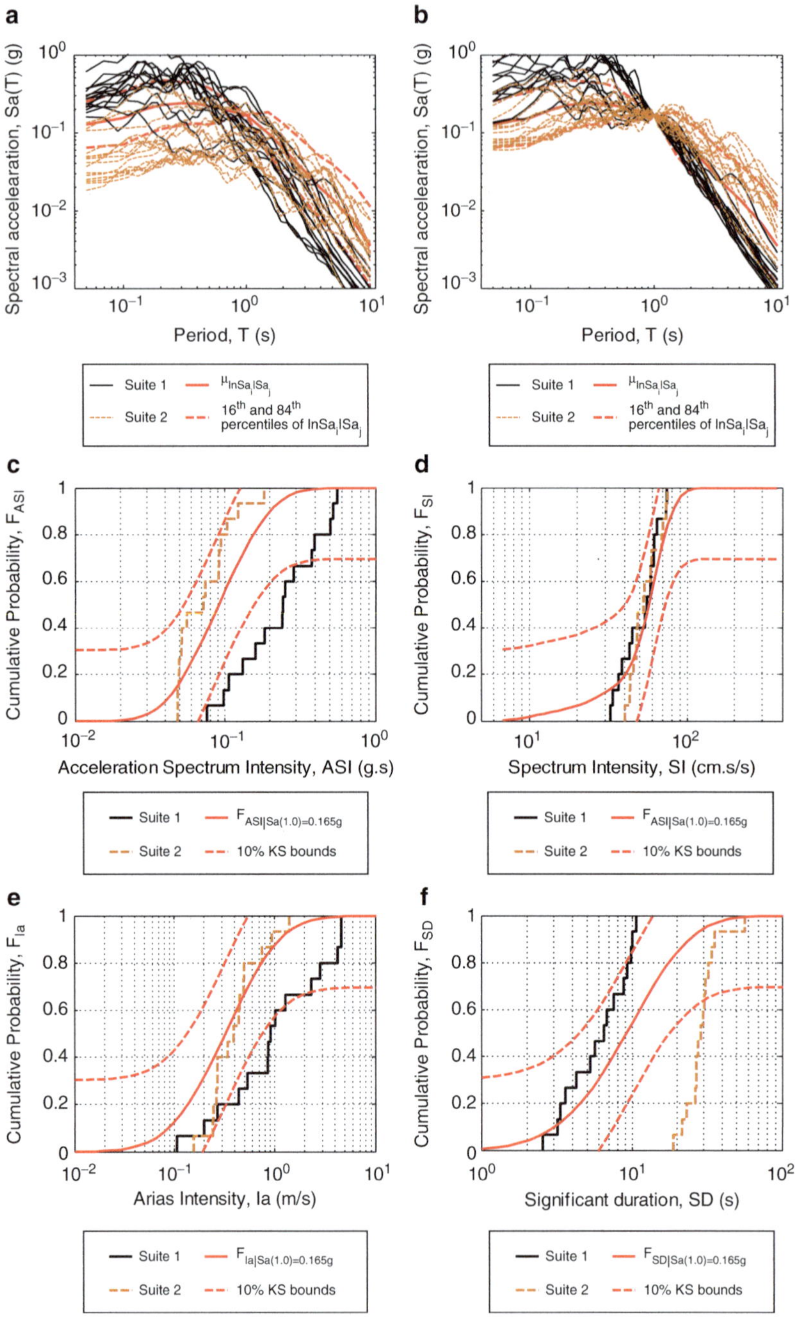

Fig. 2.3 The appropriateness of two ground motion suites for the distribution of IM|Sa (1.0) = 0.165 g: (**a**) comparison of the unscaled suite spectra; (**b**) comparison of the scaled response spectra of the suites; and the distributions of (**c**) ASI; (**d**) SI; (**e**) Ia; and (**f**) SD

greater than 1s. Despite the fact that, in their current form, Suites 1 and 2 show several departures from the theoretical response spectrum distribution at various periods, it would not be overly difficult to selectively choose other suites of ground motions (or combine these two) within these magnitude-distance ranges which provide a better fit. However, as stated previously, in the general seismic response problem, the appropriateness of a suite of ground motions is dependent on more than just their response spectrum characteristics, which is where a significant benefit of the GCIM approach can be obtained.

In addition to examination of the appropriateness of the ground motion suites in terms of their response spectra, using the GCIM approach comparisons can also be made with respect to any other arbitrary ground motion intensity measure. For example, Figs. 2.3c–f illustrate the comparisons between the empirical distribution functions (EDF's) of Suite's 1 and 2 and the theoretical distributions from the GCIM approach for Acceleration Spectrum Intensity, *ASI*; Spectrum Intensity, *SI*; Arias Intensity, I_a; and Significant Duration, *SD*, respectively. It can be seen that the ground motions of Suite 2 provide a good representation of the theoretical distribution of *ASI*, *SI* and I_a but not for *SD*, while the *ASI* and I_a distributions of Suite 1 are significantly larger (at the 10% significance level) than the theoretical distributions.

Clearly, the conditional distributions obtained for any arbitrary IM parameter from the GCIM approach can provide a significant constraint on the selection of ground motions. In fact, in many applications where as-recorded ground motions (with only amplitude scaling) are desired, it may not be possible to find a suite of N_{gm} ground motions which are not statistically different from the GCIM distributions. In such cases as above, two options are available. The first option is to reduce the number of ground motions which are desired. The second option is to explicitly neglect the statistically significant differences between the ground motion suite and theoretical distribution for one of more IM_i. Here we will only focus on the latter of these two options, but the consequences of both of these options are elaborated upon in Bradley (2010).

Neglecting certain differences between the characteristics of a ground motion suite and the theoretical characteristics of ground motions expected at the site is universally adopted in earthquake engineering research and practice. For example, selecting ground motions on the basis of their response spectra alone, as implemented in seismic design guidelines (e.g. NZS 1170.5 2004), implicitly neglects all other characteristics of a ground motion that are not directly represented through its response spectrum. The main feature of this common approach to ground motion selection is that such neglection is implicit. Two key unknowns therefore are: (i) whether the distribution of these other characteristics of the selected ground motion suite are consistent with the theoretical distribution of such characteristics; and (ii) whether the seismic response analysis problem for which such ground motions are being selected for is dependent on any of these other characteristics. The GCIM approach allows an analyst to determine the theoretical conditional distribution of any intensity measure, and therefore allows an explicit answer to (i) above. That is, for any arbitrary intensity measure, the EDF of the potential ground motions suite can be compared with the theoretical distribution (as in Fig. 2.3). The second point is addressed in the subsequent section.

2.2.4 Bias in Seismic Response Estimates from Incompatible Ground Motions

Once seismic response analyses are performed using the final selected suite of ground motions, the resulting seismic responses of interest should be examined for bias. Bias can potentially occur if the distribution of one or more intensity measures of the ground motion suite differs from the theoretical distribution. Mathematical details regarding the estimation of bias can be found in Bradley (2010).

Figure 2.4 a illustrates, for a hypothetical seismic response problem, the dependence of peak free-field ground displacement, U_g, on significant duration, SD, based on the Suite 2 ground motions used previously. The p-value of 6.9×10^{-10} for the slope of the linear regression line indicates that there is a statistically significant dependence of U_g on SD. Therefore, given that the Suite 2 ground motions have a distribution of SD which is different from the theoretical distribution obtained from the GCIM approach (Fig. 2.3f), the distribution of EDP (i.e. U_g) obtained from the seismic response analysis is potentially biased. Figure 2.4b illustrates the EDF of U_g, its lognormal approximation, and the corrected distribution which approximately accounts for this bias (Bradley 2010). It can be seen that the corrected and uncorrected distributions are notably different.

The estimation of seismic response bias due to improper ground motion selection as illustrated in Fig. 2.4, therefore provides a simple method to examine the final

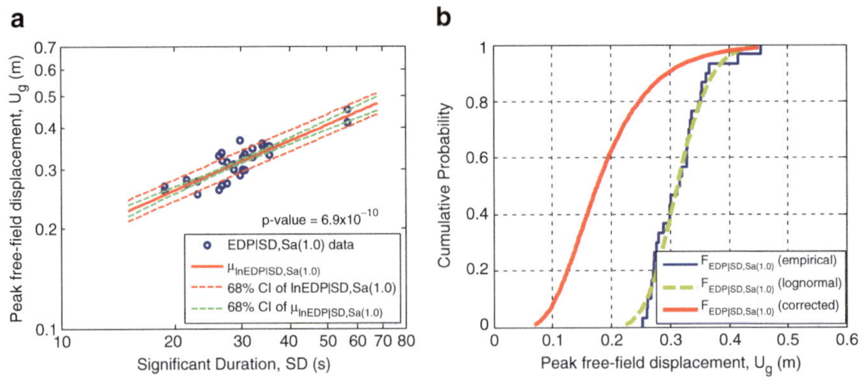

Fig. 2.4 Hypothetical results (*EDP*) of a seismic response analysis and their dependence with IMi: (**a**) Regression of peak free-field displacement, U_g, with SD; and (**b**) comparison of observed and corrected U_g distributions due to the dependence of U_g on SD and the difference between the SD statistics of the ground motion suite and theoretical distribution

consequences (in terms of bias in the estimated response) of the selected ground motion suite. Based on such information, an analyst could then decide whether the estimated bias is acceptable or whether a more suitable suite of ground motions is necessary.

2.3 Validation of and Uncertainty in Seismic Response Analyses Using Seismometer Array Recordings

In addition to the uncertainty associated with the future ground motions to affect a structure located at a particular site, there is also uncertainty associated with the prediction of the seismic performance of the considered structure even if the future ground motions were known. This uncertainty arises due to limited knowledge of the particular structure itself (which may include both the structure itself and the surficial soil layers); incomplete understanding of the physical mechanisms governing seismic response; and the limited capabilities of the mathematical models used to predict seismic response.

Quantitative data from seismometers represent one of the primary interactions between observations and computational simulation in earthquake engineering, with other interactions including: element testing, testing of subsystems, or testing of entire systems at full or reduced scales. Seismometer data offers several advantages over these other forms of quantitative data in that the instrumented facilities automatically have the correct in situ and boundary conditions which can be difficult, if not impossible, to reproduce in laboratory experiments.

This section provides an overview of a framework in which seismic response models can be validated with seismic array recordings, and consequently how uncertainties in seismic response models can be characterized. Firstly, the conventional use of seismometer arrays in validation of seismic response modelling and its limitations are discussed. The details of the proposed framework, which addresses conventional limitations, are developed and its benefits for use in seismic response prediction and uncertainty characterization are examined. Further details of the proposed framework can be found in Bradley (2011).

2.3.1 Conventional Comparison of Seismic Response Models with Array Observations

Examples of the use of seismometer arrays to examine the capabilities of (geotechnical site) response analysis include Cubrinovski et al. (2000), Bernardie et al. (2006), Elgamal et al. (1996), Finn et al. (1993), among others. Figure 2.5 illustrates a schematic example of a seismic instrumentation array which can be used to validate a one-dimensional seismic response analysis. A computational seismic response

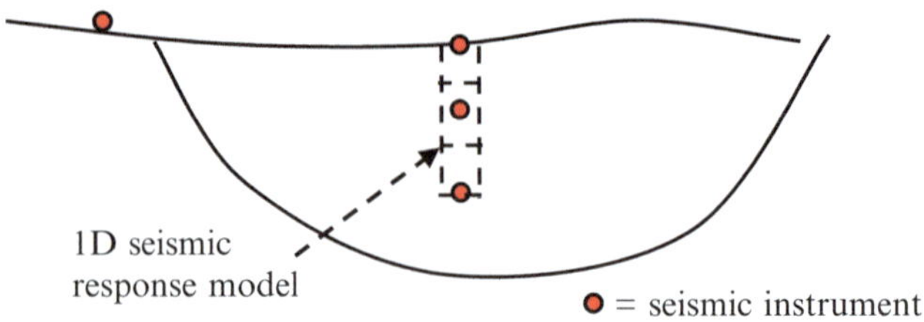

Fig. 2.5 Illustration of a site response example in which seismometer arrays can be used to provide validation of seismic site response analyses

model could be constructed, and the input excitation applied to the model can be obtained from one (or possibly more) of the seismometer recordings. Thus, in this case, the 'input' motion (i.e. that at the base of the one-dimensional computational model in Fig. 2.5) is known explicitly, such that any difference between the computational prediction and seismometer observations is due to the computational model (including how the input motion is applied as a boundary condition). This feature is a prerequisite to enable seismic arrays to be used for seismic response validation.

The conventional use of seismometer arrays for validation of seismic response computational models, as exemplified by the aforementioned references, can be regarded as deterministic in the sense that no uncertainties in the seismic response model are considered. One of the consequences of constructing a computational model of a site which exists in reality (rather than a model which is created under laboratory-type conditions) is that it is not possible to fully characterize the physical and mechanical properties of the site. Hence there exists significant uncertainty in the characterization of the problem under consideration. This consequently results in uncertainty in the parameters of the constitutive models required for computational analyses. A consequence of the failure to account for these uncertainties in the computational model is that it cannot be determined if a good agreement between a single prediction and observation is due to a capable computational model or is in fact due to 'cancellation' of errors that result from the unknown site characterization and inconsistencies in the computational model (Bradley 2011).

2.3.2 *Uncertainty Classification*

There are numerous significant uncertainties in any seismic response problem, and these must be considered in a robust validation framework. Here, such seismic response uncertainties are differentiated into four classes (Fig. 2.6): (i) site characterisation uncertainties; (ii) constitutive model parameter uncertainties; (iii) constitutive model uncertainties; and (iv) model methodology uncertainties.

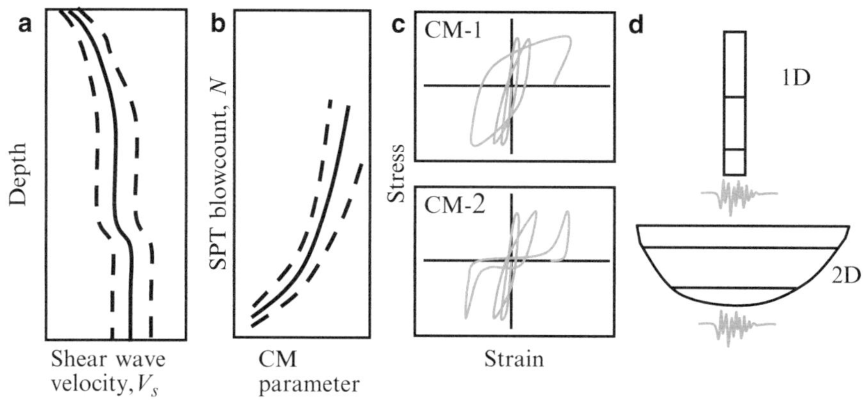

Fig. 2.6 Examples of the four different types of uncertainties in the case of a geotechnical seismic site response analysis: (**a**) site characterization uncertainty; (**b**) constitutive model (*CM*) parameter uncertainty; (**c**) constitutive model uncertainty; and (**d**) computational model methodology uncertainty

2.3.3 Consideration of Site Characterization and Constitutive Model Parameter Uncertainties

Consider initially that the computational model methodology and constitutive relations are an exact representation of the physical problem of interest. Therefore the only uncertainties in the seismic response predicted by the computational model are related to the measured values of mechanical and physical properties and the uncertain relationships between measured properties and the parameters of constitutive relationships (i.e. type (i) and (ii) uncertainties in Fig. 2.6). When type (i) and (ii) uncertainties are considered in the computational model then the resulting seismic response, measured by one of more engineering demand parameters (EDPs), will have a distribution (with each EDP having a different value for each possible realization of the uncertain parameters). Figure 2.7a illustrates this uncertainty in the form of a probability density function of a predicted EDP from the computational model. Figure 2.7a also illustrates the unique value of the seismic response quantity, $edp_{i,j,k}$, as measured from the seismometer array.

The probability density function (pdf) of the prediction of a particular demand measure, EDP_i, (e.g. peak displacement at the surface) for a single observation k, at a single site j, $f_{EDP_{i,j,k}}$, shown in Fig. 2.7a gives the likelihood that a particular value of EDP_i is observed based on the computational model. Because of the aforementioned uncertainties it is not possible to make robust inferences on the predictive capability of a particular computational model based on a single observation, and therefore multiple-prediction observation pairs are needed (Bradley 2011).

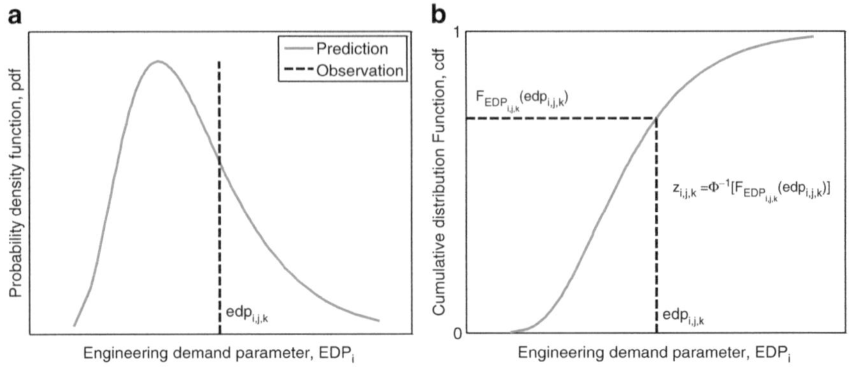

Fig. 2.7 (**a**) Schematic comparison between prediction probability density function and observation; (**b**) computation of normalized residual based on cumulative prediction distribution and observation

2.3.4 Consideration of Multiple Observations and Sites

Consider the uncertain prediction from the computational model in terms of the cumulative density function (CDF) shown in Fig. 2.7b (rather than the pdf in Fig. 2.7a). Using this CDF the actual seismometer observation (the kth observation at site j of EDP_i), $edp_{i,j,k}$, corresponds to a value $F_{EDP_{i,j,k}}\left(edp_{i,j,k}\right)$. The normalised residual of the seismometer observation for $edp_{i,j,k}$ relative to the computational model prediction can then be computed from:

$$z_{i,j,k} = \Phi^{-1}\left[F_{EDP_{i,j,k}}\left(edp_{i,j,k}\right)\right] \tag{2.6}$$

where $\Phi^{-1}[\ ]$ is the inverse normal cumulative density function. Based on its definition, $z_{i,j,k}$ represents a random observation from a standard normal distribution. In order to account for the dependence between multiple observations at a single site this normalised (total) residual is expressed as:

$$z_{i,j,k} = a + \eta_{i,j} + \varepsilon_{i,j,k} \tag{2.7}$$

where a is a constant; $\eta_{i,j}$ is the inter-site residual for EDP_i and site j, and $\varepsilon_{i,j,k}$ is the intra-site residual for the kth observation of EDP_i at site j. It is assumed that $\eta_{i,j}$ and $\varepsilon_{i,j,k}$ are independent and are characterised by a normal distribution with zero means and variances σ_s^2 and σ_o^2, respectively (Bradley 2011). Upon conducting regression to determine the unknown parameters in Eq. 2.7 (i.e. a, σ_s^2 and σ_o^2), the mean and variance of the regression model of the normalised residuals, are given by:

$$\hat{\mu}_Z = a \tag{2.8}$$

$$\hat{\sigma}_Z^2 = \sigma_s^2 + \sigma_o^2 \tag{2.9}$$

where $\hat{\mu}_Z$ is the point-estimate of the mean of Z; and $\hat{\sigma}_z^2$ is the point-estimate of the variance of Z. Based on the aforementioned assumption that the computational methodology and constitutive model are exact, each $z_{i,j,k}$ represents a random variable from a standard normal distribution. Hence, comparison of the mean and variance of the regression model for Z with that of a standard normal distribution can be used to examine the bias and precision of the computational methodology and constitutive model.

2.3.5 *Application of the Proposed Framework*

2.3.5.1 Hypothetical Observations

Figure 2.8 illustrates possible situations which may arise when comparing the predicted distribution of Z (i.e. using Eqs. 2.7–2.9), with the theoretical standard normal distribution for a particular seismic response problem. Figure 2.8a illustrates the case in which the mean and variance of Z are very similar to the standard normal distribution. It can be seen that the 90% confidence interval of μ_Z easily encompasses the theoretical value of zero, and hence the bias of the computational model methodology and constitutive model for the sites considered is relatively small. Figure 2.8b illustrates a situation where the computational model systematically over predicts the response for some EDP_i, resulting in residuals which are predominantly negative. This over-prediction bias is significant as can be seen from the 90% confidence interval for μ_Z not including the theoretical value of zero. Figure 2.8c illustrates a situation in which there is little bias in the computational model (similar to Fig. 2.8a), but that the variance of Z, σ_Z, is significantly larger than that of the theoretical value of 1, indicating that the computational model is imprecise.

2.3.5.2 Consideration of Alternative Constitutive Models

Because constitutive models used in seismic response analyses are typically empirically constructed based on direct observations, or theoretically derived based on various assumptions, then it is unlikely that a single constitutive model is perfectly representative of an engineering material. Consequently, this imperfection leads to uncertainty in the prediction of the seismic response of such a material (i.e. type (iii) uncertainty in Fig. 2.6). It is well recognised that significant differences in computational model predictions can be obtained using various commonly adopted constitutive models for certain problems (Arulanandan and Scott 1993).

The proposed seismic response analysis validation framework offers the opportunity to quantify a hierarchy of constitutive model validity based on the observed bias and precision of the alternative models, and therefore avoid problems associated with a significant reliance on expert opinion. Figure 2.9a schematically

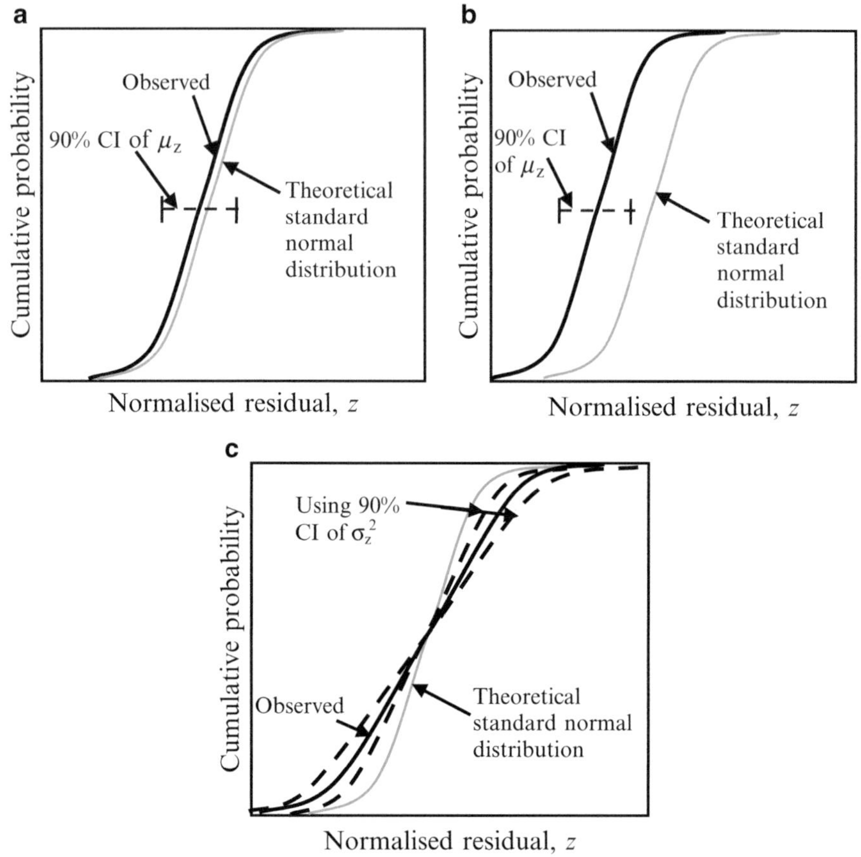

Fig. 2.8 Illustration of the resulting distribution of the normalised residuals compared to the theoretical standard normal distribution (and statistical error bounds) in the cases of: (**a**) insignificant bias; (**b**) significant bias; and (**c**) imprecision

Fig. 2.9 Use of the validation framework to assess the capability of various constitutive models (CM) based on the distribution of the normalised residuals

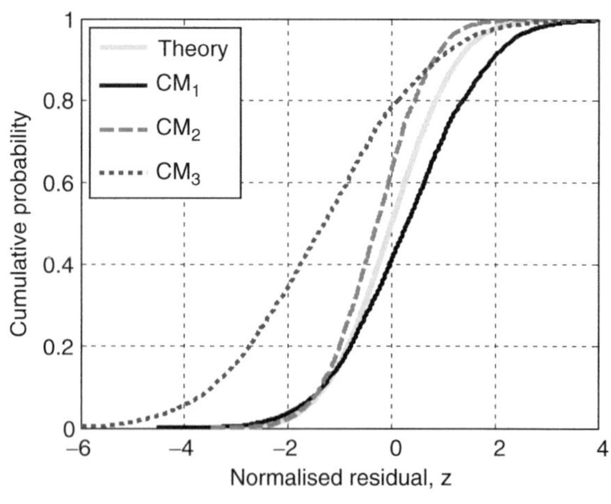

illustrates the distribution of the normalised residuals for a given computational model methodology, but using three different constitutive models. It can be seen that the use of constitutive models 1 and 2 leads to a small over-prediction and under-prediction bias as indicated by the small negative and positive mean values of the normalised residuals, respectively. It is also noted that the use of constitutive models 1 and 2 leads to an appropriate level of prediction precision (as indicated by the similarity in the variance of the normalised residuals relative to the theoretical standard normal distribution). On the other hand, the use of constitutive model 3 leads to a large over-prediction bias, as indicated by the mean value of the normalised residuals being significantly different than zero. In addition, the variance of the normalised residuals obtained using constitutive model 3 is significantly larger than one, indicating that the use of constitutive model 3 also leads to significant prediction imprecision. Hence on the basis of Fig. 2.9a, an analyst could comfortably reject the use of constitutive model 3 (for a seismic response problem which is 'within' those encompassed by the array recordings providing the observed normalised residuals), and consider only constitutive models 1 and 2 when accounting for constitutive model (type (iii)) uncertainty.

2.3.6 *Implications for Characterization of Seismic Response Uncertainty*

The proposed framework utilizing seismometer array recordings provides a direct approach in which validation of seismic response models can be achieved. However, there are also several additional benefits of the framework relating to the characterisation of uncertainties.

Firstly, the proposed framework enables analysts, researchers and practitioners to appreciate the different sources of uncertainty which affect the total seismic response uncertainty (i.e. Fig. 2.6). The framework also provides a clear motivation for the development of uncertainty models for physical quantities that characterise the problem and constitutive model parameters which are known with a relatively low precision (i.e. type (i) and (ii) uncertainties). Secondly, the proposed framework offers insight into the variability in seismic response predictions using different constitutive models and model methodology uncertainties by comparing their normalised residuals (Fig. 2.9). Thirdly, an understanding of the total uncertainty in the seismic response prediction by comparison with direct observations is made, rather than comparison of a particular seismic response model with a more 'advanced' model and the assumption that the more advanced model is a closer reflection of the true response.

It is important to understand that, in theory at least, all of the uncertainties associated with seismic response prediction (i.e. Fig. 2.6) are epistemic in nature, and thus with future research they should be able to be reduced. By characterising the total uncertainty, in the sense that one understands of the various locations where uncertainty arises, it is possible to focus resources efficiently in order to achieve the largest uncertainty reduction. The uncertainty classification and multi-tiered hierarchy adopted in the proposed framework offers such an uncertainty characterisation Bradley (2011).

2.4 Conclusions

This chapter has discussed two recently proposed frameworks which provide a rigorous and holistic characterization of uncertainties in ground motion selection and seismic response analysis. The first framework utilizes a generalised conditional intensity measure (GCIM) approach as the basis for the holistic selection of ground motions for any form of seismic response analysis, while the second framework made use seismometer arrays to validate and consequently characterise uncertainties in seismic response uncertainties.

The GCIM method is based on the construction of the conditional multivariate distribution of any set of ground motion intensity measures using PSHA results. The approach therefore allows any number of ground motion intensity measures identified as important in a particular seismic response problem to be considered. Ground motions are then selected, modified, and/or simulated based on the statistical comparison, for each intensity measure, of the empirical distribution of the ground motion suite with the 'target' GCIM distribution.

The seismic response validation framework explicitly accounts for the epistemic uncertainty related to the unknown characteristics of the 'site' and constitutive model parameters. A mixed effects model is used to allow multiple prediction-observation pairs to be utilized, enabling an improved statistical significance of inferences regarding the accuracy and precision of the computational seismic response methodology and constitutive model. Based on the results obtained by applying the validation framework it is also possible to characterise the uncertainties in computational model assumptions, constitutive models and their parameters.

References

Abrahamson NA, Silva WJ (1996) Empirical ground motion models. Report to Brookhaven National Laboratory

Arulanandan K, Scott RF (1993) Verification of numerical procedures for the analysis of soil liquefaction problems. A.A. Balkema, Rotterdam

Baker JW (2005) Vector-valued ground motion intensity measures for probabilistic seismic demand analysis. In: Department of civil and environmental engineering. Stanford University, Menlo Park, p 347

Baker JW (2007) Correlation of ground motion intensity parameters used for predicting structural and geotechnical response. In: 10th international conference on application of statistics and probability in civil engineering. Tokyo, p 8

Bernardie S, Foerster E, Modaressi H (2006) Non-linear site response simulations in Chang-Hwa region during the 1999 Chi-Chi earthquake, Taiwan. Soil Dyn Earthq Eng 26(11):1038–1048
Bradley BA (2009) Site specific and spatially distributed prediction of acceleration spectrum intensity. Bull Seismol Soc Am 100(2):792–801
Bradley BA (2010) A generalized conditional intensity measure approach and holistic ground motion selection. Earthq Eng Struct Dyn 39(12):1321–1342
Bradley BA (2011) A framework for validation of seismic response analyses using seismometer array recordings. Soil Dyn Earthq Eng 31(3):512–520
Bradley BA, Cubrinovski M, MacRae GA, Dhakal RP (2009) Ground motion prediction equation for spectrum intensity from spectral acceleration relationships. Bull Seismol Soc Am 99(1):277–285
Cubrinovski M, Ishihara K, Furukawazono K (2000) Analysis of two case histories on liquefaction of reclaimed deposits. In: 12th world conference on earthquake engineering, Auckland
Elgamal AW, Zeghal M, Parra E, Gunturi R, Tang HT, Stepp JC (1996) Identification and modeling of earthquake ground response – I, site amplification. Soil Dyn Earthq Eng 15(8):499–522
Finn WDL, Ventura CE, Wu G (1993) Analysis of ground motions at Treasure Island site during the 1989 Loma Prieta earthquake. Soil Dyn Earthq Eng 12(7):383–390
Gates W, Morden M (1995) Lessons from inspection, evaluation, repair and construction, in surveys and assessment of damage to buildings affected by the Northridge earthquake. SAC Joint Venture, Sacramento
Hanks TC, Brady AG (1991) The Loma Prieta earthquake, ground motion, and damage in Oakland, Treasure Island, and San Francisco. Bull Seismol Soc Am 81(5):2019–2047
Japanese Geotechnical Society (1998) Special issue on geotechnical aspects of the Jan 17, 1995 Hyogoken-Nambu earthquake, Soils and foundations
Luco N, Bazzurro P (2007) Does amplitude scaling of ground motion records result in biased nonlinear structural drift responses? Earthq Eng Struct Dyn 36(13):1813–1835
NZS 1170.5 (2004) Structural design actions, Part 5: earthquake actions – New Zealand. Standards New Zealand. Wellington
Shome N, Cornell CA, Bazzurro P, Carballo JE (1998) Earthquakes, records, and nonlinear responses. Earthq Spectra 14(3):469–500
Travasarou T, Bray JD, Abrahamson NA (2003) Empirical attenuation relationship for arias intensity. Earthq Eng Struct Dyn 32:1133–1155

Chapter 3
Design Earthquakes and Conditional Hazard

Iunio Iervolino

Abstract Disaggregating seismic hazard in terms of some ground motion or source parameters as magnitude and distance to the site of interest, allows to identify the earthquakes giving the largest contribution to the occurrence or exceedance of a specific value of a ground motion intensity measure (IM). If mapped, such information may be of useful to engineers in better defining the seismic threat for the structure of interest, for example in case of record selection for nonlinear seismic structural analysis. Because disaggregation results change with the spectral ordinate and return period, and more than a single event may dominate the hazard, the issues related to mapping design earthquakes (DE) are discussed in the paper with respect to Italy. Secondly, it is shown how DEs may be useful in rapidly determining conditional hazard of a secondary ground motion intensity measure with respect to a primary IM. As shown by the application for a vector comprised of peak ground acceleration as the primary IM and the *Cosenza and Manfredi Index* as the secondary IM, conditional hazard may be used for a more refined and consistent definition of design seismic actions on structures. The conditional hazard approach may be applied, in principle, to any other vector of IMs, providing the advantages of vector-valued hazard analysis requiring much less effort, and therefore rendering it ready for implementation in practice.

Keywords Disaggregation • Scenario • Performance-based earthquake engineering • Record selection • REXEL

M. Dolšek (ed.), *Protection of Built Environment Against Earthquakes*,
DOI 10.1007/978-94-007-1448-9_3, © Springer Science+Business Media B.V. 2011

3.1 Introduction

The basis for definition of seismic actions on structures in codes is the design spectrum. Because a rational design target should be representative of the seismic hazard at the construction site, the uniform hazard spectrum (UHS) or an approximation of it, is often used as the reference for design. The UHS is built entering the elastic spectral acceleration or Sa (for several structural periods, T) hazard curves with a specified probability of exceedance of (e.g., 10% in 50 years), and plotting the corresponding Sa versus T.

Given the UHS, the most of codes also require to determine the relevant scenarios for the site of interest in terms of seismic sources, and not only in terms of ground motion represented by the UHS. For example, Eurocode 8 (CEN 2003), regarding input ground motion selection for structural analysis states[1]: *in the range of periods between* $0.2T_1$ *and* $2T_1$*, where* T_1 *is the fundamental period of the structure in the direction where the accelerogram will be applied, no value of the mean 5% damping elastic spectrum, calculated from all time histories, should be less than 90% of the corresponding value of the 5% damping elastic response spectrum.* Moreover, accelerograms should be *adequately qualified with regard to the seismogenetic features of the sources* [...]. Records used in the structural analysis may be: *real, artificial or obtained by simulation of seismic source, propagation and site effects.*

It is clear that seismic source features may be required applying code prescriptions; however, it is unlikely that the engineer is able to qualify the input ground motions with respect to the seismological features of the sources. On the other hand, the most important events may be identified via *disaggregation of seismic hazard* (e.g. Convertito et al. 2009). In fact, once the UHS has been defined, it is possible to identify one or more earthquakes; i.e. the values of magnitude (M), source to site distance (R) and ε (number of standard deviations that the ground motion parameter is away from its median value estimated by the assumed attenuation relationship) providing the largest contributions to the hazard of exceeding a specified IM threshold. These events may be referred to as the earthquakes dominating the seismic hazard in a probabilistic sense, and may be used as design earthquakes (DEs) if appropriately mapped.

In the following, disaggregation of spectral acceleration hazard for all Italian sites for structural periods equal to 0s (PGA) and 1s is presented. In a lager study four different exceedance return periods (Tr) were considered (2,475, 975, 475 and 50 years) corresponding to the main limit states for civil and strategic structures, however, herein only results for Tr equal to 475 years will be shown; other results shall be available by Iervolino et al. (2010a).

The study shows how mapping DEs may be helpful, not only in ground motion record selection, but also in the low-effort definition of seismic actions for vectors of IM. In fact, vector-valued ground motion intensity measures have been recently

[1] Other minor requirements apply; see (Iervolino et al. 2008) for details.

investigated thoroughly (e.g. Baker 2007). Proposed measures are mainly function of spectral ordinates which have been shown to be useful in the assessment of structural response enhancing the confidence in estimations with respect to scalar IMs.

In a pair of IMs, it is often the case one is considered of primary importance with respect to the other. For example, it is generally believed that integral ground motion IMs, associated with duration, are less important with respect to the peak parameters of the record; nevertheless, there may be cases in which the cumulative damage potential of the earthquake is also of concern. For these IMs, it seems appropriate to develop *conditional hazard* maps; i.e., maps of percentiles of a secondary IM given the occurrence or exceedance of a primary parameter for which a design hazard map is often already available. In the presented study, this concept is illustrated and conditional hazard is developed for a parameter which may account for the cumulative damage potential of ground motion, the so-called *Cosenza and Manfredi index* (I_D), given peak ground acceleration (PGA). The study shows how easily it is possible to obtain analytical distributions of I_D conditional on PGA and on the corresponding DEs. Mapping of conditional hazard seems useful to complement design hazard maps for acceleration.

3.2 Italian Design Earthquakes

Given the characterization of seismic sources and once an IM is chosen, probabilistic seismic hazard analysis (PSHA) allows to identify, for each considered site, the probability of exceedance of different IM values in a time interval of interest. For its integral nature, PSHA, combines the contribution to the hazard from all considered sources. The event most important may be identified via disaggregation, and may be referred to as those providing the largest contributions to the hazard in terms of exceeding a specified IM threshold. Analytically, result of disaggregation is the joint probability density function (PDF) of {M,R,ε} conditional to the exceedance of the IM_0 threshold, Eq. 3.1.

$$f(m,r,\varepsilon|\mathrm{IM}>\mathrm{IM}_0)=\frac{\sum_{i=1}^{n} v_i\cdot I[\mathrm{IM}>\mathrm{IM}_0|m,r,\varepsilon]\cdot f(m,r,\varepsilon)}{\lambda_{IM_0}} \tag{3.1}$$

In Eq. 3.1 I is an indicator function that equals 1 if IM is larger than IM_0 for the specific values of r, m, and ε; n is the number of seismic sources relevant for the hazard at the site, v_i is the earthquake occurrence probability for the fault i; $f(m,r,\varepsilon)$ is the joint PDF of {M, R, ε}; and λ_{IM_0} is the hazard for IM_0. From the equation it is possible to observe that disaggregation depends on IM_0 (i.e., the hazard level being disaggregated, or the return period of the IM) and on the definition of the IM itself. If the IM of interest is Sa(T), then disaggregation, and therefore DEs, also depend on T. In fact, UHS for different return periods is characterized by different DEs, and, within a given UHS, short and long period ranges may display different M, R and ε from disaggregation (e.g. Reiter 1990).

In Italy, recent seismic code (CS.LL.PP 2008) introduced a new seismic classification of national territory, which was discretized by a grid of more than 10^4 nodes where the seismic hazard has been computed in terms of acceleration spectral ordinates from 0s to 2s, along with PGA disaggregation (see http://esse1.mi.ingv.it/). In this way, the code provides design spectra very close to the UHSs and software tools have been developed to automatically select sets of ground motion records compatible to them; i.e., REXEL (Iervolino et al. 2010b). However, PGA hazard disaggregation may be insufficient to define DEs for structures in the moderate period range. To overcome this gap, herein hazard is developed for PGA (considered as a benchmark) and Sa for T equal to 1s, and the disaggregation is computed for the whole country referring to the 475 years hazard.

Hazard values were computed for 30 values of the IMs equally distributed between 0.001 g and 1.5 g. All the analyses have been performed by a FORTRAN program specifically developed and also used in Convertito et al. (2009). The modeling of seismogenetic zones is that of Meletti et al. (2008), also adopted by the *Istituto Nazionale di Geofisica e Vulcanologia* or INGV (Fig. 3.1). Because of seismogenetic zones modeling, the hazard software assumes an uniform distribution of possible epicenters. Seismicity parameters of each zone are those used by Barani et al. (2009) (Table 3.1). The considered grid for Italy is the same of that from INGV and used in the Italian seismic code. All the analyses refer to rock site conditions according to Ambraseys et al. (1996), which is the ground motion prediction equation (GMPE) considered, magnitude is that of surface waves.

3.2.1 Disaggregation Results

The hazard results, computed in terms of PGA and Sa at T = 1s, are in fair agreement with those of INGV, and they are considered as the basis for disaggregation analyses presented in this section. The joint PDFs of M, R and ε given the exceedance of IM_0 corresponding to the four return periods considered, were computed, for each site of the grid, via simulation and using bins of M, R and ε equal to 0.05, 1.0 and 0.5, respectively. Minimum and maximum values used for ε are −3 and +3. Subsequently the first two modes of the joint PDF from disaggregation were extracted. The first mode is identified as the M, R and ε vector giving the largest contribution to the hazard, while the second mode corresponds to second higher relative largest contribution, identified if the differences between first and second mode are 5.0 km and 0.25 in terms of R and M.

In Figs. 3.2 and 3.3 modes of disaggregation distributions are shown. In the map referring to the second mode, white zones indicate that the hazard contribution of second mode is zero or negligible (i.e., providing a contribution to hazard lower than 10^{-4}). Analyses show that almost all sites are characterized by at least two DEs. This means that, from a design point of view, for each site it may be useful to

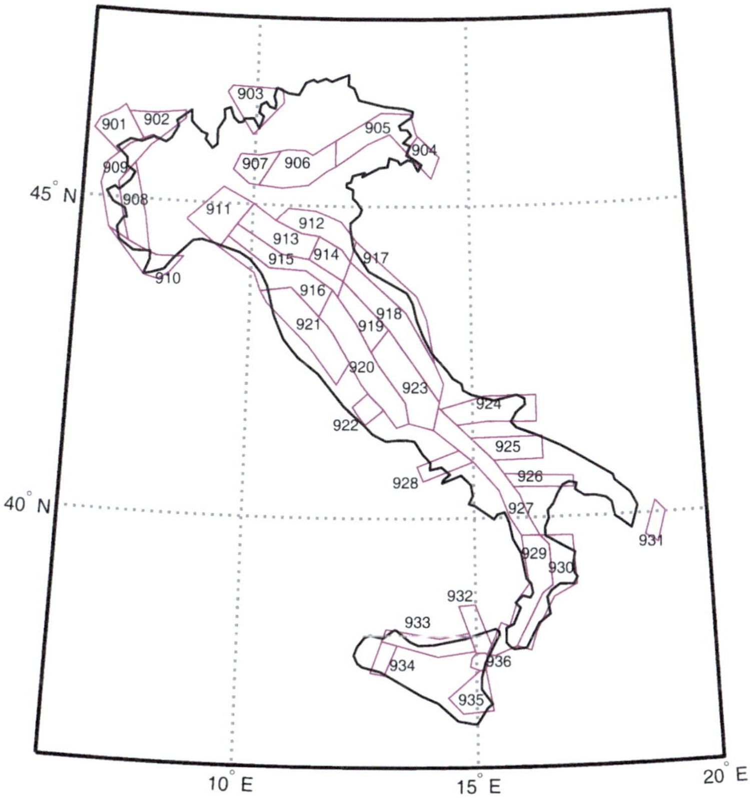

Fig. 3.1 Seismogenetic zones considered

know not only the first mode, but also the second one, in definition of seismic action on structures. Moreover it is possible to identify some general trends: (i) the first mode corresponds to an earthquake caused by the closer source (or the source the site is enclosed into) and with low-to-moderate magnitude, (ii) the second mode accounts for the influence of the more distant zones usually with larger magnitude, and (iii) moving from PGA to Sa, the number of sites with two DEs increases and the contribution of the second mode also increases (see following section). As consequence of (ii) and (iii), it can be inferred that the influence of more distant zones is higher for Sa(1s) than for PGA.

For a few sites, the particular combination of geometrical condition and seismic parameters of each source can determine an inversion of disaggregation results, and in such sites the sources influencing the first mode can be more distant than that related to the second mode. Other exceptions are represented by sites with a single mode; i.e., one DE. These sites are enclosed, or are close to, zones with high seismicity with respect to the surrounding zones which give negligible hazard contribution.

Table 3.1 Seismic sources' parameters

Zone	M_{min}	M_{max}	υ	b
901	4.3	5.8	0.045	1.133
902	4.3	6.1	0.103	0.935
903	4.3	5.8	0.117	1.786
904	4.3	5.5	0.050	0.939
905	4.3	6.6	0.316	0.853
906	4.3	6.6	0.135	1.092
907	4.3	5.8	0.065	1.396
908	4.3	5.5	0.140	1.408
909	4.3	5.5	0.055	0.972
910	4.3	6.4	0.085	0.788
911	4.3	5.5	0.050	1.242
912	4.3	6.1	0.091	1.004
913	4.3	5.8	0.204	1.204
914	4.3	5.8	0.183	1.093
915	4.3	6.6	0.311	1.083
916	4.3	5.5	0.089	1.503
917	4.3	6.1	0.121	0.794
918	4.3	6.4	0.217	0.840
919	4.3	6.4	0.242	0.875
920	4.3	5.5	0.317	1.676
921	4.3	5.8	0.298	1.409
922	4.3	5.2	0.090	1.436
923	4.3	7.3	0.645	0.802
924	4.3	7.0	0.192	0.945
925	4.3	7.0	0.071	0.508
926	4.3	5.8	0.061	1.017
927	4.3	7.3	0.362	0.557
928	4.3	5.8	0.054	1.056
929	4.3	7.6	0.394	0.676
930	4.3	6.6	0.146	0.715
931	4.3	7.0	0.045	0.490
932	4.3	6.1	0.118	0.847
933	4.3	6.1	0.172	1.160
934	4.3	6.1	0.043	0.778
935	4.3	7.6	0.090	0.609
936	3.7	5.2	0.448	1.219

For each zone it is provided: minimum (M_{min}) and maximum magnitude (M_{max}); annual rate of earthquake occurrence above M_{min}, (υ); and negative slope of Gutenberg-Richter relationship (b)

3.2.2 An Example of Multimodal Disaggregation

In Convertito et al. (2009) and in Barani et al. (2009) some interesting examples of disaggregation results for individual sites have already been presented: most of those cases are characterized by two different modal values with comparable contributions. In the wok presented in this paper the site of Lecce (southern Italy)

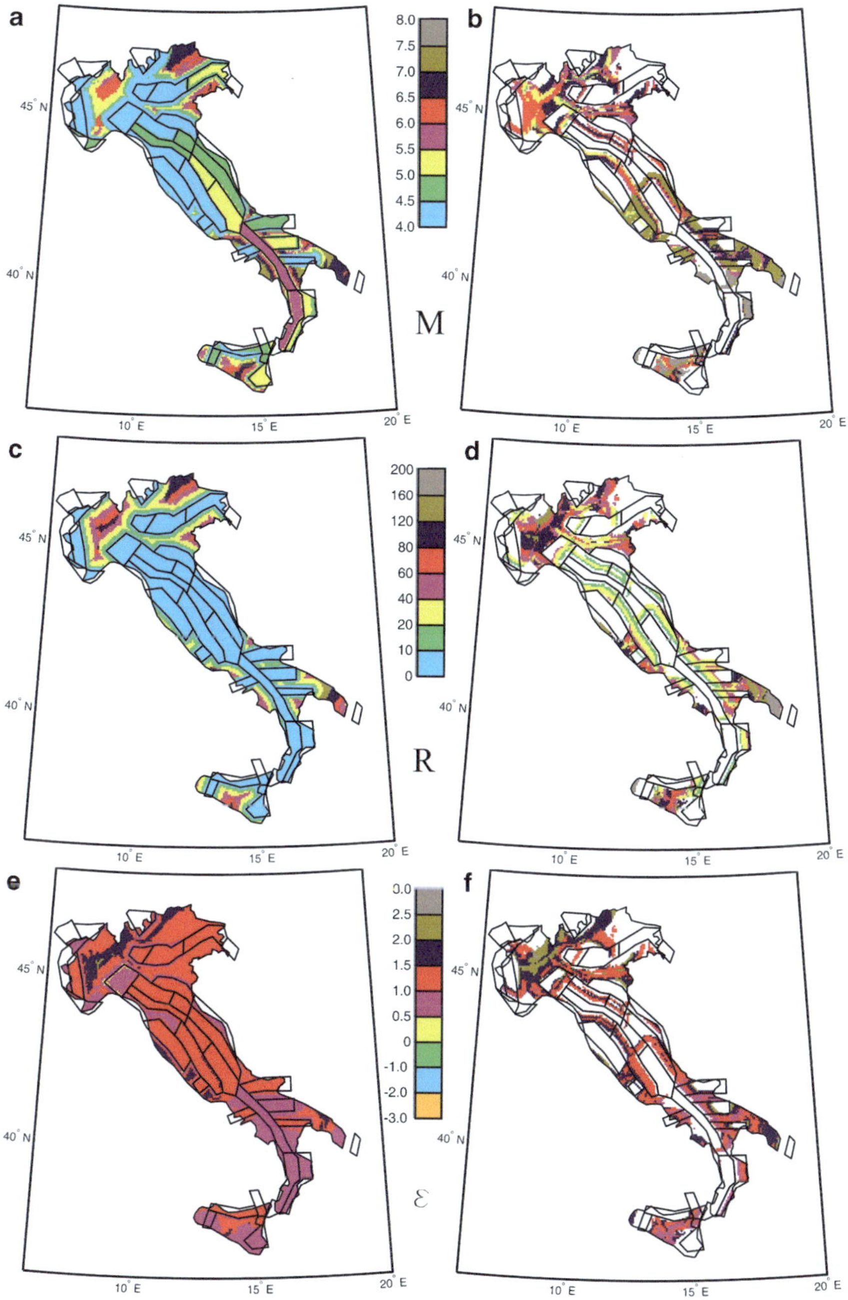

Fig. 3.2 First (*left*) and second (*right*) DEs for PGA and Tr = 475 years

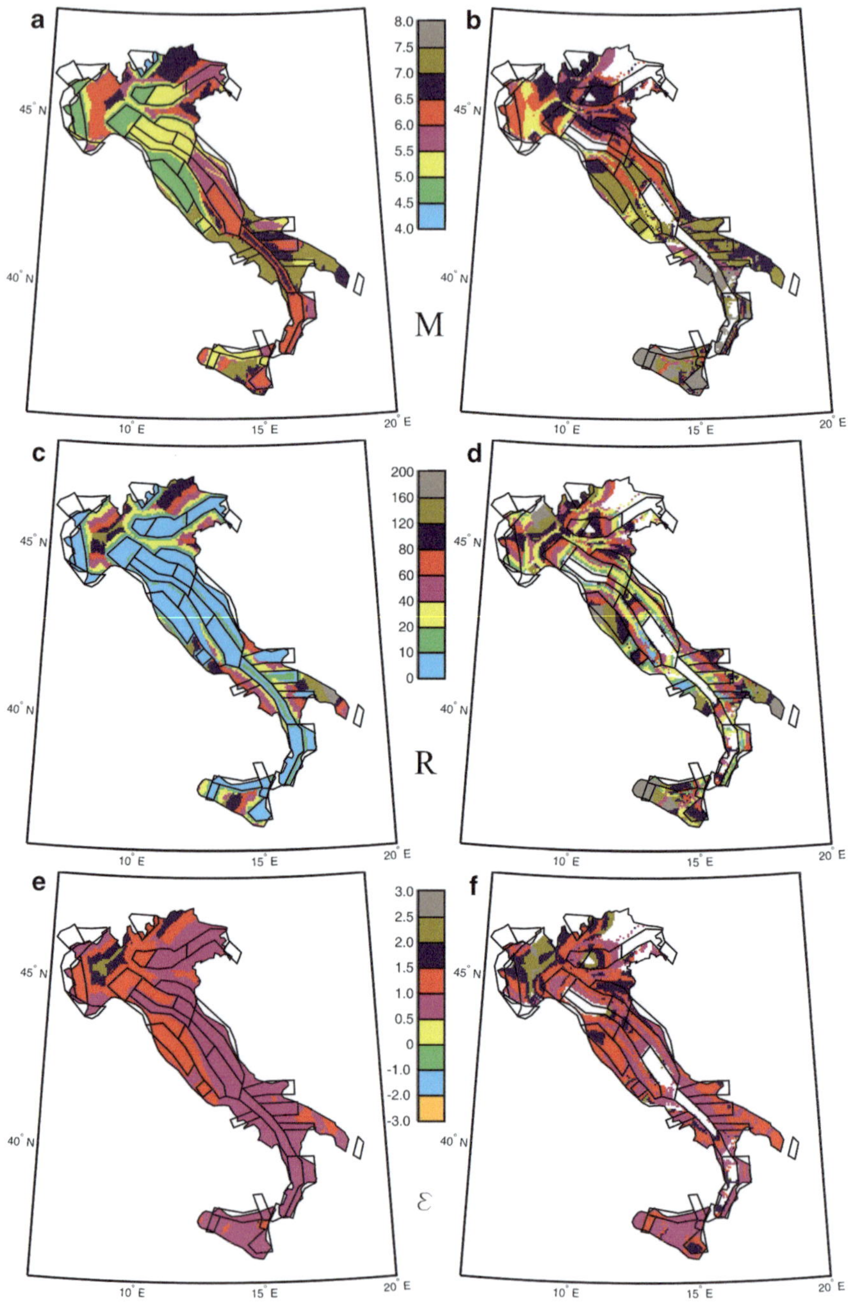

Fig. 3.3 First (*left*) and second (*right*) DEs for Sa(1s) and Tr = 475 years

is considered (latitude 40.338° N, longitude 18.147° E) and disaggregation results are shown for PGA and Sa(1 s) and for Tr = 475 years. In particular the joint PDF obtained from Eq. 3.1 is represented in Fig. 3.4 showing the marginal PDFs of R and M and of R and ε.

The considered site is not enclosed in any seismic source and its hazard is affected by sources 931 and 926 with a minimum distance lower than 100 km, and by sources 925, 927, 929 and 930 with a minimum distance between 100 and 200 km. For the combination of the characteristics of all the sources around the site, the PGA and Sa(1s) hazards for 475 years both correspond to 0.053 g.

Disaggregation of PGA hazard shows that the site is characterized by two different modal values: the first one due to R and M equal to about 80 km and 6.8, and the second one due to R equal to 180 km and M equal to 7.3. Hazard contribution of the second mode is much lower.

The same modes are computed for Sa(1s) but, as expected, the increment of spectral period determines increment of hazard contribution of more distant sources and, as consequence, the second mode becomes comparable with the first one.

These results, as noted already in Convertito et al. (2009), point out that even if hazard contribution of the second mode is comparatively low (like in the case of

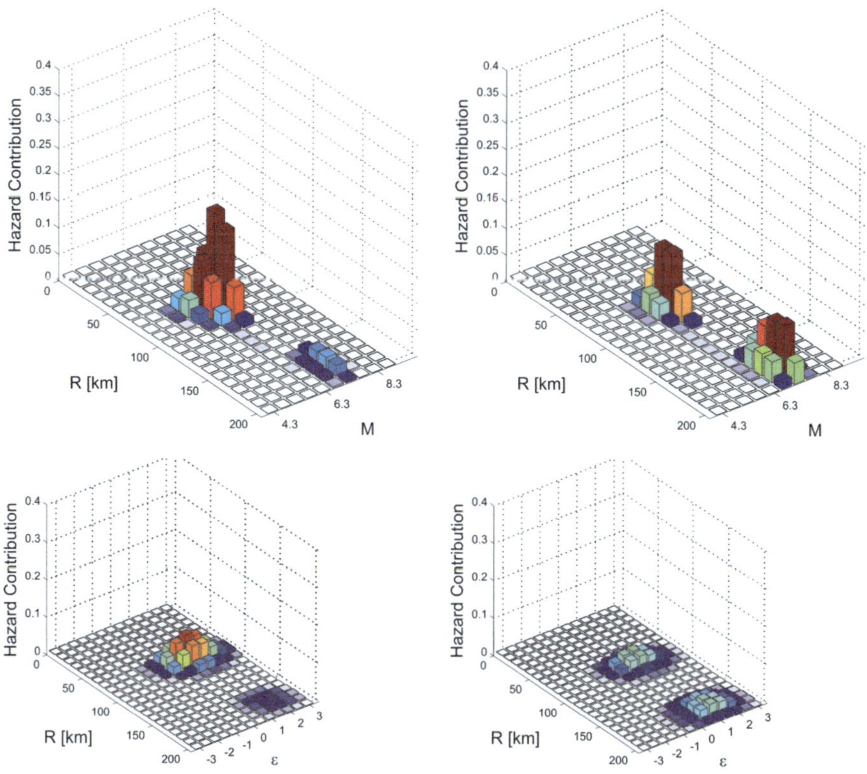

Fig. 3.4 Disaggregation results for Lecce for PGA (*left*) and Sa(1s) (*right*) and Tr = 475 years

PGA) a characterization of DEs should prudently account for it. In fact, when looking at spectral ordinates close to the fundamental period of the most common structures, the second mode may become significant. This has engineering consequences because, for example, although given a response spectrum the displacement structural response may not be very sensitive to magnitude and distance (Iervolino and Cornell 2005), ground motions characterized by different magnitudes and source-to-site distances can display different seismic demand, for example, in terms of cyclic structural response (e.g. Iervolino et al. 2006).

3.3 Conditional Hazard

Acceleration-based IMs (e.g., spectral ordinates) have been shown to be important and useful in the assessment of structural response of buildings. However, there are cases in which it is desirable to account for additional IMs while defining seismic actions. For example, although it is generally believed that, under some hypotheses, integral ground motion parameters associated to duration are less important for structural demand assessment with respect to peak quantities of ground motion, there are cases in which the cumulative damage potential of the earthquake is also of concern.

An easy yet hazard-consistent way of including secondary IMs in record selection is represented by the conditional hazard maps; i.e., maps of secondary ground motion intensity measures conditional, in a probabilistic sense, to the design hazard for the primary parameter. To illustrate the conditional hazard concept, in Iervolino et al. (2010c) the joint distribution of PGA and a parameter which may account for the cumulative damage potential of ground motion, was investigated. The chosen energy related measure is the so-called *Cosenza and Manfredi index*, the ratio of the integral of the acceleration squared to the PGA and peak ground velocity (PGV). I_D has proven to be a good proxy for cyclic structural response (Manfredi 2001). It is defined in Eq. 3.2 where a(t) is the acceleration time-history, t_E is the total duration of the seismic event. Therefore, the numerator of I_D is proportional to the *Arias Intensity* and it will be referred to as I_A.

The best candidates to be ground motion intensity measures are those for which hazard analysis is easy to compute, which requires a ground motion prediction equation (GMPE) to be available. Therefore, a GMPE was developed for I_D. The dataset used consists of 190 horizontal components from 95 recordings of Italian earthquakes used by Sabetta and Pugliese (1987). For the purposes of the present study the records were obtained by the *European Strong-motion Database* (ESD) (Ambraseys et al. 2000, Ambraseys et al. 2004). The dataset in terms of magnitude, distance, and, site conditions is given in Fig. 3.5 (left).

$$I_D = \frac{\int_0^{t_E} a^2(t)dt}{PGA \cdot PGV} = \frac{I_A}{PGA \cdot PGV} \tag{3.2}$$

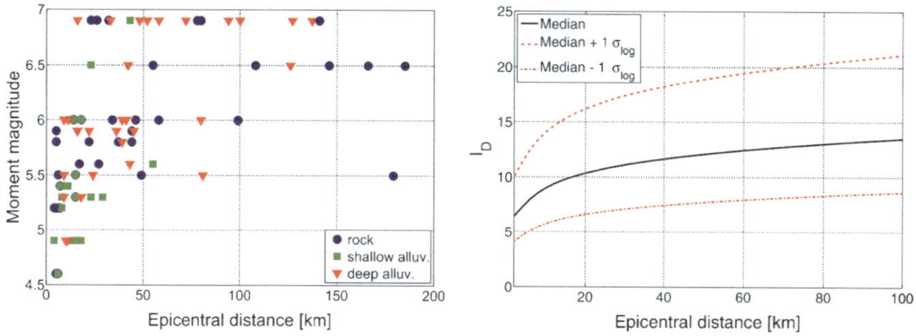

Fig. 3.5 Distribution of records with respect to moment-magnitude and epicentral distance (*left*); plot of I_D as a function of epicentral distance (*right*)

Table 3.2 Regression coefficients for PGA, PGV and I_A

Y	a	b	c	d	e	h	$\sigma_{\log_{10}Y}$
PGA [cm/s^2]	1.12	0.34	–0.89	0.16	–0.065	5.0	0.19
PGV [cm/s]	–1.27	0.55	–0.95	0.14	0.036	3.9	0.25
I_A [cm^2/s^3]	0.42	0.92	–1.69	0.24	–0.021	5.3	0.39

The empirical predictive equations for the logs of the IMs (the generic one is indicated as Y) appearing in the definition of I_D were fitted by regression using the same functional form of Sabetta and Pugliese (1996), Eq. 3.3, as a function of moment magnitude, epicentral distance (in km), and recording site geology. In this form, h is a fictitious depth, the dummy variables S_1 and S_2 refer to the site classification and take the value of 1 for shallow and deep alluvium sites, respectively, and zero otherwise. The residual, $\varepsilon_{\log_{10}Y}$, is a random variable which in ordinary least squares regressions, is implicitly assumed to be Gaussian with zero mean and a standard deviation $\sigma_{\log_{10}Y}$.

$$\log_{10}Y = a + b\,M + c\,\log_{10}\left(R^2 + h^2\right)^{\frac{1}{2}} + d\,S_1 + e\,S_2 + \varepsilon_{\log_{10}Y} \qquad (3.3)$$

The estimates for the coefficients for PGA, PGV and I_A, obtained using the ordinary least-squares regression, are given in Table 3.2. In the same table also the estimated standard deviations of the respective residuals, are also given. h values were not estimated and assumed to be coincident to those provided by Sabetta and Pugliese (1996).

In order to obtain an attenuation relation for the logs of I_D as a function of M, R and local site conditions it is possible to derive its coefficients as linear combinations of those for $\log_{10}$PGA, $\log_{10}$PGV and $\log_{10}I_A$ leading to the expression of Eq. 3.4, in which subscripts 1, 2 and 3 for c coefficient and h refer to PGA, PGV and I_A, respectively.

Table 3.3 Regression coefficients for I_D

Y	a	b	d	c_1	c_2	c_3	e	$\sigma_{\log_{10}Y}$
I_D	0.58	0.034	–0.068	0.89	0.95	1.69	0.0077	0.19

$$\log_{10}\mathrm{I_D} = a + b\mathrm{M} + \log_{10}\left(\frac{\left(\mathrm{R}^2+\mathrm{h}_1^2\right)^{c_1}\left(\mathrm{R}^2+\mathrm{h}_2^2\right)^{c_2}}{\left(\mathrm{R}^2+\mathrm{h}_3^2\right)^{c_3}}\right)^{\frac{1}{2}} + d\mathrm{S}_1 + e\mathrm{S}_2 + \varepsilon_{\log_{10}\mathrm{I_D}} \tag{3.4}$$

The coefficients of Eq. 3.4 are listed in Table 3.3. A plot of I_D versus epicentral distance is given in Fig. 3.5 (right) where the typical increasing trend with distance of duration-related measures is shown.

Because, if the vector comprised of logs of PGA and I_D can be considered jointly normally distributed, all the possible marginal and conditional distributions obtained from the joint distribution are still Gaussian, The skewness and kurtosis' tests of Mardia (1985) were used to test multivariate normality of the vector made of $\varepsilon_{\log_{10}\mathrm{PGA}}$ and $\varepsilon_{\log_{10}\mathrm{I_D}}$. With a given significance level of 0.05, the multivariate skewness and the multivariate kurtosis resulted non-significant.

The residuals of the prediction relationships for the logs of PGA and I_D have been also tested for correlation in order to compute $\mathrm{f}(\log_{10}\mathrm{I_D}|\log_{10}\mathrm{PGA})$, that is, the conditional PDF of the logs of I_D given the logs of PGA. The estimated correlation coefficient (r) between $\varepsilon_{\log_{10}\mathrm{PGA}}$ and $\varepsilon_{\log_{10}\mathrm{I_D}}$ (equal to -0.25) has been tested for statistical significance using a Student-T statistic Mood et al. (1974) and assuming as the null hypothesis $\rho = 0$ (ρ is the "true" correlation coefficient), which was rejected at 0.05 significance level.

Because of bivariate normality, the conditional PDF for one of the variables given a known value of the other, is normally distributed. The conditional mean, $\mu_{\log_{10}\mathrm{I_D}|\log_{10}\mathrm{PGA,M,R}}$, and standard deviation of $\log_{10}\mathrm{I_D}$, $\sigma_{\log_{10}\mathrm{I_D}|\log_{10}\mathrm{PGA}}$, given that $\log_{10}\mathrm{PGA} = z$, are given in Eq. 3.5 where $\mu_{\log_{10}\mathrm{I_D}|\mathrm{M,R}}$ and $\sigma_{\log_{10}\mathrm{I_D}}$ are the mean and the standard deviation from the I_D GMPE.; $\mu_{\log_{10}\mathrm{PGA}|\mathrm{M,R}}$ and $\sigma_{\log_{10}\mathrm{PGA}}$ are the mean and the standard deviation from the PGA GMPE.

Because the joint distribution of I_D and PGA depends on the I_D attenuation and from the PGA attenuation, therefore also on magnitude and distance, to obtain the conditional distribution of the logs of I_D conditional on PGA only, the marginalization in Eq. 3.6 is required.

$$\begin{cases} \mu_{\log_{10}\mathrm{I_D}|\log_{10}\mathrm{PGA,M,R}} = \mu_{\log_{10}\mathrm{I_D}|\mathrm{M,R}} + \rho\, \sigma_{\log_{10}\mathrm{I_D}} \dfrac{z - \mu_{\log_{10}\mathrm{PGA}|\mathrm{M,R}}}{\sigma_{\log_{10}\mathrm{PGA}}} \\ \sigma_{\log_{10}\mathrm{I_D}|\log_{10}\mathrm{PGA}} = \sigma_{\log_{10}\mathrm{I_D}}\sqrt{1-\rho^2} \end{cases} \tag{3.5}$$

$$\mathrm{f}\left(\log_{10}\mathrm{I_D}|\log_{10}\mathrm{PGA}\right) = \int_{\mathrm{M}}\int_{\mathrm{R}} \mathrm{f}(\log_{10}\mathrm{I_D}|\log_{10}\mathrm{PGA,M,R})\,\mathrm{f}(\mathrm{M,R}|\log_{10}\mathrm{PGA})\,\mathrm{d}m\,\mathrm{d}r \tag{3.6}$$

It is easy to recognize that the $f(M, R|\log_{10}PGA)$ term in Eq. 3.6 is the PDF of M and R given the occurrence of $\log_{10}$PGA; i.e., the result of disaggregation of seismic hazard. As an approximation of the integral, the modal values M* and R* (i.e., the first DE) may be plugged in Eq. 3.5; i.e., Eq. 3.7.

$$\mu_{\log_{10}I_D|\log_{10}PGA} \approx \mu_{\log_{10}I_D|M^*,R^*} + \rho\,\sigma_{\log_{10}I_D}\frac{z-\mu_{\log_{10}PGA|M^*,R^*}}{\sigma_{\log_{10}PGA}} \tag{3.7}$$

3.4 Illustrative Application

An example of the possible use of conditional hazard is given in Fig. 3.6 which shows the maps of seismic hazard in terms of I_D conditional to the PGA with a 10% exceedance probability in 50 years in Italy. In particular, Fig. 3.6a,b are the 50th and 90th percentiles of the conditional I_D PDF, respectively.

The conditional I_D maps were obtained using the distribution of parameters in Eq. 3.5 in which the z (log of PGA) values are those computed in the hazard analysis described in Sect. 2, while the values of magnitude and distance (M* and R*) to plug in the $\mu_{\log_{10}PGA|M,R}$ and $\mu_{\log_{10}I_D|M,R}$ terms of Eq. 3.7 are those shown in Fig. 3.2a, c; i.e., the design earthquakes[2] corresponding to the PGA on which the I_D distribution is conditional.

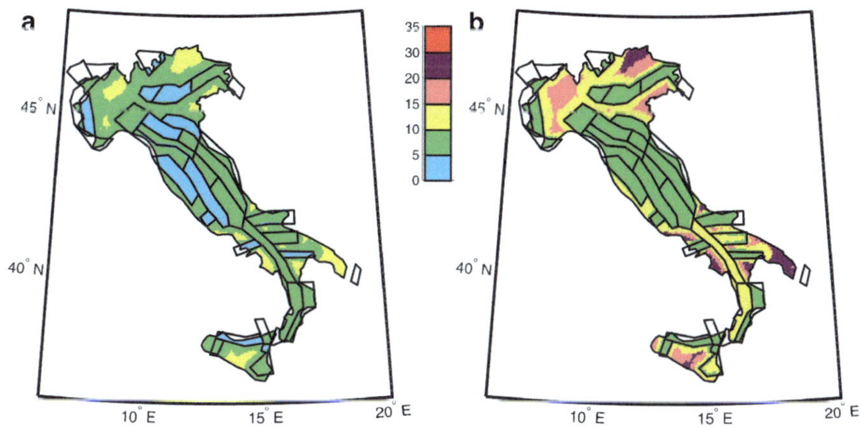

Fig. 3.6 Maps of I_D in terms of 50th (**a**) and 90th (**b**) percentiles conditional to PGA with a 475 years return period and using first DEs of Fig. 3.2

[2] For the purposes of the illustrative application it is assumed that moment magnitude can be approximated by surface-wave magnitude. Moreover, Eq. 3.5 requires disaggregation for the occurrence of PGA, while herein DEs from disaggregation of exceedance hazard are considered.

3.5 Conclusions

Disaggregation can be considered as a useful tool to address the definition of design scenarios to be used in engineering practice. In the work presented in this paper design earthquakes from disaggregation of all Italian sites for structural periods equal to 0s and 1s was presented referring to hazard with a 475 years exceedance return period. First and second modal values are used here as synthetic identifiers of DEs.

Results show that, usually, the modal value with the largest contribution to hazard corresponds to a moderate-magnitude earthquake caused by the closer source, while the influence of the more distant zones is accounted for by the second mode. Moreover, because in most cases Italian sites are located inside seismogenetic zones assumed for hazard analysis, first mode of disaggregation is characterized by a source-to-site distance lower than 10 km. For Sa at $T = 1$ s, the contribution of more distant sources is higher than in the PGA case. Finally it is to conclude that only a few sites are characterized by a single DE and this is particularly evident from disaggregation of Sa(1s) hazard, which is more representative than PGA for ordinary structures.

An immediately intelligible use of design earthquakes is ground motion record selection for nonlinear dynamic analysis of structures. However, there are other possible uses, one of which occurs when more than one ground motion parameter has to be taken into account in seismic structural assessment. For example, although it is generally believed that integral ground-motion parameters are secondary for structural demand assessment with respect to peak quantities of ground motion, sometimes the cumulative damage potential of the earthquake is also of concern. For these cases it could be useful to have a distribution of secondary intensity measures conditional on the primary parameter used to define the seismic action on structures (e.g., accelerations). Under some hypotheses, this can be carried out in close-form and was called *conditional hazard*.

This approach has the advantages of vector-valued seismic hazard analysis without the computational effort required by PSHA for vectors of IMs. To explore such a concept, in this paper the distribution of a parameter which may account for the cumulative damage potential of ground-motion, conditional to peak ground acceleration (PGA), was investigated. The chosen secondary measure is the so called *Cosenza and Manfredi index* (I_D). A ground-motion prediction relationship has been retrieved for the log of I_D on the basis of an empirical dataset of Italian records already used for well known prediction equations proposed in the past by other researchers. Subsequently, the residuals of prediction relationships have been tested for correlation and for joint normality. The study allowed to obtain analytical distributions of I_D conditional on PGA and the corresponding design earthquakes. Results of the study have been used to compute, as an illustrative example, the distributions of I_D conditional on PGA with a return period of 475 years for each node of a regular grid having about 2 km spacing and covering Italy.

The presented conditional hazard maps provide information on the values of I_D which, for example, should be taken into account along with the hazard in terms of PGA at the site. In fact, conditional hazard can complement the hazard curves or maps produced for the primary IM.

Conditional hazard may be extended, in principle, to any vector of IMs.

Acknowledgements The author wants to acknowledge professor Massimiliano Giorgio and doctors Eugenio Chioccarelli and Carmine Galasso, who have contributed to the results presented herein.

References

Ambraseys NN, Simpson KA, Bommer JJ (1996) Prediction of horizontal response spectra in Europe. Earthquake Eng Struct Dyn 25:371–400

Ambraseys N, Smit P, Berardi R, Rinaldis D, Cotton F, Berge C (2000) Dissemination of European strong-motion data (cd-rom collection). European Commission Dgxii, Science, Research and Development, Bruxelles

Ambraseys N, Smit P, Douglas J, Margaris B, Sigbjornsson R, Olafsson S, Suhadolc P, Costa G (2004) Internet-site for European strong-motion data. Boll Geofis Teor Appl 45:113–129

Baker JW (2007) Probabilistic structural response assessment using vector-valued intensity measures. Earthquake Eng Struct Dyn 36:1861–1883

Barani S, Spallarossa D, Bazzurro P (2009) Disaggregation of probabilistic ground-motion hazard in Italy. Bull Seismol Soc Am 99:2638–2661

CEN, European Committee for Standardisation TC250/SC8/ (2003) Eurocode 8: design provisions for earthquake resistance of structures. Part 1.1: general rules, seismic actions and rules for buildings, PrEN1998-1

Convertito V, Iervolino I, Herrero A (2009) Importance of mapping design earthquakes: insights for the southern apennines, Italy. Bull Seismol Soc Am 99:2979–2991

CS.LL.PP (2008) DM 14 gennaio 2008 Norme Tecniche per le Costruzioni. Gazzetta Ufficiale della repubblica Italiana, 29 (in Italian)

Iervolino I, Cornell CA (2005) Record selection for nonlinear seismic analysis of structures. Earthquake Spectra 21:685–713

Iervolino I, Manfredi G, Cosenza E (2006) Ground-motion duration effects on nonlinear seismic response. Earthquake Eng Struct Dyn 35:21–38

Iervolino I, Maddaloni G, Cosenza E (2008) Eurocode 8 compliant real record sets for seismic analysis of structures. J Earthquake Eng 12:54–90

Iervolino I, Chioccarelli E, Convertito V (2010a) Engineering design earthquakes from multimodal hazard disaggregation. Soil Dynamics and Earthquake Engineering (in press)

Iervolino I, Galasso C, Cosenza E (2010b) REXEL: computer aided record selection for code-based seismic structural analysis. Bull Earthquake Eng 8:339–362

Iervolino I, Giorgio M, Galasso C, Manfredi G (2010c) Conditional hazard maps for secondary intensity measures. Bull Seismol Soc Am 100:3312–3319

Manfredi G (2001) Evaluation of seismic energy demand. Earthquake Eng Struct Dyn 30:485–499

Mardia KV (1985) Mardia's test of multinormality. In: Kotz S, Johnson NL (eds) Encyclopedia of statistical sciences, vol 5. Wiley, New York, pp 217–221

Meletti C, Galadini F, Valensise G, Stucchi M, Basili R, Barba S, Vannucci G, Boschi E (2008) A seismic source zone model for the seismic hazard assessment of the Italian territory. Tectonophysics 450:85–108

Mood MA, Graybill FA, Boes DC (1974) Introduction to the theory of statistics, 3rd edn. McGraw-Hill Companies, New York
Reiter L (1990) Earthquake hazard analysis, issues and insights. Columbia University Press, New York
Sabetta F, Pugliese A (1987) Attenuation of peak horizontal acceleration and velocity from Italian strong-motion records. Bull Seismol Soc Am 77:1491–1513
Sabetta F, Pugliese A (1996) Estimation of response spectra and simulation of nonstationarity earthquake ground-motion. Bull Seismol Soc Am 86:337–352

Chapter 4
Orientation Dependence of Earthquake Ground Motion and Structural Response

Damian N. Grant, Diego Padilla, and Paul D. Greening

Abstract The strength of horizontal earthquake ground shaking varies with orientation, and yet is typically condensed into a single response spectrum for seismic design. This chapter investigates how this orientation-dependence ("directionality") should be considered in the specification of seismic demand in codes of practice. The sources of orientation-dependence are identified, and their implications for ground motion selection, scaling and modification are discussed. A series of nonlinear response history analyses are presented, based on simple bridge piers with circular, square and rectangular cross-sections, each subjected to a number of ground motions rotated through non-redundant orientations. It is shown that ground motions with realistic orientation dependence in their elastic spectral response are more demanding than those with elastic orientation independence, and that the variation of response with orientation is not significantly affected by the cross section shape. This suggests that design based on the geometric mean spectrum, as used in several existing codes, is unconservative with respect to real seismic demand.

Keywords Ground motion • Orientation dependence • Directionality • Spectral matching

4.1 Introduction

Earthquake ground motion is typically recorded by accelerometers in three orthogonal components – two oriented in the horizontal plane and one oriented vertically. For most building applications, it is reasonable to ignore the vertical component, and concentrate on the horizontal shaking. The amplitude of horizontal motion is not the same in all directions, and the orientation of the strongest

D.N. Grant (✉) • D. Padilla • P.D. Greening
Department of Civil, Environmental and Geomatic Engineering,
University College London, Gower Street, WC1E 6BT London, UK
e-mail: d.grant@ucl.ac.uk; padilla-diego@hotmail.com; paul.greening@ucl.ac.uk

M. Dolšek (ed.), *Protection of Built Environment Against Earthquakes*,
DOI 10.1007/978-94-007-1448-9_4, © Springer Science+Business Media B.V. 2011

acceleration will generally not coincide with one of the "as recorded" components. Although it is useful to represent ground motion in terms of orthogonal components, considering them independently can obscure the two dimensional (or three dimensional if shaking in the vertical direction is also considered) characteristics of the motion, concealing the manner in which the strength of ground shaking varies with the azimuth of measurement.

Traditionally two main approaches have been used to condense the two horizontal components into a single measure of the ground shaking (Beyer and Bommer 2007). In both cases, response spectral ordinates for a viscous damping ratio of 5% are used to characterise the motion and its effect on structural response. In the first approach, the response spectra are calculated for the two components individually and the geometric mean of the two spectral ordinates is determined at each period. In the second approach, the maximum of the two as-recorded spectral ordinates at each period are considered. Most ground motion predictive equations (GMPEs) prior to the Next Generation Attenuation (NGA) Project (Power et al. 2008) used one of these definitions, and therefore almost all current codes of practice are also based on these, at least where seismic zonation in the code is based on probabilistic seismic hazard analysis (PSHA).

The NGA Project adopted the GMRotI50 measure (Boore et al. 2006), which has the advantage that it does not vary with the arbitrary orientation of the original instrument used to record the ground motion, as installed in the field. Response spectral ordinates are calculated for all possible non-redundant orientations of the horizontal ground motion (in arbitrarily small angle increments). The GMRotI50 spectral ordinates are then those corresponding to a single orientation angle (the same for all periods) that is most representative of the overall median response. A disadvantage of the GMRotI50 measure is that it can be sensitive to the range of periods considered in its derivation, and a small change in the upper bound period can result in a disproportionately large change in the spectral ordinates (Boore et al. 2006; Grant 2011).

The recently released US standard, ASCE 7–10, adopted the maximum rotated spectral response as its measure of seismic demand. Conceptually, this is calculated for each structural period by rotating through every possible orientation in which the ground motion could be applied, and calculating the highest value of spectral response. In practice, the calculations can be carried out, for a given period and damping ratio, by evaluating the single degree of freedom response versus time for two orthogonal components of input, say X and Y, individually. The maximum rotated response is then the maximum of the square root sum of squares of the X and Y responses versus time (Grant 2011; Huang et al. 2008). This calculation is illustrated graphically for two different periods of response for a particular earthquake ground motion in Fig. 4.1a and b. Note that with this definition, each period may be governed by a different orientation angle. In this chapter, the maximum rotated response is referred to as the "major axis response", and the azimuth in which this peak is measured defines the "major axis". The orthogonal axis is referred to as the "minor axis".

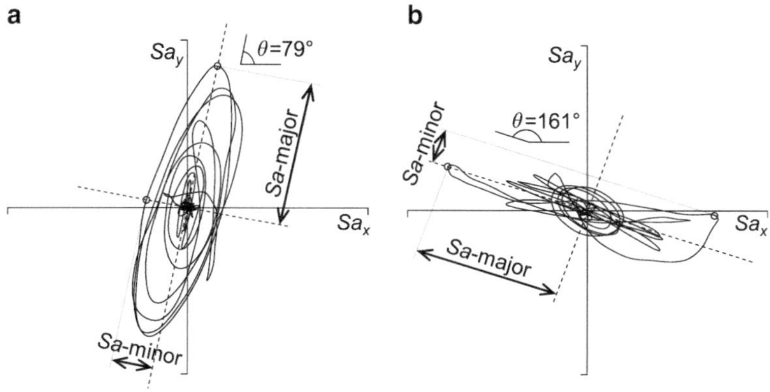

Fig. 4.1 Response orbits and definition of major and minor axes of response for PEER Record 184 from Imperial Valley event; (**a**) 1.3 s period response, and (**b**) 3.0 s period response

As long as consistent measures of ground motion are used in the specification of seismic hazard and the interpretation of analysis results, unbiased estimates of structural response can be determined (Baker and Cornell 2006). However, to increase the precision of structural analyses – especially important for response history analyses where several runs are required to take into account the uncertainty in the ground motion definition – it is important to identify areas where the traditional representations of seismic demand do not contain all the possible information about potential ground shaking, and which therefore introduce additional uncertainty into the analyses. For example, if the geometric mean spectrum is used, then realistic ground motions whose geometric mean spectra are consistent with this should be provided. Records where each component was spectrally-matched to the mean spectrum could not be considered "realistic" in this case, as they would not show an appropriate relationship between the maximum rotated response and other orientations of input (Grant 2011). Structural engineers could then carry out response history analysis for a suite of earthquake ground motions and possibly for a range of incidence angles, and evaluate the expected performance of their structure. Different representations of demand may require more or fewer ground motions and/or orientations to be used, and alternative ways of combining all the results together, but it should be possible in theory to reach an unbiased estimate of the expected performance in each case.

It can be argued that the most appropriate measure of seismic demand depends on the structural form (Grant 2011), and if a common definition is adopted in seismic design codes, then corrections may need to be introduced for different structural configurations. Many now accept that for axisymmetric structures, such as circular bridge piers or chimneys, the maximum rotated response gives a direct measure of the peak demand on the structure (Grant 2011; Grant et al. 2005a); i.e. their response may be considered "azimuth independent" (Stewart et al. 2011). The extreme of "azimuth dependent" response is a shear wall, which only carries load from one direction of ground motion. Of course, a single shear wall would not be an appropriate lateral load resisting system for a building, and shear wall building response considered holistically, with walls in orthogonal directions, is not purely azimuth dependent, as demonstrated in the next section.

The objective of this chapter is to investigate how the azimuthal variation of ground shaking amplitude (as measured by elastic spectral response), the vibration characteristics (such as period and perhaps damping), and nonlinear behaviour (strength and hysteretic response) should be taken into account in design and analysis. The chapter concludes with a summary of some analytical studies carried out by Padilla (2010), as part of a broader investigation of how structural form affects orientation dependence of structural response.

4.2 Variation of Structural Response with Orientation of Ground Motion

4.2.1 *Probability of Structural Failure*

Structural engineers are accustomed to interrogating structural models for local response quantities, such as moments, axial forces and shears in elastic analyses, and plastic deformations in nonlinear analyses. Even if multiple quantities are combined together into a single parameter, such as a utilization ratio involving moment and axial force, these quantities are local in nature. If one member fails, it may or may not lead to global collapse of the structure. In most design examples, however, failure of a member implies an inadequate design, and either that member should be strengthened, or other changes should be made to the building to reduce the demand on that member. Although the findings of this chapter apply to any design limit state, we will use the shorthand "probability of failure", or p_f, to indicate the overall probability that the structure will "fail", given a certain level of earthquake ground shaking.

When considering how the applied orientation of the ground motion affects structural response, it is tempting to consider local response quantities in isolation, and consider their variation with orientation angle. Since each orientation angle is equally likely to occur (at least in the absence of any information about how local faulting influences azimuth dependence), it is tempting to conclude that the response quantity to consider for design is the median over all 360° (Beyer and Bommer 2007; Priestley et al. 2007). The implication of this is that the value taken for each design action would be exceeded in 50% of possible orientations of the ground motion. If we are only interested in a single response quantity of the structure in the seismic load case (because, for example, we know that every other element in the structure is governed by a different load case), then we would argue that this is the correct approach – design for the expected value of the demand under any angle of incidence of the ground motion.

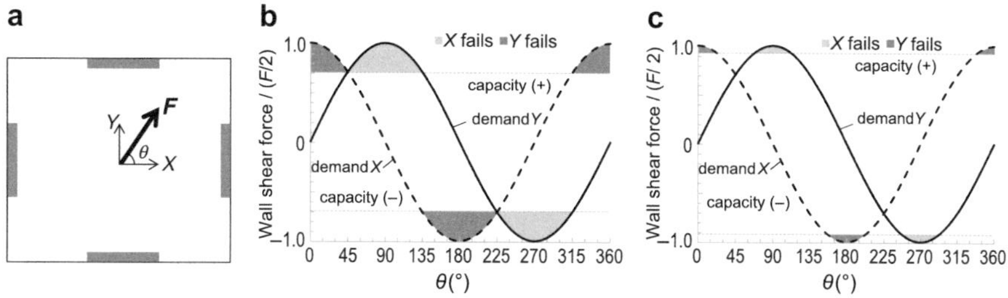

Fig. 4.2 Example of variation in wall demand for simple building plan with orientation angle of static force. (**a**) Symmetric wall layout; (**b**) probability of failure setting capacity equal to expected value of each response quantity; (**c**) probability of failure setting capacity equal to value required to give structural failure in 50% of orientations (i.e. $p_f = 0.5$)

We are seldom interested in just one quantity, however, but rather the overall performance of the structure. Different structural components will exhibit a different dependence on the orientation of the ground motion. The orientation angle required to give peak response in one component will often be 90° out of phase with the critical angle for other components. If any component fails, this will often mean overall structural failure. Rather than looking at each response quantity individually, we must consider their dependence on orientation holistically, and evaluate the demand that causes overall failure of the structure in 50% (or other known percentage) of possible ground motion orientations.

To illustrate this point, consider the building plan shown in Fig. 4.2a, with walls in two orthogonal directions, and a static force, F, applied at an unknown angle, θ. The variation in demand in the X-oriented walls and the Y-orientated walls with angle is shown in Fig. 4.2b, normalised with respect to $F/2$, which is the force in one wall when the force is parallel to the axis of the wall. We assume that the value of F has been determined by a probabilistic procedure (such as PSHA), and that we do not wish to add any additional conservatism by, for example, enveloping the demand. Considering an elastic design, if the capacity is set equal to the median absolute value of shear over all possible orientation angles, the normalised capacity is 0.71, as shown in Fig. 4.2b. However, since the response of the X and Y walls is completely out of phase, adopting this capacity would result in failure of the structure for any orientation of the static force. From 0° to 45°, X walls fail; from 45° to 135°, Y walls fail; and so on.

Arguably, a more rational distribution of capacity in the walls is as shown in Fig. 4.2c, where the capacity in each wall is set such that failure occurs for 50% of the possible orientation angles. For the sinusoidal response, the normalised capacity required to achieve this is given by cos(22.5°), or 0.92. As shown in Fig. 4.2c, we would then expect failure in X walls for angles from 0° to 22.5°, no failure from 22.5° to 67.5°, failure in Y walls from 67.5° to 112.5°, and so on.

Response of real structures to earthquake ground motion is more complicated than this simple example. Realistic earthquake ground motion imposes demand on both directions simultaneously, as well as positive and negative cycles, and therefore the demand curves shown in Fig. 4.2b and c would not pass through zero. Response quantities other than the shear force in the walls may be relevant, such as inter-storey drifts for design of facades and partitions, or overall floor movements in X and Y directions to assess the probability of pounding with adjacent buildings. Structural periods may be different in the two orthogonal directions, and therefore the absolute value of the demand imposed by the earthquake ground motion may vary with rotation. Additionally, torsion (whether due to eccentricities in the design or accidental torsional effects) may couple the responses in the plan directions. Finally, for nonlinear structural response, the simple check of demand and capacity is not sufficient, and the demand also varies as the nonlinearity modifies the structural periods.

4.2.2 Sources of Orientation Dependence in Seismic Response

In assessing the response of structures to earthquake ground motion, we can identify three main sources of orientation or azimuth dependence:

1. Input ground motion
2. Linear and/or nonlinear structural response
3. Performance criteria

The first source recognises that the ground motion is not equally strong in all directions. For example, Fig. 4.1a shows the bidirectional spectral response orbit with fundamental period of 1.3 s and damping ratio of 5% of critical, for a particular earthquake ground motion (see Grant (2011) for details). The response is heavily polarised, such that the peak response at $\theta = 79°$ is around 3.7 times the orthogonal response ($\theta = 169°$). Figure 4.1b shows the response for a period of 3.0 s and damping ratio of 5% of critical, which also shows highly polarised response, but for a major axis orientation of 161°, almost orthogonal to the 1.3 s major axis orientation. Even varying the damping ratio can modify slightly the orientation of the principal axes.

As with unidirectional structural response, the bidirectional elastic response spectrum is instructive, but does not tell us all we need to know about inelastic response. Proponents of Direct Displacement-Based Design (DDBD) (Priestley et al. 2007), or other methods that make use of equivalent linearization of nonlinear response, are accustomed to viewing structural hysteresis as an elongation of the fundamental period and an increase in energy dissipation characterised by the viscous damping ratio. However, it should be recognised that this is an approximate method of estimating nonlinear response, and equations for damping versus ductility need to be calibrated against many nonlinear analyses to give accurate results (e.g. Grant et al. 2005b). In DDBD, the inelastic response of a structure with a fundamental period of 1.3 s that undergoes a displacement ductility of 5.3, say, would be best estimated from the elastic response spectrum with an equivalent period of around $1.3 \times \sqrt{5.3} = 3.0$ seconds (the equivalent viscous damping to be used depends on the appropriate hysteresis model to calibrate the cyclic structural response). Taking the same ground motion as in Fig. 4.1, this does not necessarily imply, however, that nonlinear structural response out to this ductility level is a maximum for an orientation angle of $\theta = 161°$ (as for the 3.0-s elastic response). In this chapter, azimuth dependence of elastic response is frequently used as a proxy for inelastic response to illustrate the concepts; the extension of these conceptual arguments to inelastic response is the goal of the studies reported in Sect. 4.4.

The second source of orientation dependence is in how structural response varies with angle, again considering both elastic and nonlinear response. Simple reinforced concrete bridge piers with circular or square cross sections (as investigated later in the chapter) have identical elastic (uncracked) moment of inertia about any centroidal axis. However, for the square section (and for other shaped sections), the nonlinear cyclic response (strength and ductility capacity) varies with orientation. The response of rectangular columns is more dependent on orientation, as the elastic period and section modulus both vary with angle. In these azimuth-dependent cases, the effect of polarised ground motions on structural response will vary with the angle at which the ground motion is applied to the structural model. For example, a structure with a fundamental period of 3.0 s in one direction and 1.3 s in the orthogonal direction, subjected to the ground motion in Fig. 4.1 would be heavily shaken in both directions if the ground motion was applied to the structure at an angle of $\theta = 161°$ (measured anticlockwise from the X-axis, as shown in the figure); the demand would be much lower for $\theta = 251°$ (i.e. in the orthogonal direction), say.

Different structural configurations also differ in the way in which the ground motion is "interpreted" by the structural components. The structural wall building in Fig. 4.2 effectively decomposes the input ground motion into X and Y components, which are resisted separately by orthogonal structural systems. Polarised ground motion in which the strong axis is aligned with one of the structural axes will lead to higher demand in the walls than if the strong axis is at 45°. In the simplified models investigated in Sect. 4.4, this is equivalent to a square pier section with all the longitudinal reinforcement lumped in the middle of each side of the section. A structure with more degrees of symmetry (such as the current tallest building in the world, the Burj Khalifa, with threefold rotational symmetry in plan) decomposes the ground motion among its structural subsystems, and maximum demand is only experienced in any one of the subsystems when the strong axis of the ground motion coincides with one of the structural axes.

A subtle point is that even if structural response (linear and/or nonlinear) does not vary with orientation, the way in which we interpret the data and the performance criteria that are applied can introduce a third source of orientation dependence. As an example, consider again a simple circular bridge pier with axisymmetric structural response, subjected to a polarised earthquake ground motion. For a particular project, the strength could be governed by other considerations (say wind or traffic loading), and the seismic demand could be well under the capacity for any applied ground motion angle. However, suppose that pounding at the abutments or between adjacent spans is a concern, and that therefore longitudinal displacements of the bridge superstructure should be within a specified limit. In this case, again we must identify the percentage of possible orientation angles that would cause failure, where in this case "failure" is identified with excessive longitudinal displacements leading to pounding. For design, the required nonlinear stiffness (and therefore strength) is that which satisfies the limit in at least 50% of possible orientation angles.

Arguably, this third source of orientation dependence is an example of the second. If the full structural response up to collapse were modelled, including strength and stiffness degradation, slip and buckling of reinforcing bars, nonlinear geometric effects associated with pounding, and non-structural components, then the analyst would not need to introduce performance limits to response quantities. Unacceptable performance could readily be identified with collapse in the analysis model, or irreparable structural or non-structural damage. In practical analysis models, however, it is informative to separate the two sources.

4.3 Orientation Dependence in Ground Motion Selection, Scaling and Modification

In the past, hazard analyses were typically carried out using ground motion prediction equations based on the geometric mean of two as-recorded components of horizontal ground shaking, whereas structural analysts inconsistently applied the outputs from these studies along a single axis of the structure (Baker and Cornell 2006). Using a consistent measure of bidirectional ground motion intensity removes this source of bias. However, just as the distribution of unidirectional spectral ordinates in a suite of ground motions influences the calculated structural response quantities, for 3D analyses it is also important to control the variation in input ground motion intensity with orientation. Relationships are available in the literature that provide expected ratios of the major to minor axis (Hong and Goda 2007), or major axis to geometric mean (Huang et al. 2008) spectral response. If response history analyses are carried out for a range of input angles to measure the variation in demand with orientation, ground motions should be selected and/or modified to satisfy these literature relationships, to ensure that orientation dependence (in particular, minimum and maximum demand) is appropriately captured.

Carrying out analyses for a large suite of ground motions for multiple orientation angles is useful for research studies, but is impractical for real design projects. Therefore, for practical applications we seek a method that would reduce the number of analyses required without significantly decreasing the precision of the method or introducing bias (conservative bias is perhaps admissible for design, although this is better introduced and controlled through the reliability methods that are the basis of code material and safety factors). We concentrate here particularly on reducing the number of orientations that need to be considered in response history analyses. This is primarily motivated by the fact that the record to record variability (and the subsequent variability in the response history analysis results) is expected to be larger than the variability with orientation angle. If this assumption holds, then reducing the uncertainty in the analysis results is more efficiently accomplished by carrying out analyses for different ground motions, rather than for extra orientations of the same motions.

The variability in orientation dependence of ground motions can be reduced with response spectral matching. Spectral matching is typically used to reduce record-to-record variability of single component ground motions. A real recorded ground motion is adjusted such that its elastic response spectrum is a close match of a target spectrum, while ideally preserving the "realistic", nonstationary characteristics of the original recording (such as phase, duration, frequency content). Although several algorithms and computer programs are available, the program *RspMatch* (Abrahamson 1992) and its updated versions (Hancock et al. 2006; Al Atik and Abrahamson 2010) have become popular with practitioners.

A new version of the software, *RspMatchBi* (Grant 2011, 2010), considers two input ground motion components simultaneously. The program makes adjustments to both components, effectively modifying the bidirectional orbits (e.g. Fig. 4.1) such that the major and minor axis spectra of the record match two separate target spectra. The adjustments are carried out in such a way that the orientation of the major and minor axes for each period (and damping value) are preserved from the original record, unlike if the single component version of *RspMatch* is applied to the two components separately (Grant 2011). *RspMatchBi* was used to generate suites of ground motions for the numerical studies described in the next section.

4.4 Analytical Studies on Orientation Dependence of Seismic Response

4.4.1 Structural Models

Idealised single bridge piers – representative of a typical structural system – were modelled using the program *OpenSees* (McKenna et al. 2000). The analysis model represents a simple bridge pier, as shown in Fig. 4.3a, with superstructure mass,

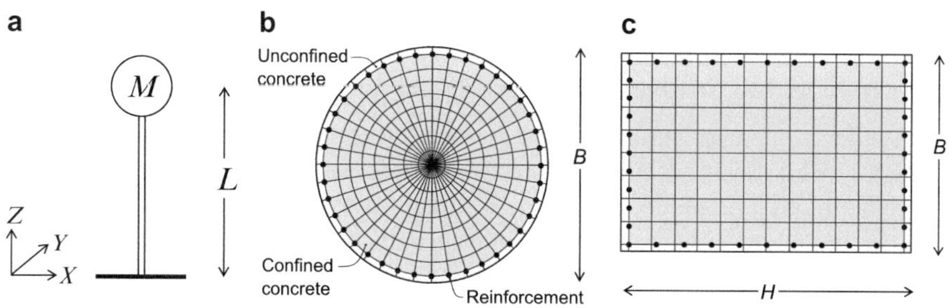

Fig. 4.3 (**a**) *OpenSees* analysis model; (**b**) circular and (**c**) square/rectangular cross section

Table 4.1 Structural model descriptions, and median scaling factors (*SF*) required to give $p_f = 0.50$ for all ground motions

	Section dimensions		Steel ratio	Median *SF* for $p_f = 0.50$		
Section description	*B* (m)	*H* (m)	ρ (%)	Suite 1	Suite 2	Suite 2/Suite 1
Circular	1.50	–	1.60	1.10	1.00	0.909
Square, $H/B = 1.0$	1.30	1.30	1.62	0.920	0.820	0.891
Rectangular, $H/B = 1.25$	1.30	1.63	1.22	1.00	0.910	0.910
Rectangular, $H/B = 1.5$	1.30	1.95	0.99	1.08	1.01	0.935
Rectangular, $H/B = 1.75$	1.30	2.28	0.82	1.17	1.08	0.923
Rectangular, $H/B = 2.0$	1.30	2.60	0.70	1.25	1.16	0.928

$M = 766$ t, and pier height, $L = 9$ m. The self-weight of the superstructure mass was applied, followed by the earthquake ground motion in the two horizontal directions. A damping ratio of 2% with tangent stiffness proportional Rayleigh damping was used.

Six cross sections were defined, with fibre cross section geometry shown in Fig. 4.3b and c, and parameters summarised in Table 4.1. Using fibre sectional analysis allows the orientation dependence of the strength, stiffness and acceptance criteria (based on material strains) to be taken into account. The *OpenSees* modelling approach, and design assumptions for the sections are described in (Padilla 2010). In each model, the total reinforcement quantity calculated from these design assumptions was divided among 40 reinforcing bars. Note that the stiffnesses and strengths of the cross sections are different.

4.4.2 *Ground Motion Input*

Two suites of ground motions, each comprising 20 two-component ground acceleration histories, were developed. The same twenty records were used as seeds for each of the suites, to allow direct comparisons between the results from each. Further information about the record selection is provided in (Padilla 2010). Summary information about the records used, and the initial linear scaling factors applied to them, in provided in Table 4.2.

RspMatchBi was used to match each record in the two suites of motions such that:

1. In suite 1, the principal axes of each record were both spectrally matched to a single mean target spectrum, based on an assumed earthquake scenario and a ground motion prediction equation (adjusted for a damping value of 2%). That is, the major and minor axis spectra of each record were approximately identical, and therefore the records were approximately orientation independent.
2. In suite 2, the major axis spectrum of each record was matched to a major axis target spectrum and the minor axis was matched to a minor axis target spectrum.

Table 4.2 Ground motions from PEER NGA database used as seeds for development of suites 1 and 2 (ranked in order of best initial fit of target spectrum)

Rank	PEER seq. No.	Event name	Recording station	M_w	R_{jb} (km)	Scale factor
1	178	Imperial Valley-06	El Centro Array #3	6.5	10.8	1.81
2	184	Imperial Valley-06	El Centro Differential Array	6.5	5.09	1.17
3	183	Imperial Valley-06	El Centro Array #8	6.5	3.86	1.30
4	802	Loma Prieta	Saratoga – Aloha Ave	6.9	7.58	1.47
5	175	Imperial Valley-06	El Centro Array #12	6.5	17.9	3.41
6	2752	Chi-Chi, Taiwan-04	CHY101	6.2	21.6	2.88
7	163	Imperial Valley-06	Calipatria Fire Station	6.5	79.4	5.34
8	176	Imperial Valley-06	El Centro Array #13	6.5	49.9	4.05
9	880	Landers	Mission Creek Fault	7.3	27.0	4.60
10	2655	Chi-Chi, Taiwan-03	TCU122	6.2	18.1	2.74
11	187	Imperial Valley-06	Parachute Test Site	6.5	12.7	3.89
12	1762	Hector Mine	Amboy	7.1	41.8	2.42
13	1521	Chi-Chi, Taiwan	TCU089	7.6	7.04[a]	1.77
14	718	Superstition Hills-01	Wildlife Liquef. Array	6.2	17.6	3.78
15	126	Gazli, USSR	Karakyr	6.8	3.92	0.993
16	832	Landers	Amboy	7.3	69.2	3.03
17	1489	Chi-Chi, Taiwan	TCU049	7.6	3.78	1.88
18	169	Imperial Valley-06	Delta	6.5	22.0	1.63
19	143	Tabas, Iran	Tabas	7.4	1.79	0.632
20	985	Northridge-01	LA – Baldwin Hills	6.7	23.5	2.69

[a] R_{jb} not given; epicentral distance reported

Target principal axis spectra were developed from the mean spectrum using the factors from references (Watson-Lamprey and Boore 2007) and (Hong and Goda 2007): the former gives factors to transform the geometric mean spectrum into the major axis spectrum, while the latter gives factors to transform from the major to minor axis spectrum. This gave a degree of orientation dependence typical for real ground motions.

The response spectra of the two suites of records are compared with the target spectra in Fig. 4.4, where the top and middle plots in each case show the major and minor axis spectra for each record, compared with the relevant major and minor target spectra, respectively. The bottom plot compares the ratio of major to minor axis spectra with the ratio of the target spectra. The difficulties in matching to identical target spectra are discussed in (Grant 2011); despite the imperfect fit to the target, the three suites of ground motions satisfy the objective of giving inputs to structural analyses with different directionality characteristics.

Ground motions were successively transformed and applied to the model at 9° increments of orientation angle, for a total of 40 orientations between 0° and 360°. Symmetry was used to reduce the number of analyses carried out; for example, for the circular cross section, only one analysis was required, and the results for each fibre were offset 9° for each 9° change of orientation angle.

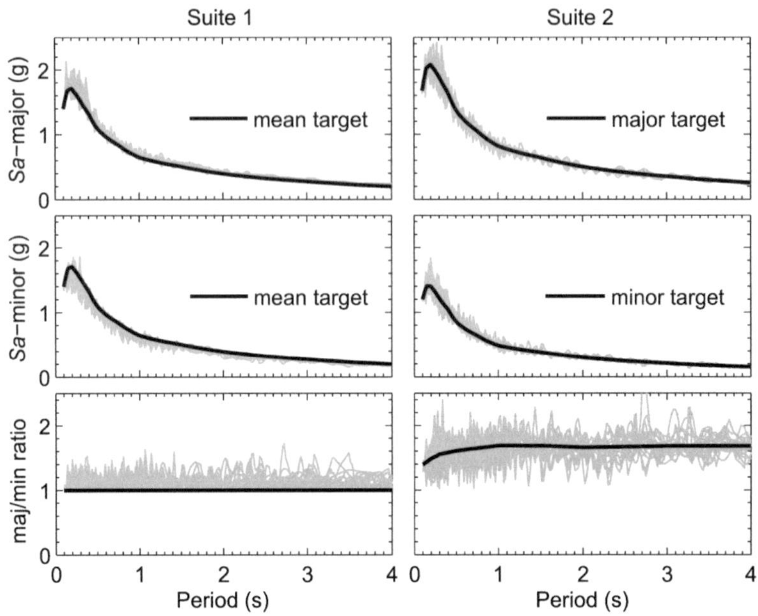

Fig. 4.4 Two suites of ground motions generated by *RspMatchBi*: response spectra for major axis (*top*), minor axis (*middle*) and ratio of major to minor (*bottom*)

4.4.3 Processing of Results

As discussed previously, the probability of failure (i.e. of exceeding a given limit state) must be considered holistically, and would depend on what response quantities are important for the design. If the pier is oversized due to high wind demand, for example, then longitudinal pounding against the abutments could be the only controlling response quantity, and the peak displacement in the X direction may be the only output of interest.

In this study, however, we considered the more common case where peak material strains control performance limits. Ultimate state strain limits of $\varepsilon_{cu} = -0.02$ for the confined core concrete, and $\varepsilon_{su} = +0.10$ for the reinforcement (where tensile values are positive) were assumed. Compression strain limits in the steel were assumed to be less onerous than the neighbouring concrete compression limits. The cross sections shown in Fig. 4.3b and c consist of hundreds of integration points (fibres). Since plane sections remain plane, it suffices to check the strain at the edge of the confined concrete core, which corresponds to the centroid of the reinforcement. Therefore, for each model, only the strains at the fibres representing reinforcing bars were evaluated, recognizing that the strain at these locations is identical to that in the neighbouring concrete (since bar slip is not modelled).

An iterative approach was adopted to determine, for each combination of ground motion and model, a linear scaling factor (*SF*) that could be applied to the ground motion that would result in a target value of $p_f = 0.50$. This implies that in 50% of the orientations, from 0° to 360°, the limit state strain values are exceeded, and in the other 50%, the strain limits are not exceeded. For each analysis, 80 response quantities were considered – maximum and minimum values of the strain, ε_{max} and ε_{min}, at each of the 40 reinforcing bar locations. Note that we consider maximum (positive) and minimum (negative) strains separately, as the limiting strain values are different in each case. For a given ground motion, the iterative approach involved guessing a *SF*, carrying out nonlinear response history analyses for all orientations, evaluating p_f, and iterating on *SF* until p_f was sufficiently close to the target value of 0.5. The objective of this approach was to be able to compare a scalar measure of the demand that the ground motion imposes on the structure, for an unknown initial value of the ground motion orientation, θ.

4.5 Results and Discussion

Analyses were carried out in *OpenSees* for the analysis models described in the previous section. For each ground motion, the iterative approach was used to determine a scaling factor that provided failure in 50% of the orientations. For each cross section and each suite of 20 ground motions, a normal distribution was fitted to the 20 scaling factors, which was found to give a good measure of the spread of data in each case. The median values of scaling factors are reported in Table 4.1; standard deviations and cumulative distribution functions are given in (Padilla 2010).

Figure 4.5 shows how the maximum and minimum strain values for the 40 monitored integration points vary with orientation angle of ground motion number 17, for the square and rectangular (aspect ratio 1.5) cross sections, for both suites. In each suite, this ground motion was found to give scale factors closest to the median values for each model, and is used here to illustrate typical variation of fibre strains vs. orientation. At the bottom of each plot is a histogram of $p_f(GM,\theta)$, which takes a value of 1 (shaded) at an orientation for which either of the strain limits are exceeded, and 0 where they are not. Since the records have been scaled to give $p_f(GM,\theta) = 0.5$, the bottom plots are half-shaded.

For the circular section (not shown in the figure), the strength, stiffness and period are axisymmetric, and in whichever orientation the ground motion is applied, the peak demand is the same. If the section remains elastic, this peak demand is related to the peak rotated response ordinate. For the square section, the initial elastic stiffness is the same in each direction, but strength varies, as do the maximum and minimum strains for a given curvature. Therefore, the peak strains vary with orientation angle, as shown in Fig. 4.5a and b. For the rectangular sections, as the aspect ratio of the rectangular section increases, the strains become more dependent on orientation angle, as shown in Fig. 4.5c and d.

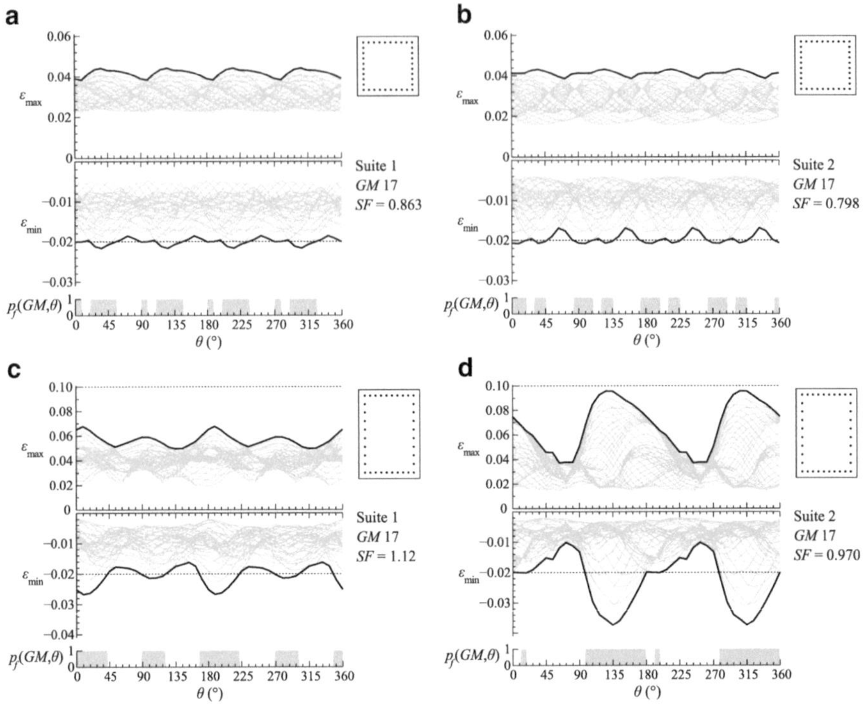

Fig. 4.5 Maximum and minimum fibre strain vs. orientation for ground motion 17 of suites 1 (**a**, and **c**) and 2 (**b** and **d**), scaled to give $p_f = 0.5$. (**a**) and (**b**) square cross section; (**c**) and (**d**) rectangular cross section with aspect ratio 1.5. Grey lines show results from individual fibres; solid black line shows strain envelope; dashed lines show strain limits ($\varepsilon_{max} = +0.10$; $\varepsilon_{min} = -0.02$). Shaded columns show orientations for which failure is indicated

Since each structural model has different strength, stiffness and vibration period, the comparison between the median scale factors for each suite is not particularly informative. More interesting is the ratio of suite 2 to suite 1 scaling factor, shown in Table 4.1. Predictably, the scaling factors required for suite 2 are lower than for suite 1, implying that suite 2 ground motions impose higher demand on the structure (a lower scaling factor is required to achieve the same probability of failure). Design carried out with a spectrum based on the geometric mean of two components, as represented by the ground motions in suite 1, is unconservative when compared to records with a realistic ratio of peak rotated to mean demand, as represented by the ground motions in suite 2. The scaling factor ratio of 0.909 for the circular pier suggests that the suite 2 records (unscaled) impose approximately 10% higher demand, on average, than the suite 1 records. Since the suite 2 records were scaled such that the elastic peak rotated demand was around 20–30% higher than the geometric mean response across the period range of interest, this is lower than expected. Referring to Fig. 4.4a, it is apparent that the orientation-independent elastic response, achieved by the spectral matching process, does not necessarily translate into orientation-independent inelastic response. For a given fibre, the maximum strains vary between around +0.03 and +0.05, and the minimum strains vary between −0.01 and −0.02. The range is slightly higher for the suite 2 data in Fig. 4.4b.

The ratio of suite 2 to suite 1 scaling factor is slightly lower for the square section than for the circular section, but generally increases with aspect ratio. Rectangular sections have stiffness, strength and period that vary with azimuth angle, and introducing this variability means that the model is in a sense less vulnerable to the peak rotated seismic demand. The difference between the highest (0.935) and lowest (0.891) scaling factor ratios is less than 5%, suggesting that there is not a statistically significant difference in the behaviour of the cross sections.

On the other hand, we would expect the ratio to lie between 0.909 (the ratio of scale factors found for the circle, which experiences the peak rotated demand for every orientation angle) and 1.0 (which would imply no sensitivity to the peak demand). Although the scale factor ratios are not significantly different in absolute terms, on the scale from purely azimuth independent to purely azimuth dependent (a single shear wall from Fig. 4.2b for example), the difference is more notable.

Another observation from Fig. 4.5 is that scaling for $p_f = 0.50$ has a very different implication for the two section shapes. For the rectangular section scaled for $p_f = 0.50$ shown in Fig. 4.5d, at certain orientations the compression strain is almost double the limiting value, whilst for the square section, the peak is about a third higher than the limit in Fig. 4.5b. This implies that if target values of p_f less than 50% were targeted, corresponding to a more conservative design assumption, the scaling factors in the rectangular section would need to be reduced, more than those for the square section. Other fractiles of the distribution of *SF* with p_f could be of interest for designers.

Finally, it is interesting to note that ASCE 7–10 contains a directionality factor for wind, K_d, which is a function of structural form and varies from 0.85 for buildings (amongst other configurations) to 0.95 for hexagonal or round chimneys, tanks and similar structures. The discussion above suggests that a similar approach could be adopted for bidirectional earthquake demand, although some of the uncertainties relating to ground motion input should be addressed, and the number of structural configurations extended, before the results from Table 4.1 can be used directly for this purpose.

4.6 Conclusions

This chapter discussed how the orientation-dependence of seismic demand should be considered in response spectrum and response history analysis. Three main sources of orientation dependence were identified – azimuth dependence of the ground motion input, variation in structural properties with orientation, and the way in which performance criteria are assessed.

To investigate this orientation dependence, a series of analyses were carried out, based on two suites of earthquake ground motions developed using the program *RspMatchBi*, with each ground motion transformed in 9° increments from 0° to 360°. An iterative approach was used, where each ground motion was scaled by a linear scaling factor, determined such that specified "failure" strains were exceeded in 50% of the orientations. The analyses investigated the difference in the response with changing orientation angle of bridge piers with circular, square and rectangular cross sections. Somewhat predictably, they showed that designs based on the geometric mean demand are unconservative when subjected to ground shaking with a realistic variation of demand with orientation. The numerical results were inconclusive in determining whether the maximum rotated response needs to be adjusted when applied to azimuth-dependent systems, such as bridge piers with rectangular sections. Further analyses, potentially with a more sophisticated failure model or more carefully controlled ground motions, would be required to establish appropriate adjustment factors, if required, similar to the directionality factors provided in ASCE 7–10 for wind loading.

References

Abrahamson NA (1992) Non-stationary spectral matching. Seismol Res Lett 63:30

Al Atik L, Abrahamson N (2010) An improved method for nonstationary spectral matching. Earthquake Spectra 26:601–617

Baker JW, Cornell CA (2006) Which spectral acceleration are you using? Earthquake Spectra 22:293–312

Beyer K, Bommer JJ (2007) Selection and scaling of real accelerograms for bi-directional loading: a review of current practice and code provisions. J Earthquake Eng 11:13–45

Boore DM, Watson-Lamprey J, Abrahamson NA (2006) Orientation-independent measures of ground motion. Bull Seismol Soc Am 96:1502–1511

Grant DN (2010) RSPMatchBi: a program for spectral matching two-component ground motions. SECED Young Engineers' Conference, London

Grant DN (2011) Response spectral matching of two horizontal ground motion components. J Struct Eng 137:289–297

Grant DN, Fenves GL, Auricchio F (2005a) Modelling and analysis of high-damping rubber bearings for the seismic protection of bridges. Research Report ROSE-2005/01, IUSS Press, Pavia

Grant DN, Blandon CA, Priestley MJN (2005b) Modelling inelastic response in direct displacement-based design. Research Report ROSE-2005/03, IUSS Press, Pavia

Hancock J, Watson-Lamprey JA, Abrahamson NA, Bommer JJ, Markatis A, McCoy E, Mendis R (2006) An improved method of matching response spectra of recorded earthquake ground motion using wavelets. J Earthquake Eng 10:67–89

Hong HP, Goda K (2007) Orientation-dependent ground-motion measure for seismic-hazard assessment. Bull Seismol Soc Am 97:1525–1538

Huang Y, Whittaker AS, Luco N (2008) Maximum spectral demands in the near-fault region. Earthquake Spectra 24:319–341

McKenna F, Fenves GL, Scott MH, Jeremić B (2000) OpenSees: open system for earthquake engineering simulation. http://opensees.berkeley.edu/. Accessed July 2010

Padilla D (2010) The effect of directionality of ground motion on structural response. MSc Project, University College London

Power M, Chiou B, Abrahamson N, Bozorgnia Y, Shantz T, Roblee C (2008) An overview of the NGA project. Earthquake Spectra 24:3–21

Priestley MJN, Calvi GM, Kowalsky MJ (2007) Displacement-based seismic design of structures. IUSS Press, Pavia

Stewart JP, Abrahamson NA, Atkinson GM, Baker JW, Boore DM, Bozorgnia Y, Campbell KW, Comartin CD, Idriss IM, Lew M, Mehrain M, Moehle JP, Naeim F, Sabol TA (2011) Representation of bi-directional ground motions for design spectra in building codes. Earthquake Spectra (in press)

Watson-Lamprey JA, Boore DM (2007) Beyond SaGMRotI: conversion to SaArb, SaSN, and SaMaxRot. Bull Seismol Soc Am 97:1511–1524

Part II
Seismic Performance Assessment, Risk Analysis and Design of Structures

Chapter 5
Some Thoughts on Methods to Compare the Seismic Performance of Alternate Structural Designs

Dimitrios Vamvatsikos

Abstract The process of structural design ultimately hinges upon the selection of the top alternative designs from a group of viable choices, ideally choosing the one that best satisfies the requirements, as set by codes or guidelines. Comparing structural configurations to find the best candidate has thus remained a favorite subject of researchers and engineers alike, especially in the case of seismic loads. With the emergence of performance-based earthquake engineering, such comparisons now need to be performed on the basis of the seismic performance, preferably at several limit-states. Such a direct evaluation can become cumbersome, requiring seismic hazard information. Therefore, shortcuts and simpler techniques have been introduced that are generally based on the concept of system fragility, as estimated through the various methods of structural analysis. Still, there is no general consensus on the metrics that can be used for such an evaluation; some researchers adopt force or displacement response quantities derived from static or dynamic methods, while others prefer to compare capacities in terms of intensity or response measures. In order to even out the field, we perform a comparative evaluation of the available choices and point out the pros and cons of each, showing some of the common fallacies that plague the results of such comparisons.

Keywords Performance-based earthquake engineering • Nonlinear static analysis • Nonlinear dynamic analysis • Probabilistic methods • Design

---

D. Vamvatsikos (✉)
Metal Structures Laboratory, School of Civil Engineering,
National Technical University of Athens, Athens, Greece
e-mail: divamva@mail.ntua.gr

M. Dolšek (ed.), *Protection of Built Environment Against Earthquakes*,
DOI 10.1007/978-94-007-1448-9_5, © Springer Science+Business Media B.V. 2011

5.1 Introduction

The performance comparison of different structural designs, be they alternate structural configurations or simply differently proportioned versions of the same type, is a common, yet little-understood operation in earthquake structural engineering, both in practice and in research. Although it may not explicitly appear in typical engineering calculations, it is a fundamental task that every professional engineer sooner or later encounters. It is essentially the basic premise of seismic design, needed to rationally select, e.g., one structural system or rehabilitation strategy over another, especially when little relevant experience is available. It is also an important tool in research where the need for comparison arises in many situations. For example, when performing automated optimal design, or ascertaining the influence of structural properties and configurations or even the improvement brought on by innovative structural systems such as dampers, base isolation, new connection details, versus simpler systems. The concept of comparison may even enter the picture from a quite deceptive angle whenever we are performing simple parametric studies that seek to discover the influence of various structural variables as when seeking the corresponding sensitivity or variability in response, often in search of the error of estimation or the propagated epistemic uncertainty.

Thus, comparisons are becoming ubiquitous but also numerous. While they may often be done on a one-to-one basis, the recent dramatic increase in the computational power available to researchers and professional engineers alike has led to a proliferation of performance comparisons. Unfortunately, this has also led to a proliferation of different metrics for comparison that may or may not lead to objective results.

For example, in many cases such comparisons are being performed on the basis of testing for a higher or lower response at a given design seismic level. It seems logical that an objective test may be achieved by subjecting the structures, either statically or dynamically, to their corresponding 10% in 50 years seismic loads and proclaiming the one with the lower global or local response(s) the winner. While deceptively intuitive, such approaches can only provide some evidence in favor of one structure but no conclusive proof. The recent emergence of the performance-basis concept within a probabilistic framework has shown that the prevalence of one system over another can only be proven satisfactorily in terms of their actual seismic performance which is not perfectly correlated with the response at a single intensity level. Thus, seemingly intuitive methods that lack one-to-one connection to performance lack in resolving power and will often mislead us.

The proper question then arises of how to quantify performance and what metric to use to enable reliable comparisons. In the sections that follow we will adopt a top-down approach for our investigation. Starting from the very ideal of what a proper performance-based comparison should be (or become in the future), we will slowly start hacking away at our elaborate ideal to arrive at the bare minimum basis that one can use to achieve a meaningful comparison. At the same time we will also elaborate on several techniques that have appeared in the literature, trying to tie them into a hierarchy of comparison methods and assessing the true extent of their resolving power by using simple practical examples or even counter-examples to illustrate any weaknesses.

5.2 Life Cycle Cost, Annualized Losses and Limit-State MAFs

Perhaps there is no better way to enable a performance-based comparison than using risk engineering concepts to estimate a total ownership cost, i.e., a lifecycle cost, for the structures involved. This involves estimating the initial costs plus any insurance, repair, inspection and maintenance costs, including time-to-repair and any associated rental costs in case of a forced relocation. If casualties can also be monetized, a difficult but still common concept in risk engineering, then we can add up all the costs and convert them to a net present value. This will actually become a probabilistic distribution of a single scalar variable that will accurately characterize the building performance for financial decision-making. Typically, taking the expected value as the sole metric means that the entire performance is reduced to a single scalar.

While such an estimate will allow a simple comparison of scalar values, e.g. see Lagaros et al. (2006), it will come at the expense of a complex effort that is by no means trivial, even for academic research. Thus, on a somewhat simpler level than full lifecycle costing, one may choose to adopt a slightly less holistic approach to performance metrics. This typically entails the assumption of a Poisson model for earthquake events and foregoing any initial, maintenance and operational costs to allow expressing the structural performance directly via annualized earthquake-related losses. These may be quantified, e.g., by the triptych of repair costs, downtime and casualties that has been adopted by the Pacific Earthquake Engineering Research (PEER) Center (Yang et al. 2009). Such a methodology is best exemplified by the PEER Center equation (Cornell and Krawinkler 2000):

$$\lambda(D_V) = \iiint G(D_V|D_M) \cdot |\mathrm{d}G(D_M|E_{DP})| \cdot |\mathrm{d}G(E_{DP}|I_M)| \cdot |\mathrm{d}\lambda(I_M)| \tag{5.1}$$

D_V is a single or a vector of decision variables, such as cost, time-to-repair or human casualties that are meant to enable decision making by the stakeholders. D_M represents the damage measures, typically discretized in a number of Damage States (e.g. red/yellow/green) of structural and non-structural elements and even building contents. E_{DP} contains the engineering demand parameters such as interstory drift or peak floor acceleration and I_M is the seismic intensity, for example represented by the 5%-damped first-mode spectral acceleration $S_a(T_1,5\%)$. The function $\lambda(y)$ provides the mean annual frequency (MAF) of exceedance of values of its operand y, while $G(x)$ is the complementary cumulative distribution function (CCDF) of variable x.

By adopting the three D_V's of repair costs, downtime and casualties, this seemingly becomes a multi-criteria approach where the performance of each candidate design should be tested against the other designs for three different values which come with their own distributions. Nevertheless, a safer structure will in general offer improvements in all three categories: the more the repairs, the costlier they will be, the longer they will take and the more casualties they are likely to cause. This apparently strong correlation will probably make any such D_V-based comparison a relatively simple matter indeed. What remains problematic is the complexity of estimating each of these three quantities. While being a lesser problem than estimating the lifecycle cost, it remains a highly difficult operation that is not foreseen to become common practice in the near future.

Defining performance without involving any decision variable D_V or the closely related damage measure D_M makes sense for many engineers. Engineering quantities may be much preferable, especially when working at the level of a design office, to discern which structure is outperforming the rest. This may be best achieved by moving to the familiar territory of limit-states by appropriately modifying the PEER framework, as shown by Vamvatsikos and Cornell (2004) (see also Vamvatsikos and Dolšek (2011) for an alternative derivation from first principles). Defining D_V and D_M to be simple indicator variables that become one when a given limit-state (LS) is exceeded, transforms Eq. 5.1 to estimate λ_{LS}, the MAF of violating LS:

$$\lambda_{\mathrm{LS}} = \int_0^{+\infty}\int_0^{+\infty} F(E_{DPc}\,|\,E_{DP}) f(E_{DP}\,|\,I_M)\,\mathrm{d}E_{DP} \left|\frac{\mathrm{d}\lambda(I_M)}{\mathrm{d}I_M}\right| \mathrm{d}I_M \tag{5.2}$$

where F is the cumulative distribution function (CDF), f the probability density function (PDF) and E_{DPc} is the limit-state capacity expressed in terms of the engineering demand parameter. Further simplification may be achieved by integrating out the E_{DP}, to offer a fundamental result where the CDF of I_{Mc}, i.e., of the limit-state capacity expressed in terms of the intensity measure, is multiplied by the absolute slope of the hazard curve $\lambda(I_M)$ and integrated to produce the limit-state MAF:

$$\lambda_{\mathrm{LS}} = \int_0^{+\infty} F(I_{Mc}\,|\,I_M) \left|\frac{\mathrm{d}\lambda(I_M)}{\mathrm{d}I_M}\right| \mathrm{d}I_M \tag{5.3}$$

This MAF is going to be our engineering-level indicator, or metric, of performance. While deeply rooted into a performance-basis it retains a natural connection with the familiar concept of the probability of violating a certain performance level or limit-state. For example, the well-known 10% probability of exceedance in 50 years for a Life Safety limit-state will, via the Poisson assumption for seismic events, corresponds directly to $\lambda = -\ln(1 - 0.1)/50 = 0.21\%$, a threshold MAF value that can be compared to results derived via Eqs. 5.2 or 5.3.

Under the premise of multiple limit-state design, such operations will naturally have to become multi-criteria comparisons, essentially requiring testing across a range of different limit-states and with varying performance results. Weighing such criteria together is an issue best left to the individual situation at hand: It is difficult to say what is preferable, e.g., better performance in an Immediate Occupancy setting or in a Life-Safety one. Unless there is a clear superiority of one design over the others (i.e., it bests them in all limit-states) there is going to be a trade-off when choosing one building over another. One way to resolve this would be to move back to the higher level of a D_V basis. Alternatively, it can also be decided by directly selecting the desirable compromise, e.g., of a better performance for frequent events versus a slightly worse performance for rarer ones, based on the intended use of the structure. This has been implied by several guidelines, e.g., SEAOC Vision 2000 (1995) and has been actively performed by engineers and researchers alike since the very beginning of performance-based earthquake engineering.

Thus, in our opinion, if a single value is sought to perform engineering-level alternate design comparisons for a given limit-state, then the current best compromise between ease-of-computation and relevance for performance is using the MAF of the limit-state. For example, such comparisons were used effectively by Vamvatsikos (2009) where Pareto optimal design of a typical highway overpass bridge was performed using the MAF of global collapse to judge the seismic performance of each candidate design and achieve a desirable compromise with initial cost. Also, Tagawa et al. (2008) used the drift hazard concept (Cornell et al. 2002) or equivalently the MAF of a continuum of limit-states defined on the basis of interstory drift to compare one-way versus two-way designs of steel moment-resisting frames.

Naturally, when using more than one limit-state it is not possible to resolve such multiple comparisons in a general sense; we must set apart the multi-criteria problem to be resolved with the specific information available in each individual case. From here and on we will focus instead on the simplest underlying problem that lies behind each such comparison: Comparing at a single limit-state in a manner that is directly comparable to the corresponding MAFs.

5.3 Comparisons in IDA-Space

While using the MAF of the limit-state is a very powerful way to achieve a robust comparison, there are significant disadvantages that may preclude the wider use of such a method, the most important being the need for seismic hazard information. While it is conceivable, many people would not agree that a comparison of two different designs might shift one way or another based on the site characteristics. It may happen for two relatively close candidates when comparing them at two very different sites, but that is probably not the issue that will trouble most comparisons. Therefore the logical question arrives of whether we can drop the hazard info and focus just on fragility-level information, i.e., on the CDF of the limit-state I_M -capacity in Eq. 5.3.

This is further motivated by a simplification of the integrals of Eqs. 5.2 and 5.3 into the closed-form solutions developed for SAC/FEMA by Cornell et al. (2002). Thus, if $H(.)$ is the hazard curve function of the scalar I_M which, for reasons of conforming to the original and without loss of generality, is chosen to be S_a , then it can be approximated as

$$H(S_a) \cong k_0 (S_a)^{-k}. \tag{5.4}$$

We need also assume that the relationship of I_M and E_{DP}, which is represented by the maximum interstory drift θ (but can be any response parameter), is approximately a power law:

$$S_a \cong a \cdot \theta^b. \tag{5.5}$$

Then, if θ_c is the median E_{DP} capacity and $\beta_{RU\theta}$ is the corresponding total dispersion (standard deviation of the logarithm of the data) due to both aleatory randomness and epistemic uncertainty, Eq. 5.2 becomes

$$\lambda_{\mathrm{LS}} \cong H\left[\left(\frac{\hat{\theta}_c}{a}\right)^{\frac{1}{b}}\right] \exp\left(\frac{k^2}{2b^2}\beta^2_{RU\theta}\right). \tag{5.6}$$

Similarly, by employing only the hazard curve approximation of Eq. 5.4, if S_{ac} is the median I_M capacity, and β_{RUSa} is the corresponding total dispersion, Eq. 5.3 turns to

$$\lambda_{\mathrm{LS}} \cong H(\hat{S}_{ac}) \exp\left(\frac{k^2}{2}\beta^2_{RUSa}\right). \tag{5.7}$$

Thus, a MAF comparison may be performed either by numerically integrating Eqs. 5.2 or 5.3 or by using the approximate SAC/FEMA closed-form solutions. Still, some care needs to be exercised when doing so because of the assumptions included. Some, like homoscedasticity and lognormality of the I_M and E_{DP} capacities are easy to satisfy in the area of interest. On the other hand, the functional approximations employed are more difficult to justify. These closed-form solutions are sensitive to the way that the fitting is performed for deriving the constants in the power law approximations of Eqs. 5.4 and 5.5. Especially regarding the hazard curve, small changes in the capacities may produce disproportionately large changes in the MAF estimated, especially if a point-fit is performed at the median capacity, as proposed in Cornell et al. (2002). A local fit within [0.25, 1.25] of the median S_a capacity will in general improve the robustness of the fit (see Dolšek and Fajfar 2008 or Vamvatsikos and Dolšek 2011).

In general, if we trust the approximation of Eqs. 5.6–5.7, then it all depends on the few variables present in the equations, most notably the median I_M or E_{DP} capacities, the associated dispersions and the constants k and b related to the hazard curve and

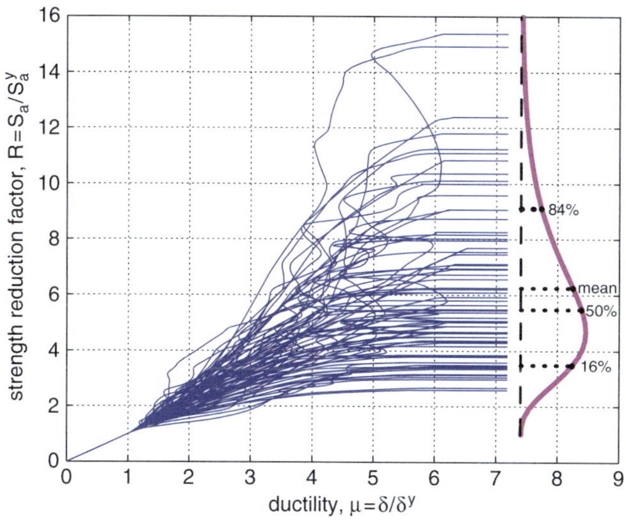

Fig. 5.1 IDA curves and corresponding collapse I_M capacity distribution for an oscillator in terms of the strength ratio R given the ductility μ (I_M given E_{DP}) for a single-degree-of-freedom system (From Vamvatsikos et al. 2009)

the $I_M - E_{DP}$ relationship. In other words, it all depends on the distributions of capacity and one or two constants depending on whether we employ Eqs. 5.7 or 5.6, respectively. Assuming that the two candidate designs allow us to use the same intensity measure (e.g. S_a on the same period) we can envision making our comparisons directly on the $I_M - E_{DP}$ plane.

Considering that incremental dynamic analysis (Vamvatsikos and Cornell 2002) is arguably the prime method for establishing in detail the complex $I_M - E_{DP}$ relationship for any given structure, we will call such a methodology a comparison in IDA-space, as envisioned in Fig. 5.1. Invariably, this should serve as a reminder that despite the simplifications achieved coming down from the top of the comparison pyramid that has been established, such a comparison may still entail numerous nonlinear dynamic analyses under a multiply-scaled suite of ground motion records. Simpler methods do exist that based on Eq. 5.3 are able to approximate the limit-state MAF without using full IDA. Instead, they employ only a few nonlinear dynamic analyses at one or two levels (stripes) of a given I_M (Jalayer and Cornell 2009) but at the heavy cost of reduced accuracy. Therefore, if uncompromised resolving power is what we are after, IDA or an equivalent multi-stripe analysis (Jalayer and Cornell 2009) is a one way street.

5.3.1 Comparing in IM Terms – Theory and Practice

In general, comparing in terms of the capacity under a given I_M is a multi-faceted issue. First of all, while recent work has shown that vector or composite I_M's can be very useful for reasons of efficiency and sufficiency (Baker and Cornell 2005; Vamvatsikos and Cornell 2005; Luco and Cornell 2007) such complex I_M's are problematic for our purposes. For example, when dealing with vectors, they cannot be compared by definition. Thus, options like $S_a(T_1)$ and epsilon (Baker and Cornell 2005) or a set of two S_a's at different periods (Vamvatsikos and Cornell 2005) become impossible to use. In addition, we cannot easily compare structure-dependent I_M's, like the inelastic spectral displacement $S_{di}(T_1)$ (Tothong and Cornell 2008), since it heavily depends on the equivalent single-degree-of-freedom (SDOF) force-deformation relationship that is building-specific. Actually, even the simple $S_a(T_1)$ is structure-dependent and needs a common period selection, rather than the actual first-mode period of each structure, to become a standard for comparison.

Perhaps one possibility is to fall back to the use of peak ground acceleration (PGA) or velocity (PGV). While this may seem as a good idea it can easily become a thorny sufficiency issue and thus remove any accuracy from our approach on the basis of invalidating even a MAF approach on such an I_M, unless of course the structure is in a period/response range were their sufficiency might be adequate. As a balance between the difficult requirements of comparability and sufficiency (i.e., ease versus validity of the comparison) we propose the use of $S_a(T)$ that can be made to work if we shift both structures to a common period T that lies between their actual fundamental periods (Vamvatsikos and Cornell 2005). Of course it would be best if we avoid large period changes, as other sufficiency questions will again come into play. Barring such issues, any I_M-capacity-basis comparison is essentially a fragility-based comparison, and can be easily resolved by comparing the CDFs of capacity.

On the premise of a common $S_a(T)$ description, we can use the exact Eq. 5.3 to compare the performance of structures 1 and 2 at any given limit-state LS. If there is a large difference between the median S_{ac} of the two structures, say $S_{ac1} > S_{ac2}$ and assuming that the dispersion β_{RUSa1} of S_{ac1} is not extremely larger than β_{RUSa2}, or in other words the CDF of S_{ac1} is nearly always to the right of the CDF of S_{ac2}, then it is quite obvious from Eq. 5.3 that due to the monotonically decreasing nature of the hazard curve (and the monotonically increasing nature of its absolute derivative), the exceedance probabilities of S_{ac2} will always be multiplied by higher slope values and always lead to higher MAFs. Therefore, in this situation the inequality of the median capacities directly translates to an inverse inequality of the corresponding MAFs.

If there is only a small difference between the median capacities of the two structures, then the above argument cannot be used any more. Nevertheless, we can now take advantage of the approximate Eq. 5.7, as the proximity of the capacities allows the same hazard curve approximation to be employed. Since there is now a common description of the hazard via Eq. 5.4 for both structures, the following result is easily derived:

$$\lambda_{\mathrm{LS1}} < \lambda_{\mathrm{LS2}} \Leftrightarrow \hat{S}_{ac,1} > \hat{S}_{ac,2} \cdot \exp\left[\frac{k}{2}\left(\beta^2_{RUSa1} - \beta^2_{RUSa2}\right)\right] \tag{5.8}$$

Table 5.1 Values of the I_M exponential factor in Eq. 5.8 for a hazard slope of $k = 3$

		β_{RUSa2}										
		0.30	0.35	0.40	0.45	0.50	0.55	0.60	0.65	0.70	0.75	0.80
β_{RUSa1}	0.30	1.00	0.95	0.90	0.84	0.79	0.73	0.67	0.61	0.55	0.49	0.44
	0.35	1.05	1.00	0.95	0.89	0.83	0.76	0.70	0.64	0.58	0.52	0.46
	0.40	1.11	1.06	1.00	0.94	0.87	0.81	0.74	0.67	0.61	0.55	0.49
	0.45	1.18	1.13	1.07	1.00	0.93	0.86	0.79	0.72	0.65	0.58	0.52
	0.50	1.27	1.21	1.14	1.07	1.00	0.92	0.85	0.77	0.70	0.63	0.56
	0.55	1.38	1.31	1.24	1.16	1.08	1.00	0.92	0.84	0.75	0.68	0.60
	0.60	1.50	1.43	1.35	1.27	1.18	1.09	1.00	0.91	0.82	0.74	0.66
	0.65	1.65	1.57	1.48	1.39	1.30	1.20	1.10	1.00	0.90	0.81	0.72
	0.70	1.82	1.74	1.64	1.54	1.43	1.32	1.22	1.11	1.00	0.90	0.80
	0.75	2.03	1.93	1.83	1.72	1.60	1.48	1.35	1.23	1.11	1.00	0.89
	0.80	2.28	2.17	2.05	1.93	1.79	1.66	1.52	1.39	1.25	1.12	1.00

Therefore the hazard curve may influence the results only if the dispersions are different. If testing for structure 1 being better than structure 2, as implied by the above equation, then the higher the dispersion of the capacity in structure 2 relatively to 1, the easier it is to get the inequality we are seeking. If, on the other hand, the dispersions are relatively close together, say within 5–8%, e.g. 0.45 versus 0.4, then Table 5.1 shows that the influence of the hazard disappears for all practical purposes, being at most equal to 10%. Higher dispersions for the high-capacity structure actually heavily favor the seemingly inferior structure, as they disproportionately increase the probability of failure for the former.

Summing up, the only doubt cast on a pure I_M-based comparison appears when one of the structures combines a slightly higher median capacity (say up to 20% higher) with significantly higher capacity dispersion than its competition. Then, one needs to resort back to MAFs to make sense of the comparison. Otherwise, the inequality of the median capacities is directly linked to a performance inequality. Since relatively similar structural configurations at nearby periods are generally expected to have very similar dispersions, there is every reason to support the validity of an I_M-basis for comparisons.

This approach is best exemplified by the work of Fragiadakis et al. (2006) where the ratios of the median S_a-capacities were used to qualitatively and quantitatively measure the influence of mass, strength and stiffness irregularities along the height of a 9-story steel frame (Fig. 5.2). Their application also included the use of bootstrapped confidence intervals (Fig. 5.3) on the capacity ratios to take into account the influence of sample size, i.e., of the limited number of ground motion records employed for IDA. While this factor often tends to be neglected, it is a standard in statistics. Its inclusion is heavily recommended as many close comparisons are often rendered completely inconclusive due to sample size effects.

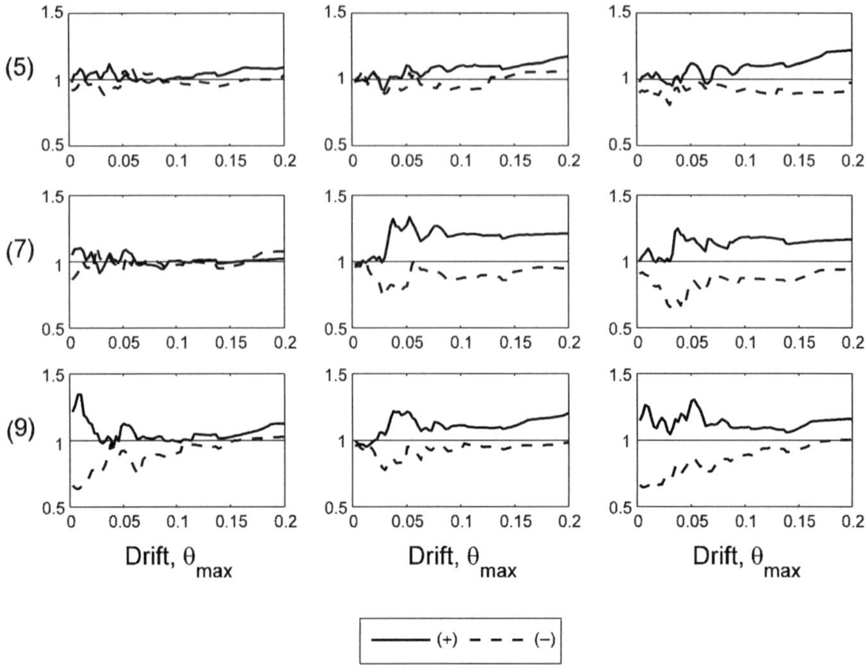

Fig. 5.2 Irregularities: Changes in stiffness (*left*), strength (*center*), or both (*right*), for stories 5, 7 and 9 of a 9-story steel moment-resisting frame. Showing the ratio of I_M capacities relative to the unmodified building for both upgraded and degraded cases (From Fragiadakis et al. 2006)

5.3.2 Comparing in EDP Terms – Capacity Basis

Prehaps the most important problem with an I_M comparison is the apparent difficulty in obtaining the distribution parameters on a "required I_M" basis. From one point this necessitates the use of some clever postprocessing (Vamvatsikos and Cornell 2004) to extract the needed information. It may also be argued that the concept of an I_M capacity is still foreign to engineers, while an E_{DP} capacity is far more intuitive. It thus makes sense to ask whether it would be possible to make comparisons on the basis of the median E_{DP} capacity.

The concept of the E_{DP} capacity was introduced through SAC/FEMA (2000) as a convenient way to represent the capacity of structures via Eq. 5.6. Through this closed-form solution it is obvious that E_{DP} capacity is related to the MAF, albeit needing one more constant than an I_M basis, namely the parameter b that relates to the shape of the median I_M–E_{DP} relationship. Unfortunately there are many problems plaguing such a comparison than just this constant.

One major issue in this approach is the definition of the E_{DP} capacity itself. Even according to SAC/FEMA (2000), for most limit-states the E_{DP} capacities are essentially fixed values, same for a whole class of structures. For example, for an Immediate Occupancy limit-state there is a uniform 2% maximum interstory drift capacity for

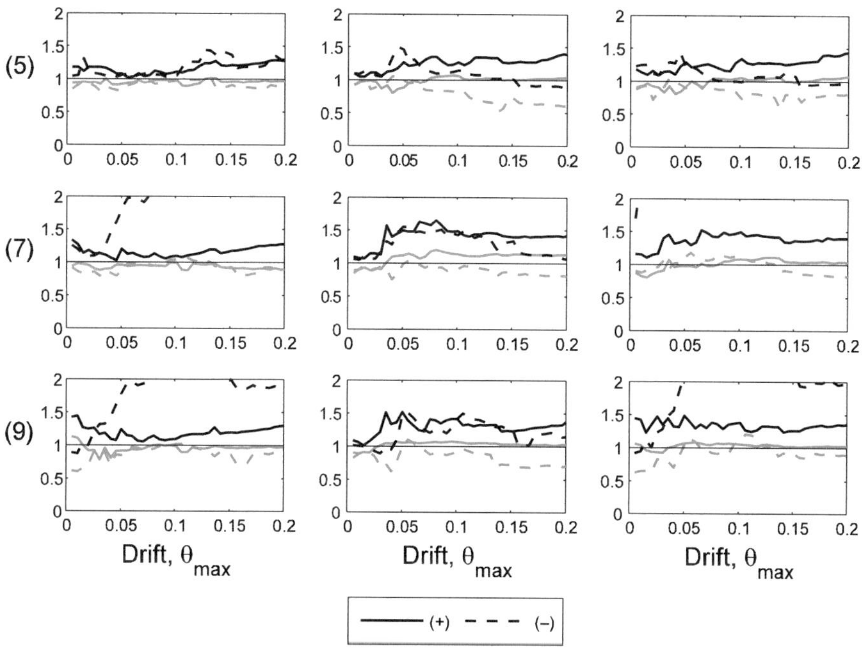

Fig. 5.3 Changes in stiffness (*left*), strength (*center*), or both (*right*) for stories 5, 7 and 9 of a 9-story steel moment-resisting frame. Light gray lines indicate the lower bound and darker ones the upper bound of the 90% confidence interval (From Fragiadakis et al. 2006)

all steel moment-resisting frames. Similarly, more recent guidelines specify fixed beam or column plastic rotation capacities that only differ among different classes of buildings, typically based on a ductility classification (e.g. CEN 2005). Clearly no comparison can be made on the basis of such constant-value definitions. Still, it may seem attractive to use them if they vary from building to building.

There are indeed cases where building-specific E_{DP} capacities have been defined in a way that is characteristic of the performance of structures. That is the case of the SAC/FEMA Collapse Prevention (CP) limit-state, where the "drift capacity" is defined as the point where the IDA curve softens to less than 20% of its initial elastic slope with a maximum allowable capacity value of 0.10. On this basis, capacity-points can be defined either on each IDA curve individually, as done in Fig. 5.4 for one record and Fig 5.5 for 20, or on the median curve itself, as shown in Fig. 5.6 for two idealized median curves. The differences between the two modes of application are insignificant and the final results are quite similar in both cases. Despite the various issues with such a definition itself (Vamvatsikos and Cornell 2004), it may nevertheless open a door for a useful, period-free measure of performance that can be used for comparison. Actually, although not explicitly used for this purpose, such values of drift capacities are catalogued in the work of Liao et al. (2007) and Huang and Foutch (2009), where they may be construed as

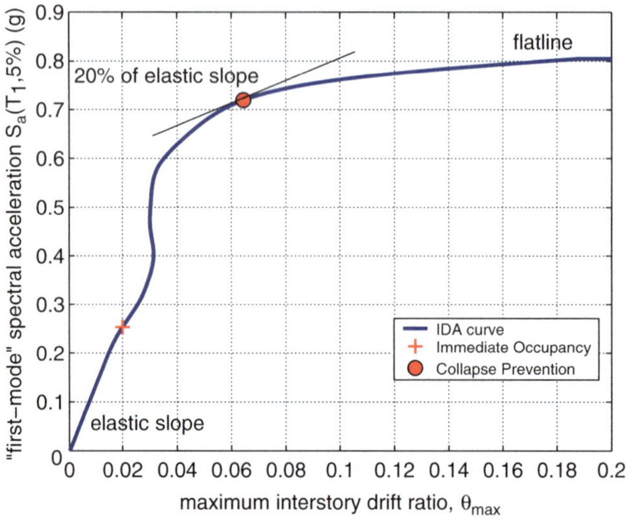

Fig. 5.4 The definition of limit-state capacities for IO and CP according to SAC/FEMA on a single IDA curve. Both the I_M and E_{DP} capacity values can be extracted from this graph (From Vamvatsikos and Cornell 2004)

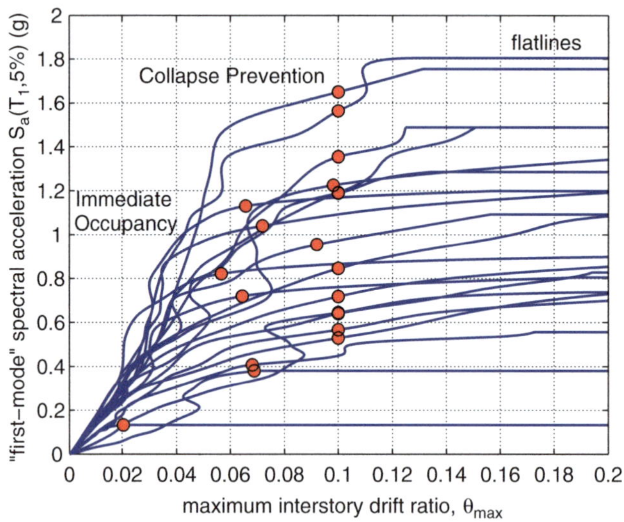

Fig. 5.5 The definition of limit-state capacities CP according to SAC/FEMA for multiple IDA curves of a single building (From Vamvatsikos and Cornell 2004)

indicative of the collapse potential of different structural systems or configurations of the same system.

Unfortunately, as Fig. 5.6 shows, an E_{DP}-capacity basis for comparison may prove to be misleading. Therein we show the idealized median IDA curves of two

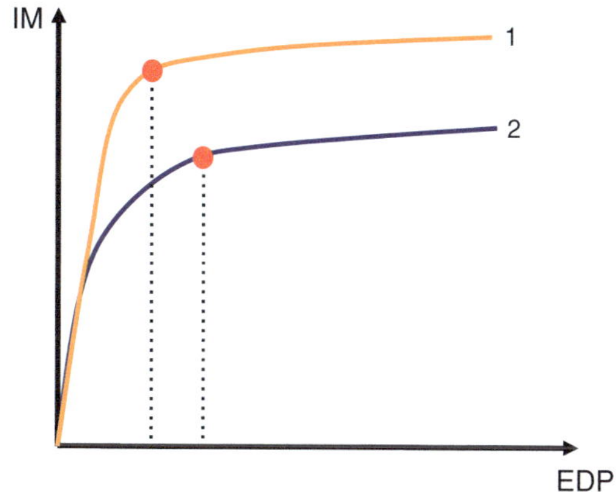

Fig. 5.6 Two idealized median IDAs with different I_M and E_{DP} capacities

different systems that have the same period and initial elastic slope but quite different flatlines, i.e., different I_M-levels where global dynamic instability manifests itself. These IDA curves are shown to have slight differences in the curvature as they rise towards their flatlines that force the E_{DP} capacity of system 1 to be lower than that of 2, although the relation of their median I_M capacities is reversed. Therefore, based on our previous discussion on I_M-comparisons, system 1 will clearly outperform system 2. A comparison based on their E_{DP} capacities would erroneously indicate otherwise. In other words, using this simple counter-example, we can show that E_{DP}-based comparisons can be severely flawed.

Let us now seek the theoretical reason behind this discrepancy. Following the same route we took to establish the validity of the I_M-comparison, we can assume that as shown in Fig. 5.6 the systems have the same period and their capacities are close enough to allow us using the same hazard curve approximation via Eq. 5.4. Then, using the SAC/FEMA approximation of Eq. 5.6, and if we let a_i, b_i $(i = 1{,}2)$ represent the IDA-shape parameters of Eq. 5.4 for each of the two systems, we arrive at the following requirement:

$$\lambda_{\mathrm{LS}1} < \lambda_{\mathrm{LS}2} \Leftrightarrow \hat{\theta}_{c,1} > \hat{\theta}_{c,2} \cdot \left(\frac{a_1}{a_2}\right) \exp\left[\frac{k}{2}\left(\frac{\beta^2_{RU\theta 1}}{b_1} - \frac{\beta^2_{RU\theta 2}}{b_2}\right)\right] \tag{5.9}$$

The above relation clearly shows that the outcome of the comparison does not depend only on the E_{DP} capacities but also on the parameters that represent the IDA curve. In other words, the shape implied by a_i, b_i is a very important consideration that, as exemplified by Fig. 5.6, cannot be discarded, even if we assume the very simple power-law shape implied by Eq. 5.5. Furthermore, Eq. 5.6, and consequently Eq. 5.9, has another important limitation: it is not able to take into account the probability of collapse which may become important when close to the flatline. For all of the above reasons, unless under quite restrictive assumptions, any comparison on an E_{DP} basis should not be considered indicative of the actual performance and it should be avoided in general.

5.3.3 Comparing in EDP Terms – Demand Basis

Another intuitive way of comparing two different structures involves performing a single stripe analysis, for example at the design level I_M corresponding to, e.g., 10% in 50 years for a Life Safety comparison, and comparing the statistics of the E_{DP} response of structure 1 versus those of structure 2. This does away with the need for a cumbersome IDA that the previous approaches dictate, restricting the computational complexity to a single intensity level rather than multiple ones, making this a very attractive proposition.

While such a comparison may seem simple, especially if the dispersions are similar where it would point towards a comparison of mean or median E_{DP} responses, it only provides some evidence that may inconclusively favor one of the two structures. It does not allow for a definitive comparison unless the difference of the responses is disproportionately large and some severe constraints and assumptions are in place. There are two important reasons for this.

One problem is the issue of collapse, i.e. some ground motions may be found to produce "infinite" E_{DP} results due to non-convergence of the dynamic analysis that signals, on a well executed analysis and a numerically robust model, the onset of global dynamic instability. Then, the envisioned "simple" comparison of two mean or median EDP responses may become a difficult multi-criteria approach: Structure 1 may show a 3% drift and 20% probability of collapse (e.g. 4 out of 20 records have not converged), while structure 2 has 3.5% drift and 10% probability of collapse. Which one is preferable? One cannot answer such a question without resorting to an I_M-based technique. Therefore, when the probability of collapse cannot be ignored, such a single-I_M-level comparison may not even be feasible.

The most important reason, though, is that information at only a single level of intensity cannot provide much intuition on what is happening at other intensities. There is indeed some correlation to be expected among responses at nearby intensities but this does not change the fact that our information is restricted to a point-estimate of the relationship between the two structures. As even the simple Eq. 5.6 shows, that is not enough to allow us to determine the performance even for a single limit-state without some information about the shape of the complete IDA curves. It is easier to understand the problem if we realize that we cannot determine the probability of exceedance of a given E_{DP}-level only by testing at a single level of intensity, simply because of the record-to-record variability. Due to such variability there are both lower and higher I_M levels than the one tested that may cause high enough E_{DP} responses that will significantly contribute to the exceedance of the limit-state. Actually, considering that the lower I_M's are associated with disproportionately higher exceedance frequencies, they often contribute much more to the overall performance. In other words, the comparative shapes of the IDA curves, or, as discussed in the previous section, the terms a_i and b_i, remain equally important as before and it is not possible to remove them from the comparison. Unless we can make certain assumptions about their comparative values, which can be quite hazardous in many cases, we should refrain from using such narrow-range comparisons. Unless very large differences in response are involved, they are completely unreliable.

5.4 Comparisons in SPO-Space

Considering the computational load incurred by IDA, it becomes desirable to scale it down to more manageable levels. The static pushover (SPO) curve is an essential ingredient of the Nonlinear Static Procedure that has found its way in various codes in the past decade, most recently ASCE/SEI 41–06 (ASCE 2007) and Eurocode 8 (CEN 2005). Thus it has become quite familiar to engineers, and it additionally carries within itself three major advantages: The uniqueness of the curve (assuming a single load pattern is used), the relative simplicity of its generation, compared to selecting ground motion records and performing dynamic analyses, and its relatively intuitive nature. It seems natural to compare in terms of the base shear reached at a given limit-state. This can be defined at the pushover space (on the curve) in similar ways as for the IDA, e.g., when certain values of one or more E_{DP}'s are exceeded. When thus operating in the base shear V versus E_{DP} plane we will use the term SPO-space regardless of how the relevant information displayed therein was acquired (i.e. statically or dynamically). Let us now see what can and cannot be done in such coordinates.

5.4.1 Using Static Pushover Results

The main argument in favor of the SPO curve is that it clearly represents capacity. Hence it is often called the capacity curve, e.g., in FEMA-440 (ATC 2005) or the capacity boundary in FEMA-P440A (ATC 2009). Actually, engineers and researchers alike often intuitively compare SPO curves together for ductility and overstrength, thus providing a long history of such use. Unfortunately, such comparisons should be made with a lot of care.

First of all, it should be recognized that SPO-based comparisons are haphazard when there is a difference in the periods. In each Nonlinear Static Procedure (NSP) framework there is a duality of demand and capacity, and for similar base shear capacities, any difference in periods will render the demands quite different. Thus, unless we are facing very small differences in periods we cannot reliably use a SPO-based comparison. For larger period differences, some adjustment schemes have appeared, most notably by Al-Ali and Krawinkler (1998). These are cumbersome to apply as they necessitate careful manipulation of the mass and/or stiffness of the systems to bring them to a common period and allow a reasonable comparison. Nevertheless, even such meticulously prepared comparisons, while being qualitatively useful they may still lead to quantitative errors as shown by Fragiadakis et al. (2006).

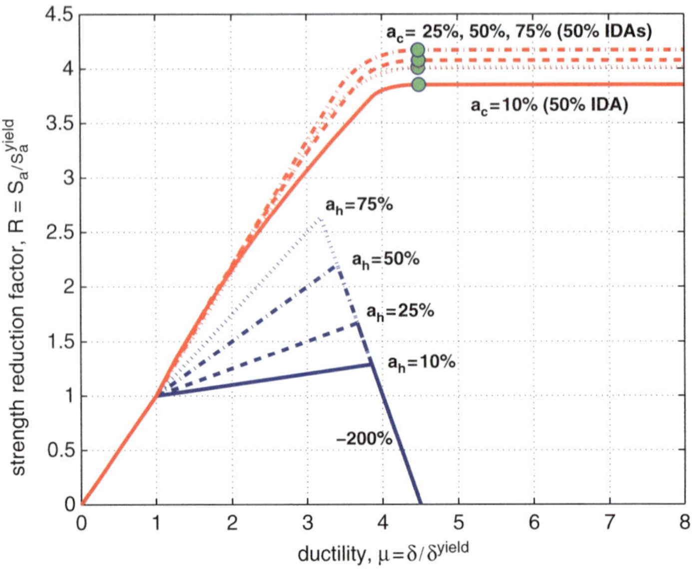

Fig. 5.7 The median IDAs produced by SPO2IDA for four different SDOF systems having the same period and coincident negative stiffness segments. Despite the large difference in the positive stiffness segments their dynamic response is practically the same (From Vamvatsikos and Cornell 2006)

Even if there is no period difference, we should be careful of SPO-based comparisons as they carry all the disadvantages associated with the SPO itself. For example, irregular structures or structures significantly influenced by higher modes cannot be captured reliably by the SPO curve, therefore making any such comparisons essentially non-robust.

To top it all, even when discussing SDOF systems where many disadvantages of the SPO completely disappear, the backbone does not always make for a perfect predictor of the dynamic response. That is not to say that the pushover cannot provide good qualitative info, but the correspondence between what appears in SPO space and how this manifests itself in IDA space can be quiet surprising. The resolving power of the SPO is not as good as many engineers might believe, and while large differences in strength and ductility in SPO space can generally be held to imply similar large differences in I_M-capacity in IDA-space, many relatively close cases are impossible to resolve via the pushover. And by close we mean cases that may appear to have a relatively wide margin among their backbone curves but often in places where such differences do not really matter, thus placing many SPO-based comparisons on a shaky ground. A prime example appears in Fig. 5.7, where four different SDOF systems having the same period and coincident negative stiffness segments are shown together with their median IDA curves. Despite the large differences in the positive stiffness segments their dynamic response is practically the same in all cases.

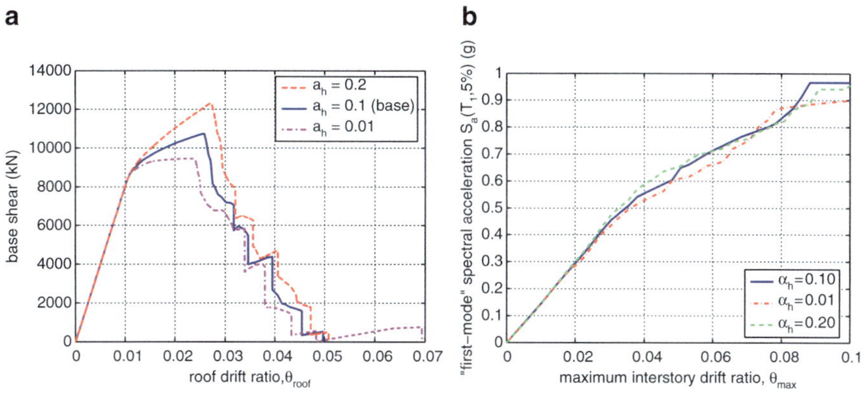

Fig. 5.8 The (**a**) SPO curves and (**b**) corresponding median IDAs from three different realizations of the same building using different properties for the beam plastic hinges. Despite the obvious differences in the post-yield region their dynamic response is indistinguishable (From Vamvatsikos and Fragiadakis 2010)

In order to help us distinguish with relative reliability any such cases, be they close together or wider apart, we need to leave the SPO space and establish a direct link at least to the reliable I_M -based comparison in IDA-space. Actually, such a link is readily available through IN2 (Dolšek and Fajfar 2005) or SPO2IDA (Vamvatsikos and Cornell 2006). The latter is a powerful R-μ-T (reduction factor-ductility-period) relationship that has been encoded in an interactive tool and can help us to make most such distinctions with relative safety. By providing the 16/50/84% IDA curves it enables a comparison of both the median and the dispersion of I_M capacity, thus enabling an I_M-based comparison as discussed in previous sections. Actually, Fig. 5.7 has been produced via SPO2IDA, exactly as a test of its power to resolve tricky cases involving close SPO curves.

Another example appears in Fig. 5.8, where the obvious differences in the maximum base shear capacity of three realizations of the same structure do not translate to IDA-space, where their median I_M capacities are practically indistinguishable. Still, the most surprising case is shown in Fig. 5.9, based on the work of De Luca et al. (2011) where a clearly enveloping piecewise-linear capacity curve is up to 25% lower in I_M-capacity than the seemingly inferior curvilinear SPO. In conclusion, while SPO-space comparisons can provide valuable information, they should be interpreted with caution.

5.4.2 *Using Dynamic Pushover Results*

An idea born out of the need to reduce the dispersion of IDA curves and somehow correlate them with the static pushover (SPO) curve, was the representation of IDA results in the SPO space (Mwafy and Elnashai 2001; Pinho et al. 2007).

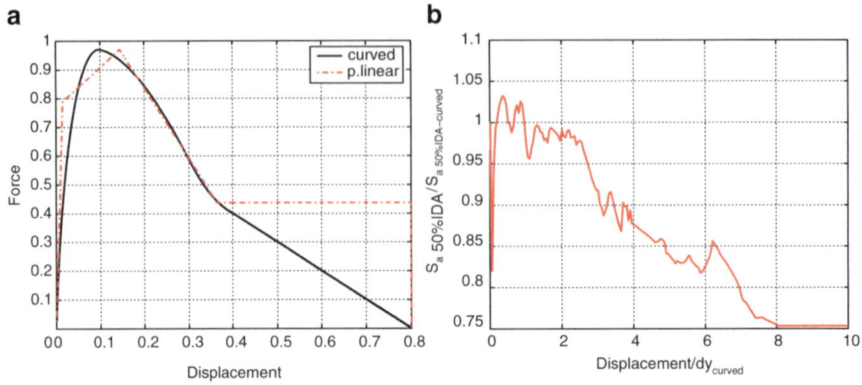

Fig. 5.9 Two SDOF systems with the same period have (**a**) a piecewise linear capacity curve enveloping the curvilinear one, but (**b**) the median I_M capacity ratios show that the "inferior" curvilinear system has up to 25% higher performance

Thus, the dynamic pushover (DPO) was born that, although based on the same underlying machinery as IDA, it comes with its own quirks. This dynamic version of the pushover obviously does away with some of the advantages of the SPO, namely the "uniqueness", assuming a single load pattern, and the absence of ground motion records and dynamic analyses. Still, it carries with it the weight of a higher level method based on dynamic results. Could this representation be used as an intuitively simpler way to compare competitive designs, retaining the self-explanatory nature of SPO combined with the accuracy of IDA?

First of all there are some issues with the definition of the dynamic pushover, regarding the selection of the base shear V versus E_{DP} (typically roof drift θ) pair of values that will be selected at each intensity level from an entire timehistory. For example one can define DPOs using the peak base shear V and the peak roof drift θ, referred as a max$V$-maxθ approach, or use the corresponding peak θ that occurs within an arbitrarily selected time window around the occurrence of max$V$. This would thus become the max$V$-corrθ approach. Similarly one can define the corr$V$-maxθ case, with the latter being the best one in terms of better matching the expected pushover curve in the post-peak region as discussed at least by Ferracuti et al. (2009).

An example appears for an SDOF system in Fig. 5.10. Therein four different SPO curves appear, having different post-peak negative stiffness, plotted together with their corresponding median IDAs. The derived DPO curve depends on its definition. If the max$V$-maxθ or max$V$-corrθ concept are employed, then the DPO curve retains the same value for base shear, equal to the maximum strength of the system, for any displacement beyond the peak of the system's backbone. On the other hand, if a corr$V$-maxθ definition is used, which is the suggested one by Ferracuti et al. (2009), then the DPO faithfully coincides with the corresponding SPO curve of each system. Even under such a definition, the SPO will differ from the DPO curve for an MDOF structure depending on the width of the time-window

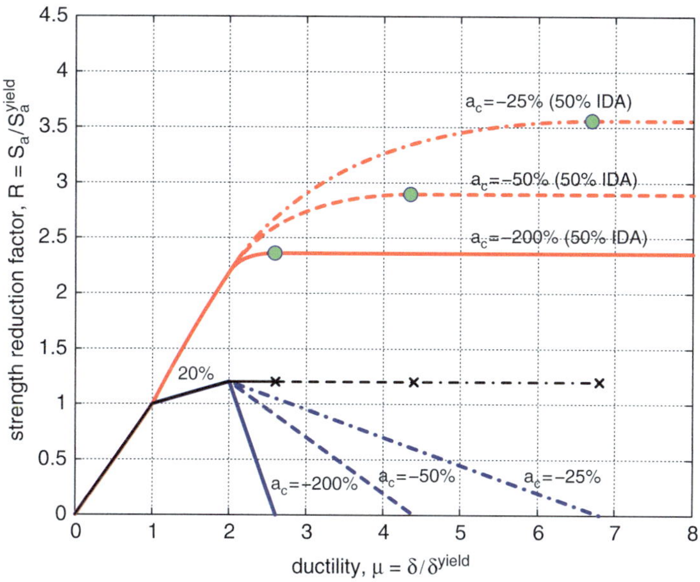

Fig. 5.10 Static pushover curves (*blue*), max$V$-maxθ dynamic pushover curve (*black*) and the median IDA curves (*red*) shown in reduction factor (normalized intensity) R versus ductility μ coordinates

employed to determine the corresponding base shear and on the load pattern used for the static pushover, as shown, e.g., by Mwafy and Elnashai (2001). The latter property has been exploited in several cases as a way to compare the validity of different pushover methods or load patterns in matching to the dynamic results.

Even if we disregard all the little issues with defining the DPO curve, its main feature is also its worst drawback: Using an SPO-space representation may bring back familiarity and remove any issues of the asymmetry or higher mode influence making the original static pushover inaccurate but it also lets in practically all the problems that plague the SPO-based comparison. It is actually worse in some sense as it needs IDA-grade data (together with the associated effort) to offer only an improved validity within the limited SPO-space.

Actually, while DPO may have indeed found its niche for comparing together different pushover patterns, this probably makes sense only at a global level. It would be equally or even more useful to perform such comparisons in IDA space or in terms of interstory drift story-wise profiles (see for example Chap. 6 by Fragiadakis et al. 2006). These are a more reliable indicator of how a load pattern impacts the structure, clearly pointing to the mechanism that forms. Dynamic pushover comparisons are not inherently wrong, but they have a low resolving power that is on par with the simple static pushover at the expense of much heavier computational load. Therefore, they had better be avoided for general applications.

5.5 Conclusions

Comparing design alternatives on the basis of performance is not straightforward, as there are a lot of metrics, each with its own implementation issues and sometimes plagued by many pitfalls and fallacies. Generally, it can be said that top level, holistic approaches such as using an all-encompassing lifecycle cost or even annualized losses in terms of one or more decision variables like repair cost, downtime or casualties, is the only sure way. While comprehensive, such approaches are very cumbersome and still remain impractical for most situations.

Thus it becomes a much needed simplification to move to an engineer-friendly structural limit-state basis, where all seismic hazard, cost and damage information is removed to leave only a clean fragility-based comparison. Such an approach can take us back to the familiar ground of seismic intensity versus structural response and it comes with many benefits but also some costs. The best compromise is using a common intensity measure, for example the spectral acceleration at a common period, to allow a comparison in terms of the median I_M-capacity. This needs IDA-level information, i.e., multiple nonlinear dynamic analyses under multiple earthquake records spanning different intensity levels, but retains all the robustness of a performance basis by being directly linked to the mean annual frequency of exceeding a limit-state.

The seemingly attractive approach of using a common E_{DP} as the basis of comparison, either in the form of E_{DP} capacity or as statistics of E_{DP} response at a single intensity level, should be avoided as it tends to be unreliable. The only simpler alternative is moving to a static pushover but it should be done with caution. Keeping in mind the issues of inaccuracy surrounding the pushover, it should not be used for comparisons at the familiar coordinates of base shear versus response but should be instead upgraded to the IDA-space of intensity versus response, for example via the SPO2IDA/IN2 approach or the IDA analysis of an equivalent oscillator. This is the only way that it can present a viable, simpler, albeit less accurate, alternative to an IDA-level comparison.

References

Al-Ali AAK, Krawinkler H (1998) Effects of vertical irregularities on seismic behaviour of building structures. Report No. 130, John A. Blume Earthquake Engineering Center, Stanford University, Stanford

ASCE (2007) Seismic rehabilitation of existing buildings, ASCE standard ASCE/SEI 41–06. American Society of Civil Engineers, Reston

ATC (2005) Improvement of nonlinear static seismic analysis procedures. Report No. FEMA-440, prepared for the Federal Emergency Management Agency, Washington, DC

ATC (2009) Effects of strength and stiffness degradation on seismic response. Report No. FEMA-P440A, prepared for the Federal Emergency Management Agency, Washington, DC

Baker JW, Cornell CA (2005) A vector-valued ground motion intensity measure consisting of spectral acceleration and epsilon. Earthq Eng Struct Dyn 34:1193–1217

CEN (2005) Eurocode 8: design of structures for earthquake resistance. EN 1998–3, European Committee for Standardisation, Brussels

Cornell CA, Krawinkler H (2000) Progress and challenges in seismic performance assessment. PEER Center News 3(2). URL http://peer.berkeley.edu/news/2000spring/index.html

Cornell CA, Jalayer F, Hamburger RO, Foutch DA (2002) Probabilistic basis for 2000 SAC federal emergency management agency steel moment frame guidelines. J Struct Eng ASCE 128 (4): 526–533

De Luca F, Vamvatsikos D, Iervolino I (2011) Near-optimal bilinear fit of capacity curves for equivalent SDOF analysis. Proceedings of the CompDyn 2011 conference on computational methods in structural dynamics and earthquake engineering, Corfu

Dolšek M, Fajfar P (2005) Simplified non-linear seismic analysis of infilled reinforced concrete frames. Earthq Eng Struct Dyn 34(1):49–66

Dolšek M, Fajfar P (2008) The effect of masonry infills on the seismic response of a four storey reinforced concrete frame–a probabilistic assessment. Eng Struct 30:3186–3192

Ferracuti B, Pinho R, Savoia M, Francia R (2009) Verification of displacement-based adaptive pushover through multi-ground motion incremental dynamic analyses. Eng Struct 31: 1789–1799

Fragiadakis M, Vamvatsikos D, Papadrakakis M (2006) Evaluation of the influence of vertical irregularities on the seismic performance of a 9-storey steel frame. Earthq Eng Struct Dyn 35(12):1489–1509

Huang Z, Foutch DA (2009) Effect of hysteresis type on drift limit for global collapse of moment frame structures under seismic loads. J Earthq Eng 13:939–964

Jalayer F, Cornell CA (2009) Alternative non-linear demand estimation methods for probability-based seismic assessments. Earthq Eng Struct Dyn 38(8):951–1052

Lagaros ND, Fotis AD, Krikos SA (2006) Assessment of seismic design procedures based on the total cost. Earthq Eng Struct Dyn 35:1381–1401

Liao K-W, Wen Y-K, Foutch DA (2007) Evaluation of 3D steel moment frames under earthquake excitations. I. Model J Struct Eng ASCE 133(3):462–470

Luco N, Cornell CA (2007) Structure-specific scalar intensity measures for near-source and ordinary earthquake ground motions. Earthq Spectra 23(2):357–392

Mwafy AM, Elnashai AS (2001) Static pushover versus dynamic collapse analysis of RC buildings. Eng Struct 23:407–424

Pinho R, Casarotti C, Antoniou S (2007) A comparison of single-run pushover analysis techniques for seismic assessment of bridges. Earthq Eng Struct Dyn 36:1347–1362

SAC/FEMA (2000) Recommended seismic design criteria for new steel moment-frame buildings. Report No. FEMA-350, prepared for the Federal Emergency Management Agency, Washington, DC

SEAOC Vision 2000 (1995) A framework of performance-based seismic engineering of buildings. Structural Engineers Association of California, Sacramento

Tagawa H, MacRae G, Lowes L (2008) Probabilistic evaluation of seismic performance of 3-story 3D one- and two-way steel moment-frame structures. Earthq Eng Struct Dyn 37:681–696

Tothong P, Cornell CA (2008) Structural performance assessment under near-source pulse-like ground motions using advanced ground motion intensity measures. Earthq Eng Struct Dyn 37:1013–1037

Vamvatsikos D (2009) Optimal multi-objective seismic design of a highway bridge by selective use of nonlinear static and dynamic analyses. Proceedings of the 9th international conference on structural safety and reliability (ICOSSAR), Osaka

Vamvatsikos D, Cornell CA (2002) Incremental dynamic analysis. Earthq Eng Struct Dyn 31(3):491–514

Vamvatsikos D, Cornell CA (2004) Applied incremental dynamic analysis. Earthq Spectra 20(2):523–553

Vamvatsikos D, Cornell CA (2005) Developing efficient scalar and vector intensity measures for IDA capacity estimation by incorporating elastic spectral shape information. Earthq Eng Struct Dyn 34(13):1573–1600

Vamvatsikos D, Cornell CA (2006) Direct estimation of the seismic demand and capacity of oscillators with multi-linear static pushovers through incremental dynamic analysis. Earthq Eng Struct Dyn 35(9):1097–1117
Vamvatsikos D, Dolšek M (2011) Equivalent constant rates for performance-based assessment of ageing structures. Struct Saf 33(1):8–18
Vamvatsikos D, Fragiadakis M (2010) Incremental dynamic analysis for estimating seismic performance uncertainty and sensitivity. Earthq Eng Struct Dyn 39(2):141–163
Vamvatsikos D, Miranda E, Akkar S (2009) Estimating the dynamic instability of oscillators with non-trivial backbones. Proceedings of the COMPDYN2009 conference on computational methods in structural dynamics and earthquake engineering, Rhodes
Yang TY, Moehle J, Stojadinovic B, Der Kiureghian A (2009) Seismic performance evaluation of facilities: methodology and implementation. J Struct Eng ASCE 135(10):1146–1154

Chapter 6
Static Versus Dynamic Methods of Analysis for Estimating Seismic Performance

Michalis Fragiadakis, Dimitrios Vamvatsikos, and Mark Aschheim

Abstract Nonlinear static methods are evaluated and compared with nonlinear dynamic methods for estimating the seismic performance of structures. Emphasis is given on assessing the applicability of nonlinear static methods for RC buildings, and on comparing the building's capacity obtained using nonlinear static and nonlinear response history analysis. The first task refers to the ability of alternative static pushover-based methods to estimate the response at the level of a member or of a story. Plain as well as more elaborate pushover methods such as the Modal Pushover Analysis method and the Consecutive Modal Pushover method are included in our evaluation. The second task refers to the qualitative comparison at the global level between static pushover and nonlinear response history analysis when either the static pushover or the Incremental Dynamic Analysis (IDA) setting is adopted. When the static pushover setting is adopted, we show that nonlinear static methods can be compared with the IDA curve when the base shear instead of spectral acceleration is plotted on the ordinates, while the dispersion among the single-record IDAs is considerably reduced. Alternatively, the comparison can be performed within the IDA setting if appropriate R-C_1-T relationships, simplified or more advanced (e.g. SPO2IDA), are adopted. Each setting shows different qualitative characteristics of the two seismic performance estimation approaches and has different practical applications.

M. Fragiadakis (✉)
Department of Civil and Environmental Engineering, University of Cyprus, Nicosia, Cyprus
e-mail: mfrag@mail.ntua.gr

D. Vamvatsikos
Metal Structures Laboratory, School of Civil Engineering,
National Technical University of Athens, Athens, Greece
e-mail: divamva@mail.ntua.gr

M. Aschheim
Department of Civil Engineering, Santa Clara University, Santa Clara, CA, USA
e-mail: maschheim@scu.edu

M. Dolšek (ed.), *Protection of Built Environment Against Earthquakes*,
DOI 10.1007/978-94-007-1448-9_6,

Keywords Seismic performance assessment • Static pushover • Nonlinear response history analysis • Incremental Dynamic Analysis • Modal pushover analysis • Dynamic pushover

6.1 Introduction

Although Nonlinear Response History Analysis (NRHA) is arguably the most rigorous analysis method available, Nonlinear Static Procedures (NSPs) are still popular for the analysis and design of building structures. This stems from several reasons which include the computing cost of NRHA, the difficulty of choosing appropriate ground motions and the fact that NSP is closer to the simplified elastic methods traditionally relied upon by seismic design codes worldwide.

The scope of this chapter is to provide a qualitative comparison between static and dynamic seismic performance assessment approaches. Initially we compare common NSP methods, such as those suggested in the ASCE/SEI 41-06 (ASCE 2007) guidelines and also more elaborate NSP methods that attempt to enhance the capability of the typical NSP approach in order to provide results closer to those from NRHA. This comparison has been part of a study carried out in the framework of ATC 76-6 (Consultants Joint Venture 2010) guidelines where various NSP methods were compared with respect to their ability to provide estimates of the demand at the member level, as measured using different Engineering Demand Parameters (EDPs). Given the inability of simple nonlinear static methods of analysis to reflect the diversity of response apparent in NRHA, the accuracy of NSP methods is evaluated relative to the central tendency (mean or median) of the chosen EDPs.

Furthermore, our discussion extends to comparing the results of NSP and NRHA on a global level, i.e. comparing the structure's capacity curves. NRHA is considered in the form of Incremental Dynamic Analysis (IDA) in order to obtain a measure of the "dynamic" capacity of the structure. It is demonstrated that the comparison can be performed either in the framework of static methods or alternatively in the framework of IDA. In the former case the comparison can performed if the IDA capacities are presented using an appropriate parameter on the ordinate, while in the latter case an R-C_1-T (or R-μ-T) relationship (such as SPO2IDA (Vamvatsikos and Cornell 2006) or IN2 (Dolsek and Fajfar 2005)) is necessary to bring the results of the NSP analysis to the setting of IDA.

6.2 Archetype Models

The structural systems considered are a two-story, a four-story and an eight-story reinforced concrete moment resisting frame (MRF) buildings. All three buildings have three bays and are completely regular and symmetric, thus, offering an ideal setting for the NSP methods to reproduce the results of NRHA. The buildings

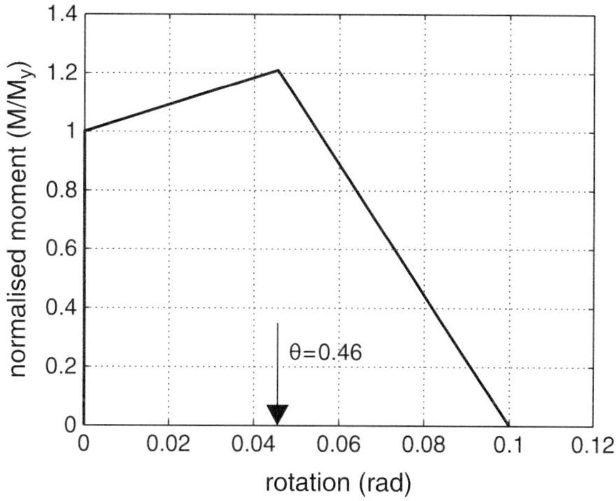

Fig. 6.1 Typical moment-rotation relationship used for the plastic-hinge part of the beam-column component models

were designed as special RC moment frames following the provisions of the 2003 IBC (International Code Council 2002). The member sizes were determined by minimum size requirements and the required column-to-beam strength ratio compatibility, in addition to joint shear requirements. The column strengths were determined on the basis of the strong-column-weak-beam philosophy. The selection of the beam stirrups was controlled by shear capacity design, while column transverse reinforcement was based on confinement requirements. The design of the frames is described in FEMA P695 (FEMA 2009).

Models of the buildings were created for structural analysis. The planar 2D models incorporate one-dimensional line-type elements, as discussed in FEMA P695 (FEMA 2009). Component models are used to simulate the nonlinear degrading response of beams, columns and joints. Typical component models are shown in Fig. 6.1. The hysteretic model used for modeling the cyclic behavior in NRHA is the "bilin" material available in the Opensees software platform McKenna et al. (2008). This is a peak-oriented model with a multinear envelope that is able to incorporate stiffness and strength degradation. According to Fig. 6.1, the model consists of three branches, an initial elastic (cannot be seen in Fig. 6.1 due to the scale of the horizontal axis), a strain hardening and a descending branch that terminates to an ultimate chord rotation equal to 0.1 rad. Due to lack of sufficient data, a conservative ultimate rotation value has been considered (Haselton et al. 2010). The gravity loads were considered following the 1.05(Dead Load)+0.25(Live Load) combination and remained constant throughout the loading history. The flexural strengths of the beam to column component models were based on calibration with test data from columns and beams with low-to-moderate axial load and ductile detailing (Haselton et al. 2010). The models do not have the capability to represent shear failure, meaning that the shear capacities are not represented in the structural models. More details about the models can be found in (FEMA 2009; Haselton et al. 2010; Lignos and Krawinkler 2009). The first mode period was found equal to 0.625, 0.855, and 1.80 s, for the 2-, 4-, and 8-story frames, respectively.

In addition to seismic framing, a column is used to capture P-Delta effects caused by the gravity load on the internal gravity frames, using a leaning column element. Rayleigh damping amounting to 5% of critical damping is assigned to the 1st and the 3rd mode of vibration, at all beams and all columns, but not at the joints. To compensate for the absence of damping in the joints, the stiffness proportional damping coefficient was increased by 10%.

6.3 Nonlinear Static Analysis Methods

6.3.1 Problems and Limitations

The limitations of the nonlinear static methods are related to the fact that their theoretical background is not robust. The assumption that the response of a multi-degree of freedom system is directly related to the response of an equivalent single-degree of freedom (SDOF) system often is not accurate enough, as higher modes may contribute significantly to some EDPs, and peak values of different EDPs (or the same EDP at different locations) may occur at different times in the NRHA in a way that depends on the ground motion excitation. Conventional NSPs are not well-suited to represent this variation. In the simplest NSPs, the lateral load pattern is applied without taking into consideration member yielding and the resulting modification of building properties as the lateral forces are incremented. Improved NSPs such as multi-modal methods attempt to improve the prediction of nonlinear static analysis methods at the expense of complicated computations (Fragiadakis et al. 2007). However, these approaches are usually based on engineering judgment and intuition instead of being conceptually robust and numerically justified.

The accuracy of pushover methods of analysis, usually assessed relative to the results of nonlinear dynamic analyses, is sensitive to the type of inelastic mechanism that develops and to the modeling of the structural components. Figure 6.2 illustrates estimates of floor overturning moments determined for a 9-story steel moment frame using two multimode pushover methods and baseline results obtained from nonlinear response history analyses. Results for a regular model (Fig. 6.2a) suggest that relatively good estimates can be obtained using an energy-based approach, but this finding is not sustained in the presence of a weak story (Fig. 6.2b). Also worth noting is that even a small number of nonlinear dynamic analyses provided more consistent estimates of the response than could be obtained with static pushover analyses. This observation is conditioned on the ground motions being scaled to produce the same peak roof displacement (equal to 4% of the building height). Similar results would be expected for ground motions scaled to produce the same peak displacement of the equivalent SDOF model of the building.

Kunnath and Erduran (Kunnath and Erduran 2008) have shown that not only are the results of NRHA affected by modeling choices, but the evaluation of adequacy of an approximate method is a function of these modeling choices. Divergence in the story drifts determined by different analysis methods were observed depending

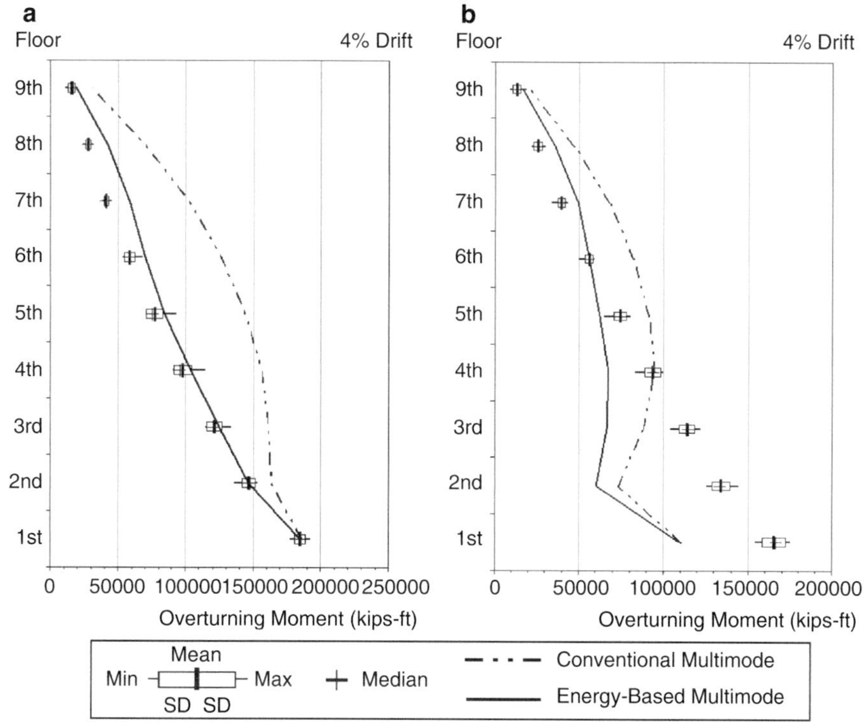

Fig. 6.2 Accuracy of floor overturning moment estimates relative to results from NRHA for (**a**) a regular 9-story steel moment frame, and (**b**) a weak-story version of the same frame (taken from (ASCE 2007))

on whether P-delta effects were modeled or not. Similarly, record-to-record variability was shown to affect story drift demands determined by NRHA differently from those determined by pushover analysis. One explanation for these observations might be that these modeling and ground motion choices affected the inelastic mechanism that developed during response. Consequently, an evaluation of the accuracy of pushover methods relative to results obtained by NRHA is conditioned on assumptions made in modeling.

Nonlinear dynamic response of structures involves fairly complex interactions among the evolving modes of the structure (Haselton and Deierlein 2007). The development of different inelastic mechanisms will increase dispersion in the values of at least some engineering demand parameters (EDPs) of interest. Nonlinear static methods generally are incapable of representing the development of multiple inelastic mechanisms and the variety of modal interactions and timing that produce maxima in the NRHAs. Often, nonlinear static methods tend to exaggerate deformation demands where mechanisms are determined to occur, while underestimating deformation demands that are observed to occur at other locations in nonlinear dynamic analysis.

6.3.2 NSP Methods Considered in the Framework of ATC-76-6

In recent years, fairly complicated methods have been proposed, involving (i) single-run pushover analyses using load patterns that represent multiple modes, (ii) multiple modes considered in step-by-step analyses, and (iii) progressive changes in the load (or displacement) pattern(s) applied to the structure. A detailed literature review on existing pushover methods can be found in the ATC-76-6 document (Consultants Joint Venture 2010). We compare the current state of practice NSP methods and enhanced multi-modal methods. The enhanced methods aim to improve the accuracy of nonlinear static analysis, ultimately aiming to avoid NRHA. Generally, experience with these methods is quite limited, and their relative complexity is a barrier to implementation. However, one procedure was selected for further exploration: the Consecutive Modal Pushover (CMP), as put forth by Poursha et al. (Poursha et al. 2009). Thus, the NSPs evaluated in this chapter are:

- *ASCE/SEI 41-06*: This is the basic procedure contained in the standard. There can be variations in the resulting force-displacement curve depending on component modeling assumptions and load patterns, as discussed below. The buildings are "pushed" with a first-mode lateral load pattern. The target displacement is obtained as:

$$d_t = C_0 C_1 C_2 C_3 S_a(T_e) \frac{T_e^2}{4\pi^2} g \tag{6.1}$$

  where C_0, C_1, C_2 and C_3 are modification factors and T_{e} is the effective fundamental period of the building. More specifically, C_0 is here taken equal to the modal participation factor of the equivalent SDOF system and C_3 was assumed equal to 1. The C_1 coefficient is obtained with the improved ASCE/SEI 41-06 relationship:

$$C_1 = 1 + \frac{R-1}{aT_e^2} \tag{6.2}$$

  where R is the strength reduction factor $R = C_{\mathrm{m}} S_{\mathrm{a}} W / F_{\mathrm{y}} \mathrm{g} \geq 1$. In Eq. 2 we assume: $a = 130$ (for site class B), $C_m = M_n^* / W$ is the modal mass participation ratio. C_3 is adopted to include the P-Δ effects and is here taken equal to 1. The coefficient C_2, takes into consideration the effect of hysteresis rules, and is calculated with the formula:

$$C_2 = 1 + \frac{1}{800} \left(\frac{R-1}{T_e^2} \right)^2 \tag{6.3}$$

- *Eurocode/N2 method*: The N2 method was initially proposed by Fajfar and Fischinger (Fajfar and Fischinger 1988) and was later expressed in a displacement-acceleration format (Fajfar 1999). Recently, the method was included in the Eurocode 8 (Eurocode 8 2004) standards. Conceptually, the method is a variation of Capacity Spectrum Method that instead of highly damped spectra uses an R-C_1-T relationship. The method, as implemented in Eurocode 8 (EC8), consists of the following steps: (i) Perform pushover analysis and obtain the capacity curve in V_b-u_r terms, (ii) Convert the pushover curve of the MDOF system to the capacity diagram of an equivalent SDOF system and approximate the capacity curve with an idealized elasto-perfectly plastic relationship to determine the period T_e of the equivalent SDOF system, (iii) the displacement of the MDOF system is simply calculated as $d_t = C_0 d_t^*$, where d_t^* *is* the target displacement of the corresponding inelastic SDOF system. Different expressions are suggested for short and for medium-to-long period ranges, for the latter case d_t^* is equal to the displacement of the corresponding elastic SDOF system, calculated as:

$$d_{et}^* = S_a(T_e)\left[\frac{T_e}{2\pi}\right]^2 \tag{6.4}$$

where $S_a(T_e)$ is the elastic acceleration response spectrum at the period T_e.
- *Modal Pushover Analysis (MPA)*: This procedure, initially proposed by Chopra and Goel (Chopra and Goel 2002), combines two or more pushover curves generated by assuming load patterns based on the first mode and one, or more, higher modes. Although subsequently Chopra and Goel (Goel and Chopra 2005) recommended a second analysis phase to determine member forces, this investigation follows the process more common among practitioners and researchers of determining all response quantities in a single application.

The steps of the Modal Pushover Analysis (MPA) method (Chopra and Goel 2002), are summarized as follows: (*i*) Calculate the natural frequencies, the mode shapes and the lateral load patterns $\mathbf{s}_n = \mathbf{m}\boldsymbol{\varphi}n$, (*ii*) For the nth mode, develop the base shear-roof displacement curve, $V_{bn}-u_{rn}$, for the $s_n$distribution of lateral forces. (*iii*) Idealize the pushover curve as a bilinear curve and compute the target displacements δ_t for every mode using the ASCE/SEI 41-06 R-C_1-T relationships. (*iv*) From the pushover results (Step ii), extract values of desired responses r_{n+g} due to the combined effects of gravity and lateral loads at roof displacement equal to $u_{rn} + u_{rg}$. (*v*) Repeat steps ii–iv for as many modes as required for sufficient accuracy, thus 2 modes for the 2-story RCMRF and 3 for the 4 and the 8-story RCMRF buildings. (*vi*) Compute the dynamic response due to the nth mode: $r_n = r_{n+g} - r_g$, where r_g is the contribution of gravity loads alone. Determine the total response (demand) by combining gravity response and the peak modal responses using the SRSS rule: $r \approx \max[r_g \pm (\Sigma(r_n)^2)^{1/2}]$.
- *Consecutive Modal Pushover (CMP)*: This procedure, initially proposed by Poursha et al. (Poursha et al. 2009) uses invariant load patterns for up to three

modes, applied consecutively in stages in a single pushover analysis after application of the gravity loads. This procedure is of interest because: (*i*) interaction of multiple modes is considered in a way that may cause different inelastic mechanisms to form, and (ii) the member forces resulting from the analysis are consistent with member capacity limits (e.g. beam shears do not exceed the shears associated with development of a plastic mechanism). Other methods reported in the literature (of which there are many), including those that combine multimode effects at each pushover step, such as Incremental Response Spectrum Analysis (Aydinoglou 2003) and Displacement Adaptive Pushover (Antoniou and Pinho 2004) were judged to be too complex for routine use unless specialized software is available.

The CMP considers up to three modes, applied consecutively in stages in a single pushover analysis. In this way, it may come closer to representing the higher mode responses that take place when the peak displacement response is realized dynamically. The first pushover analysis uses an inverted triangular load pattern for medium-rise buildings and a uniform force distribution for high-rise buildings. The second pushover analysis consists of a sequence of first and second mode forces. The first mode forces are applied until the roof displacement equals $a_1\delta_t$, where a_1 is the first mode modal mass ratio and δ_t is the target displacement determined for the first mode. Upon reaching $a_1\delta_t$, incremental forces are applied that follow a second mode pattern. The incremental displacement used for this analysis stage is $(1-a_1)\delta_t$. The third pushover analysis, required only for buildings with periods of 2.2 s or higher, consists of a sequence of first, second, and third mode forces. As before, the first mode forces are applied until the roof displacement equals $a_1\delta_t$. Upon reaching $a_1\delta_t$, incremental forces are applied that follow a second mode pattern until the roof displacement increases by $a_2\delta_t$. At this point, incremental forces that follow a third mode pattern are applied until the roof displacement increases by $(1-a_1-a_2)\delta_t$. The peak value of any EDP of interest obtained in the separate stages is retained and the final EDP estimate is obtained from the envelope of the three stages.

- *Modal Response Spectrum Analysis (MRSA)*: When applied to structures in the nonlinear range of response, Modal Response Spectrum Analysis (MRSA) relies on simple extrapolations of linear behavior and thus approximately represents the equal displacement rule for deformation-related quantities. The demands are calculated performing linear-elastic analysis using lateral load patterns proportional to the modes of vibration, which are similar to those of the MPA procedure. Target displacements are calculated using the C_1 and C_2 relationships of ASCE/SEI 41-06 and C_3 was taken equal to 1.0. The EDP values obtained using every mode-proportional lateral load pattern are then combined with the SRSS rule to obtain the final response estimates.

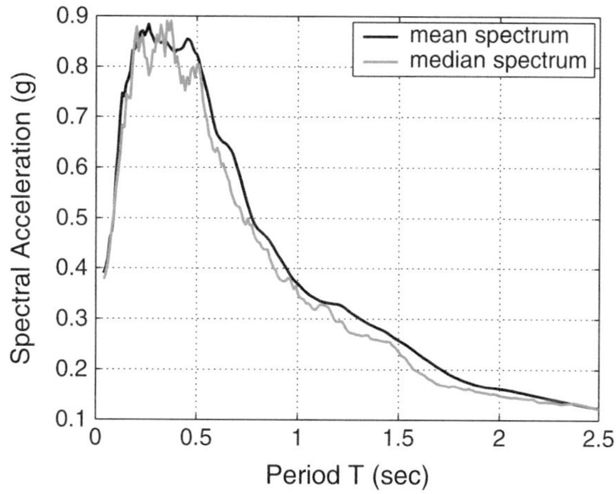

Fig. 6.3 5%-damped unscaled mean and median response spectra of the normalized 44 ground motions used in this study

6.4 Ground Motions

Nonlinear response history analysis is performed using the suite of 44 ground motion records utilized in the FEMA P-695 (FEMA 2009) far-field data set. The ground motions are normalized according to the geometric mean of their peak ground velocity (PGV) in two orthogonal directions, as discussed in FEMA P-695. The scale factors (SF) considered for all buildings of this study equal to 0.5, 1 and 2. These scale factors correspond to ground motion at a Los Angeles, California site with mean recurrence intervals of approximately 100, 400, and 2,475 years. Figure 6.3 shows the mean and the median design spectra for a 400-year mean recurrence interval (scale factor of 1.0).

6.5 Numerical Results

6.5.1 Nonlinear Response History Analysis

Figures 6.4 and 6.5 present selected results from nonlinear response history analysis (NRHA). The curves shown correspond to median response quantities plotted as function of the scale factor. The results for these structures indicate:

- Maximum story drift demand is localized at the lower stories and usually at the second story, while the minimum demand appears at the top story. Maximum story drift is defined as the maximum difference of the horizontal displacement of adjacent stories, normalized by the height of the story.
- Inspection of story drift demands indicates that higher modes generally have an appreciable effect on story drift and story shears, as discussed in reference

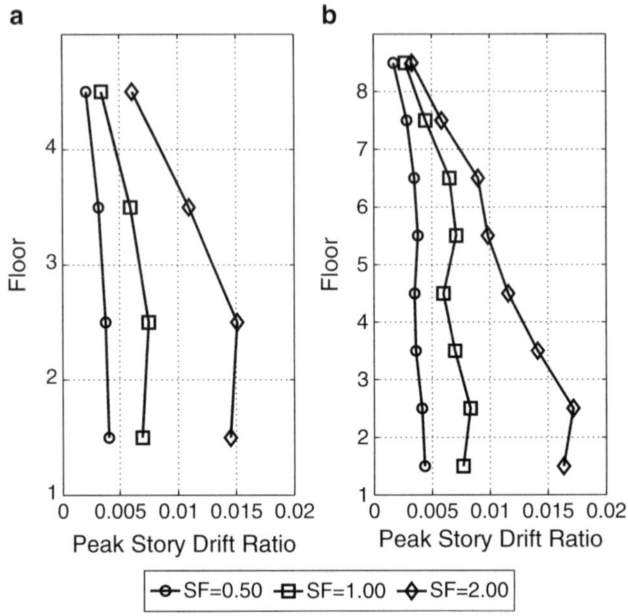

Fig. 6.4 Profiles of the peak story drift ratio: (**a**) 4-Story RCMRF, and (**b**) 8-Story RCMRF

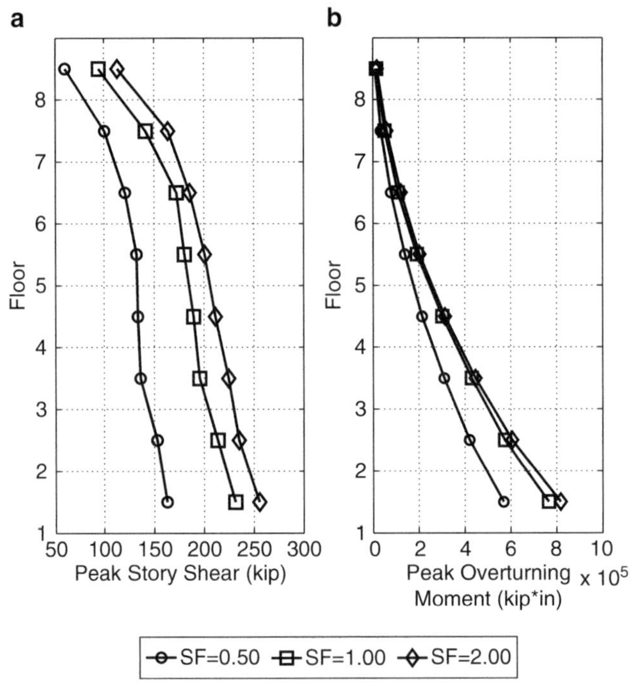

Fig. 6.5 8-Story RCMRF, profiles of: (**a**) the peak story shears, and (**b**) maximum overturning moments

(Aschheim et al. 2007). In Figs. 6.4 and 6.5 this can be seen in the drift and the shear profiles by the curviness in the mid-height stories which makes the profiles to differ from those of a response dominated by the first-mode.

- Median peak story drifts tend to concentrate in the lower stories with increasing scale factor.
- Story shears do not follow the lateral load pattern used to design the structure or the pattern resulting from lateral forces applied in a first-mode based pushover analysis. Story shears in the upper stories are significantly larger than would be expected from these load patterns. Story shears in the upper stories increase disproportionately with an increase in scale factor.
- Peak overturning moments tend to follow the concave pattern associated with the lateral load patterns used in design or first-mode pushover analysis.
- Story shears and maximum overturning moments tend to "saturate" as the scale factor increases. This happens because the capacity is bounded by the maximum moment capacity of the component models and thus any further increase of the scale factor cannot increase the corresponding story shears.

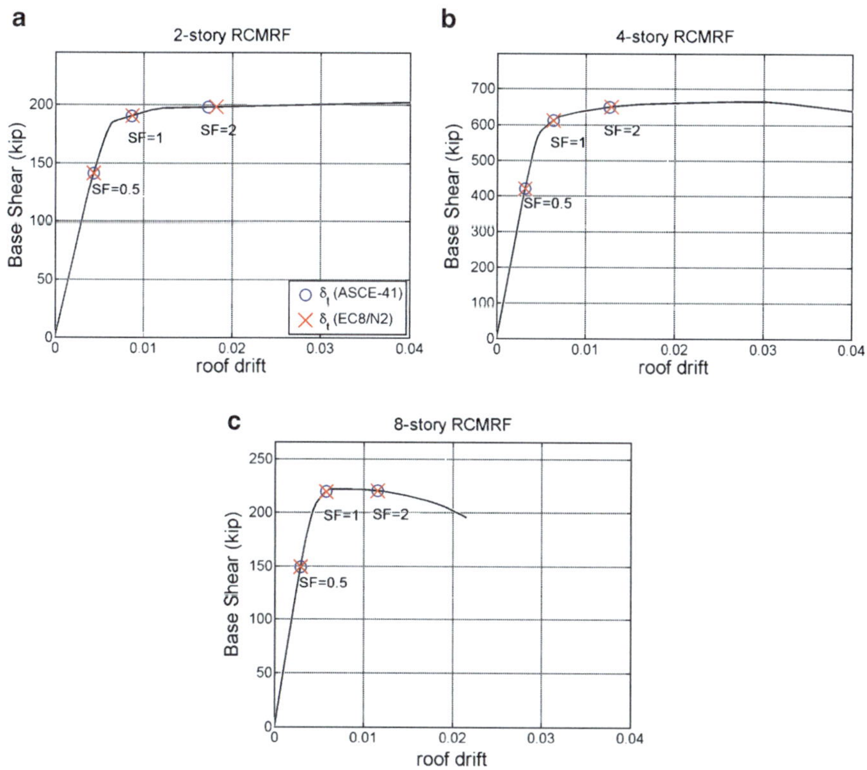

Fig. 6.6 First mode static pushover curves and target displacement estimates for: (**a**) 2-story RCMRF, (**b**) 4-story RCMRF, and (**c**) 8-story RCMRF

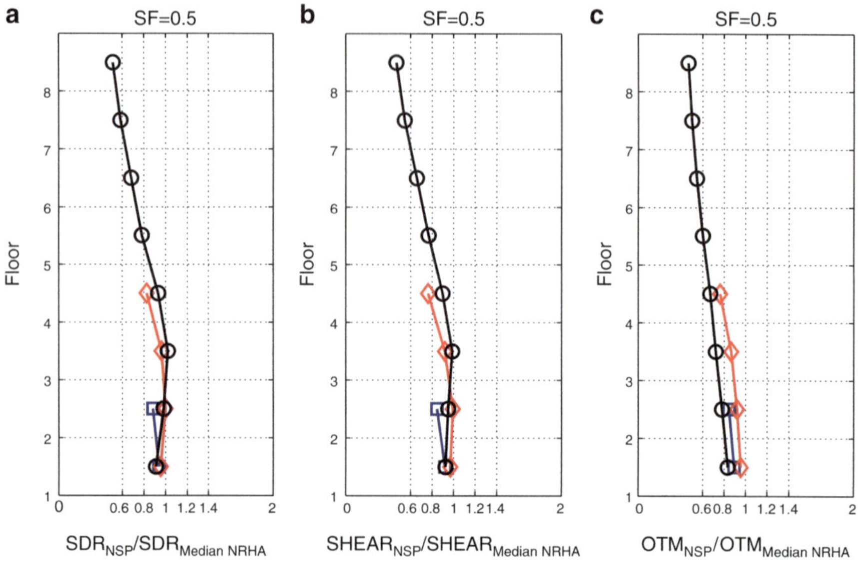

Fig. 6.7 All buildings: ratios of ASCE-41 NSP and NRHA. (**a**) peak story drifts (SF = 0.5), (**b**) peak story shears (SF = 0.5), and (**c**) peak overturning moments (SF = 0.5)

6.5.2 *Single Mode Nonlinear Static Analysis*

Figure 6.6 shows the capacity curves obtained in first-mode pushover analyses of the three RC moment frames considered. Also shown on this figure are the target displacements, δ_t, determined using the formulas of ASCE/SEI 41-06 (ASCE 2007) and Eurocode 8(2004). Roof displacements at yield are observed to occur at target displacements about 0.5 to 0.6% of the building's height. Both methods produce nearly identical target displacement estimates for the three moment frames and therefore the ASCE-41 target displacements were used to estimate response quantities using first-mode pushover analyses. Target displacements for a scale factor of 0.5 are in the elastic regime, those for the records scaled by 1.0 are nearly elastic, and those for a scale factor of 2.0 cause moderate inelastic response, developing system ductilities of 2-3.

Ratios of estimated values and nonlinear response history analysis medians are plotted over the height of each frame in Figs. 6.7 and 6.8 at scale factors of 0.5 and 2.0, respectively. The accuracy is good for all three EDPs examined for the 2-story RCMRF, where the error is less than 20%. For the 4-story frame, the accuracy of story drifts, story shears and overturning moments degraded as the scale factor increased from 0.5 to 2.0. This is caused by the tendency of equivalent SDOF systems to overestimate peak displacements of MDOF systems with increasing severity of nonlinear response as has been recognized previously. This also happens in the predicting equations of ASCE/SEI 41/06 which we used to estimate the target

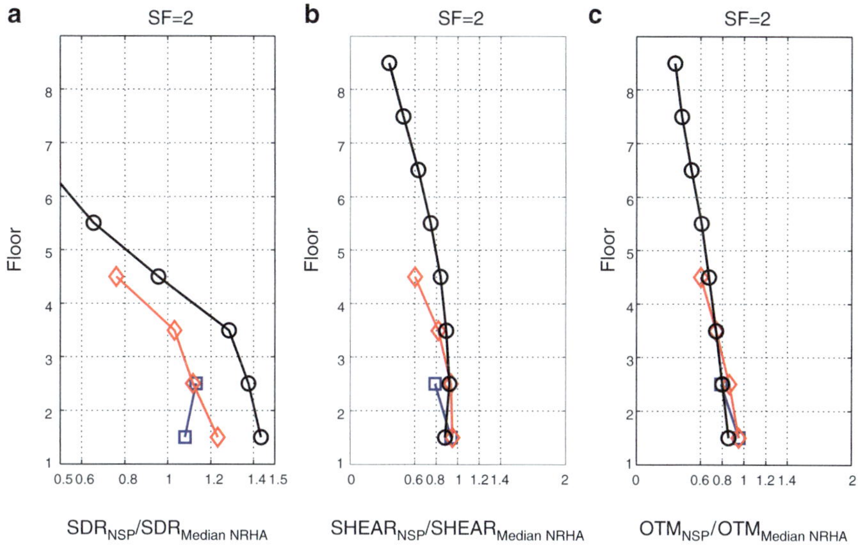

Fig. 6.8 All buildings: ratios of ASCE-41 NSP and NRHA. (**a**) peak story drifts (SF = 2.0), (**b**) peak story shears (SF = 2.0), and (**c**) peak overturning moments (SF = 2.0)

displacements of our NSP analyses. The degradation in accuracy with increase in scale factor was small for story shears and overturning moments and greater for the drift profiles. The accuracy of NSP estimates also degraded with increase in scale factor for the 8-story frame, where the error in the story shears and overturning moments of the top stories was of the order of 60%. Story drifts were grossly underestimated at the top stories and overestimated at the bottom at higher scale factors. This is attributed to the constant shape of the lateral loading scheme.

6.5.3 Modal Pushover Analysis

As already discussed, the Modal Pushover Analysis (MPA) procedure was applied taking SRSS combinations of the individual modal contributions for any response quantity. Therefore, the target displacements are determined using higher mode pushover analyses to determine inelastic response independently in each mode considered (Chopra and Goel 2002). The target displacements were determined using the coefficients suggested in ASCE/SEI 41-06 (ASCE 2007).

Figure 6.9 shows the pushover curves for the 2-, 4-, and 8-story reinforced concrete moment frames, when lateral load patterns based on the first and higher modes are used. The target displacements determined using the mean elastic response spectrum (Fig. 6.3) for scale factors of 0.5, 1.0, and 2.0 are shown. Two

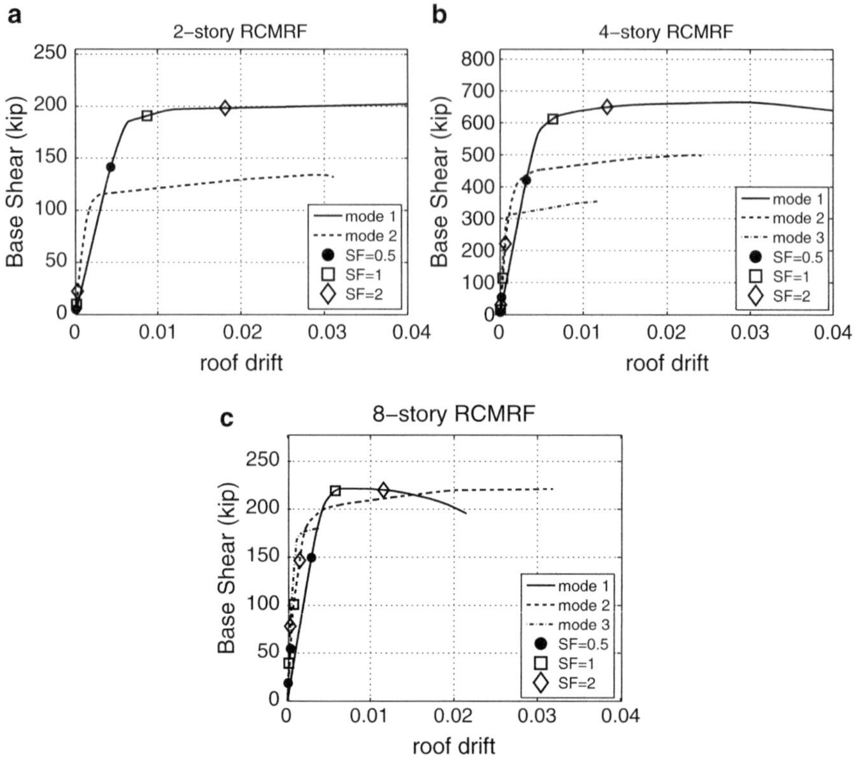

Fig. 6.9 Static pushover curves and target displacement estimates for the 2-, 4-, and 8-story reinforced concrete moment frames, for ground motion scale factors of 0.5, 1.0, and 2.0

modes were used for the 2-story frame, and three modes for the 4- and the 8-story frames. It is shown that the target displacements corresponding to the second and the third mode lie on the initial elastic branch for all three buildings.

Selected results for the three buildings are shown in Figs. 6.10, 6.12 and 6.13. Apart from the NSP estimates, the peak responses for the individual records and their mean and dispersion are also shown. Furthermore, ratios of NSP estimated values and NRHA medians are plotted over the height of each frame in Figs. 6.14 and 6.15 at scale factors of 0.5 and 2.0, respectively.

For the 2-story frame, second mode contributions to story drifts, and overturning moments were negligible and reasonably accurate estimates of these quantities were obtained with the first mode estimates (Fig. 6.10). While story shears were estimated accurately for a scale factor of 0.5 (Figs. 6.7a and 6.10a), the inclusion of second-mode contributions in the MPA procedure did not sufficiently increase the story shears to result in an accurate estimate at a scale factor of 2.0 (Fig. 6.10b).

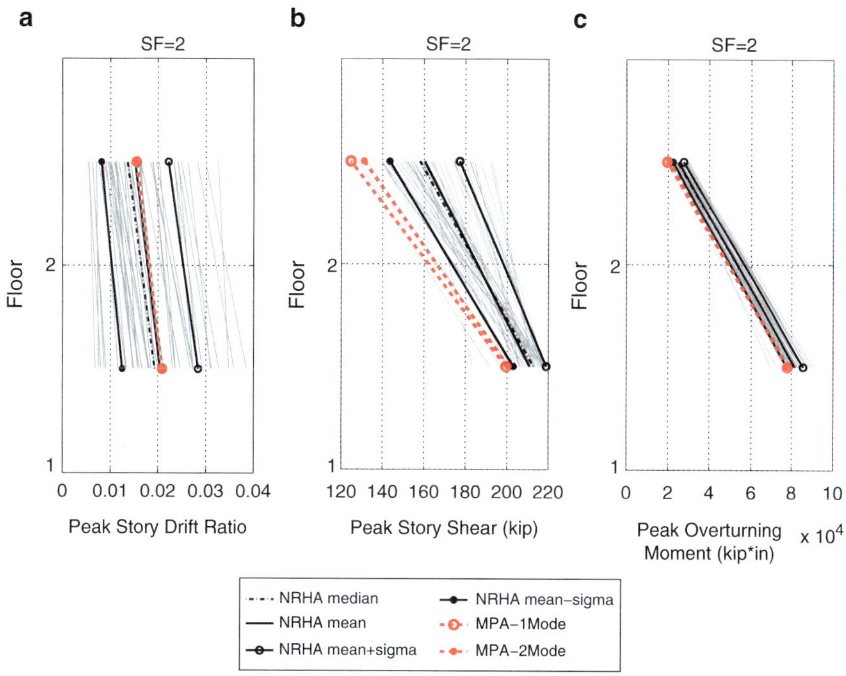

Fig. 6.10 2-story RCMRF: profiles of (**a**) peak story drift (SF = 2), (**b**) of peak story shears (SF = 2), and (**c**) peak overturning moments (SF = 2)

For the 4-story frame, second and third mode contributions to floor displacement and story drift were negligible and reasonably accurate estimates of these quantities were obtained with first-mode estimates (Fig. 6.11). In contrast to the 2-story frame, the accuracy of modal pushover analysis estimates of story shears improved as the scale factor increased from 0.5 to 2.0 (Fig. 6.12). Whereas second-mode shears had a significant and beneficial effect, third mode story shears, while beneficial, had a relatively small contribution. Relatively accurate estimates of overturning moments were made, and these benefitted from inclusion of the second mode contribution.

For the 8-story frame, second mode contributions to story drift were not negligible and improved the story drift estimates, but accuracy varied with location and scale factor (Fig. 6.13). Moreover, as shown in Figs. 6.14 and 6.15, the accuracy of story drift estimates improved with an increase in scale factor at the upper stories, while story drift estimates became less accurate as the scale factor increased at the lower stories. The reason for the overestimation observed at the lower stories is probably the constant shape of the lateral load pattern. As for story drifts, the accuracy of story shear estimates varied with location and scale factor. Both second and third mode contributions to story shears were appreciable. At the lower stories, story shear

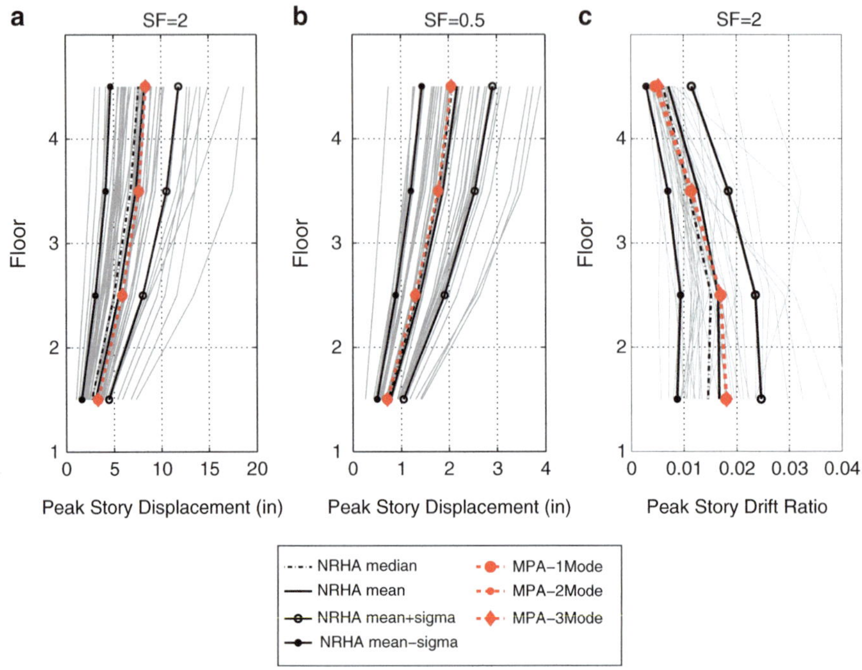

Fig. 6.11 4-story RCMRF: profiles of (**a**) peak displacement (SF = 2), (**b**) of peak story drift (SF = 0.5), and (**c**) peak story drift (SF = 2)

estimates were most accurate at a scale factor of 0.5 (Fig. 6.14); an increase in scale factor led to significant overestimates (Fig. 6.15). In contrast, at the upper stories, the story shears were underestimated at a scale factor of 0.5, were estimated with reasonable accuracy at a scale factor of 1.0, and were significantly overestimated at a scale factor of 2.0. Second mode contributions to overturning moments were not negligible and improved the estimates; overturning moments tended to be underestimated at a scale factor of 0.5, and were overestimated at a scale factor of 2.0.

6.5.4 *Consecutive Modal Pushover*

Selected results obtained by application of Consecutive Modal Pushover (CMP) analysis to the three RCMRFs are shown in Figs. 6.16 and 6.17. In addition to the single-record and NRHA curves, the figures also show with dashed lines the profiles of the two CMP stages and with solid line their envelope. Ratios of estimated values and nonlinear response history analysis medians are plotted over the height of each frame in Figs. 6.18 and 6.19 at scale factors equal to 0.5 and 2.0.

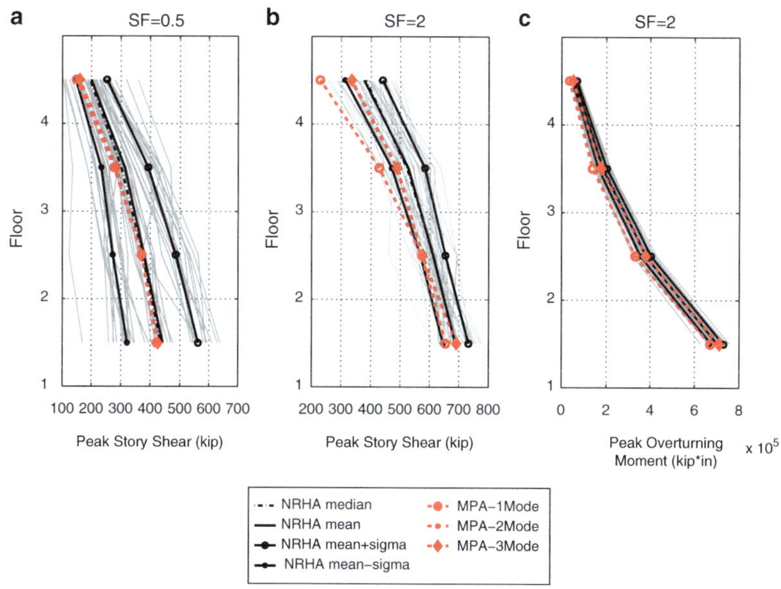

Fig. 6.12 4-story RCMRF: profiles of (**a**) peak story drift (SF = 0.5), (**b**) peak story shears (SF = 2), and (**c**) peak overturning moments (SF = 2)

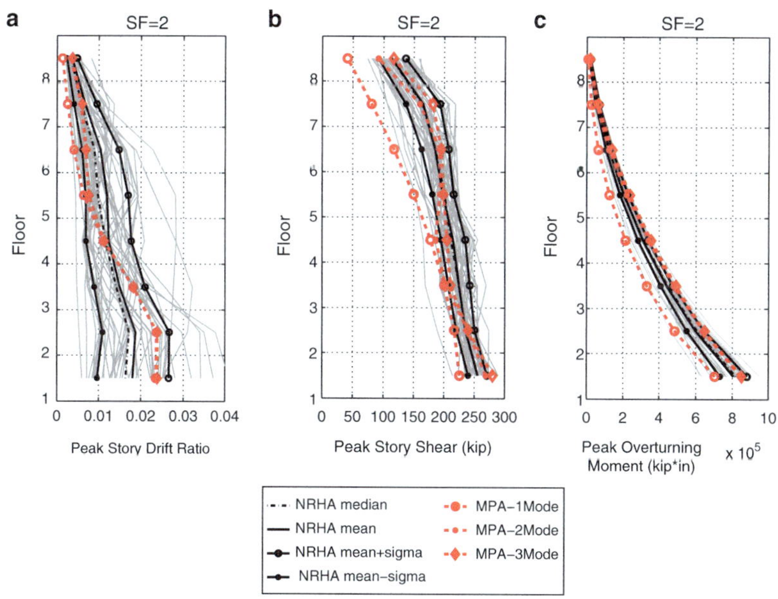

Fig. 6.13 8-story RCMRF, profiles of: (**a**) peak story drift (SF = 2.0), (**b**) peak story shears (SF = 2), and (**c**) peak overturning moments (SF = 2)

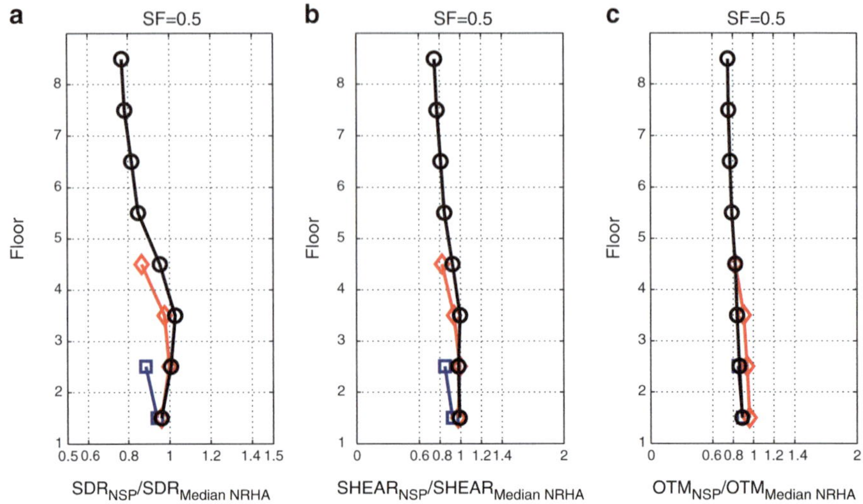

Fig. 6.14 All buildings: ratio of MPA and NRHA. (**a**) peak story drifts (SF = 0.5), (**b**) peak story shears (SF = 0.5), and (**c**) peak overturning moments (SF = 0.5)

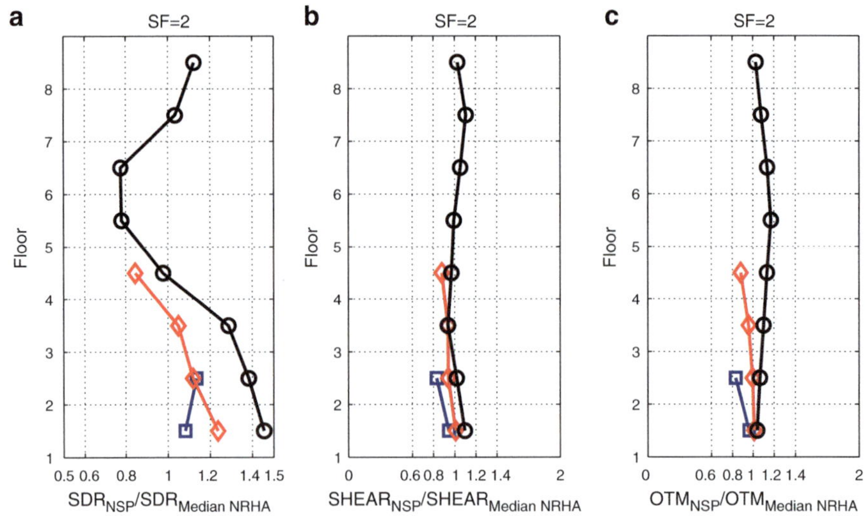

Fig. 6.15 All buildings: ratio of MPA results and NRHA. (**a**) peak story drifts (SF = 2.0), (**b**) peak story shears (SF = 2.0), and (**c**) peak overturning moments (SF = 2.0)

For the 2-story frame, story drifts were estimated with reasonable accuracy (Fig. 6.18a). Story shears were accurately estimated at a scale factor of 1.0, were significantly overestimated at a scale factor of 0.5, and were significantly underestimated at a scale factor of 2.0 (Figs. 6.16a and 6.17a).

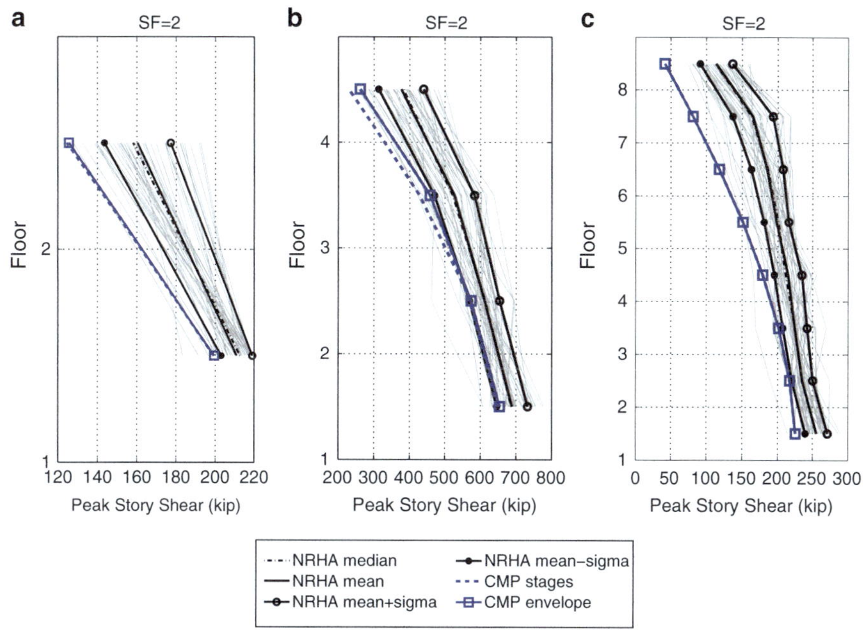

Fig. 6.16 Peak story shears obtained with Consecutive Modal Pushover method: (**a**) 2-story RCMRF (SF = 2), (**b**) 4-story RCMRF (SF = 2), (**c**) 8-story RCMRF (SF = 2)

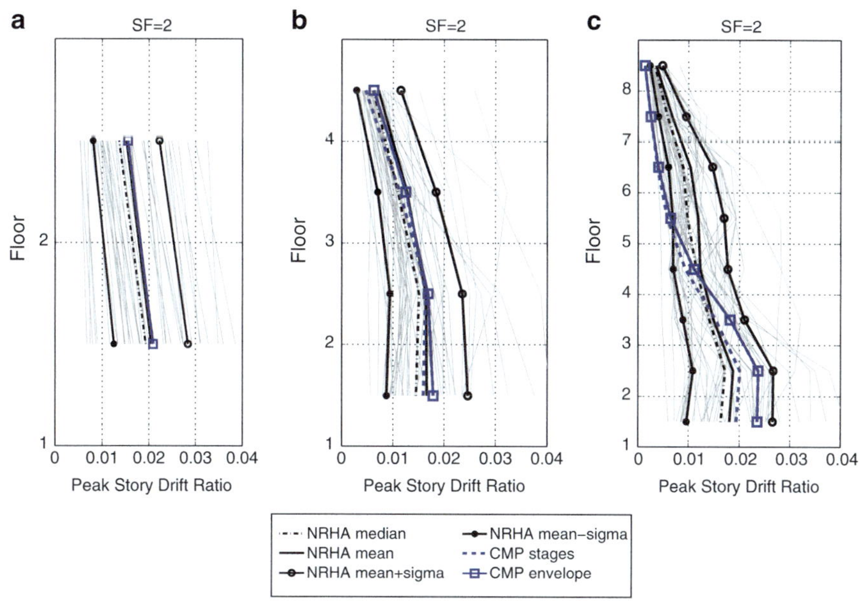

Fig. 6.17 Peak drift profiles for the Consecutive Modal Pushover method: (**a**) 2-story RCMRF (SF = 2), (**b**) 4-story RCMRF (SF = 2), (**c**) 8-story RCMRF (SF = 2)

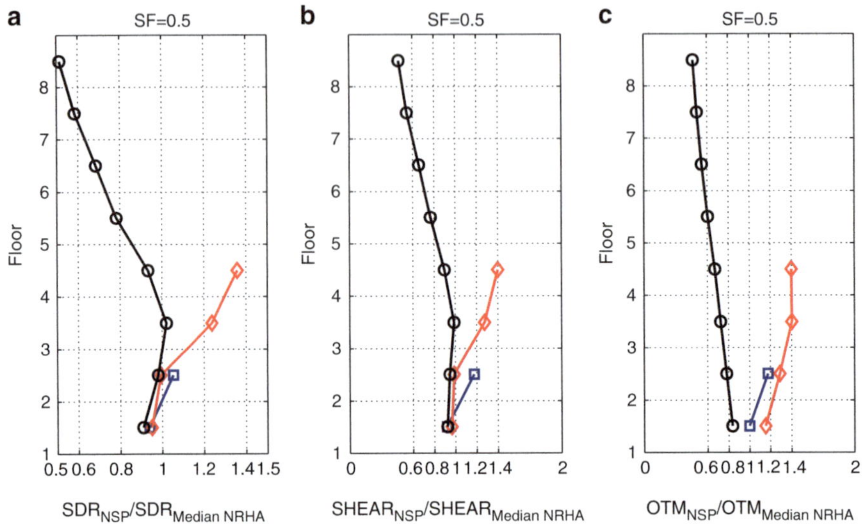

Fig. 6.18 All buildings: ratio of CMP and NRHA. (**a**) peak story drifts (SF = 0.5), (**b**) peak story shears (SF = 0.5), and (**c**) peak overturning moments (SF = 0.5)

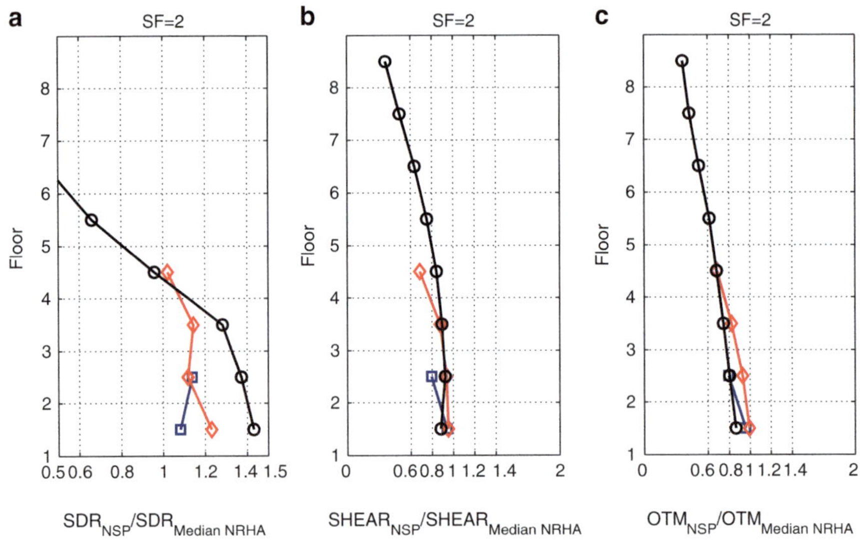

Fig. 6.19 All buildings: ratio of CMP and NRHA. (**a**) peak story drifts (SF = 2.0), (**b**) peak story shears (SF = 2.0), and (**c**) peak overturning moments (SF = 2.0)

For the 4-story frame, the accuracy of estimates of story drift, story shear, and floor overturning moments varied with location and scale factor. For example, while story drifts at the upper stories were overestimated at a scale factor of 0.5, relatively accurate estimates of story drift were obtained over the height of the building at higher scale factors. As well, story shears in the upper stories were significantly overestimated at a scale factor of 0.5, and story shears over the height of the building were significantly underestimated at a scale factor of 2.0.

For the 8-story frame, peak displacements were overestimated at a scale factor of 2.0, just as occurred with the first mode and multiple mode pushover analyses. As for the 4-story frame, the accuracy of estimates of story drift, story shear, and floor overturning moments varied with location and scale factor.

6.5.5 *Elastic Modal Response Spectrum Analysis*

Estimates of response quantities were also made by elastic modal response spectrum analysis as previously discussed. When this procedure is followed, response estimates are made by taking the SRSS combinations of the modal values made on the basis of linear elastic response extrapolated to the spectral accelerations obtained using scale factors of 0.5, 1.0, and 2.0. Drift profiles over the height of the buildings are proportional to the elastic distributions. In cases where drift patterns obtained in modal pushover analyses resemble the elastic distributions, displacement and story drift estimates made using elastic modal response spectrum analysis will resemble those obtained with modal pushover analysis. Story shear and overturning moment distributions at low scale factors will also resemble those obtained with modal pushover analysis. Of course, at high scale factors, these force-related quantities will be significantly overestimated.

Selected results of the elastic modal response spectrum analysis are shown in Figs. 6.20 and 6.21. Ratios of estimated values and nonlinear response history analysis medians are plotted over the height of each frame in Figs. 6.22 and 6.23 for selected response quantities at scale factors of 0.5 and 2.0.

For the 2-story frame peak story drifts were estimated accurately just as occurred with first mode and multiple mode pushover analysis. While story shears and overturning moments were estimated accurately at a scale factor of 0.5, the assumption of linear elastic response in every mode led to overestimation of story shears and overturning moments at higher scale factors. For elastic response and inelastic response where the equal displacement rule applies ($C_1 = 1.0$) elastic modal response spectrum analysis and multiple mode pushover analysis provide identical estimates of peak displacements and drifts.

For the 4-story frame, accurate estimates of floor displacements and story drifts were obtained at all scale factors. Story shears were underestimated at the upper stories at a scale factor of 0.5, and were overestimated at scale factors of 1.0 and 2.0. Overturning moments were slightly underestimated at the lower floors at a scale factor of 0.5, and are grossly overestimated at scale factors of 1.0 and 2.0.

For the 8-story frame, story drifts were underestimated at a scale factor of 0.5, estimated reasonably accurately at a scale factor of 1.0, and overestimated at the

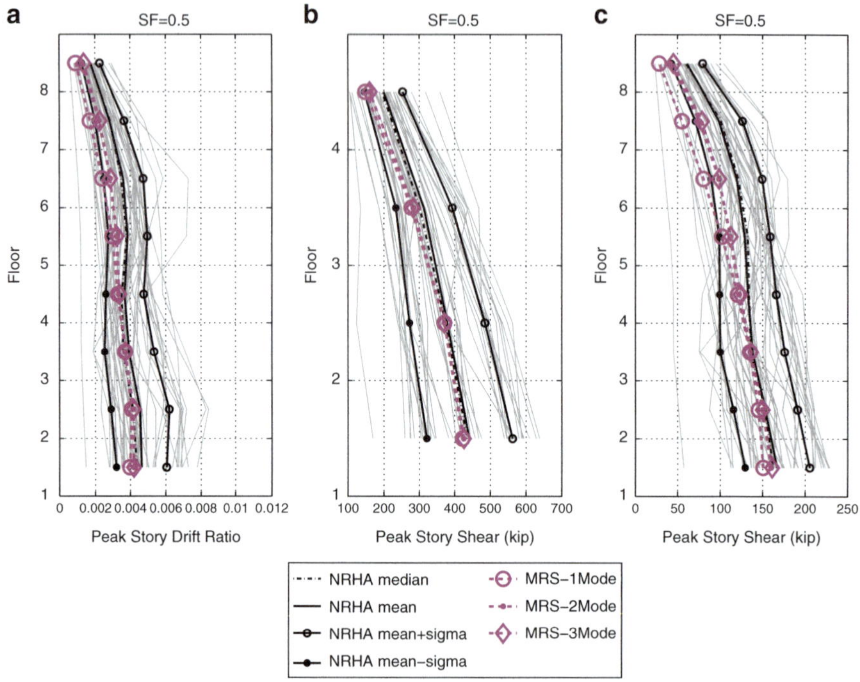

Fig. 6.20 Selected results for the modal response spectrum analysis: (**a**) 8-story RCMRF peak story drift (SF = 0.5), (**b**) 4-story RCMRF peak story shear (SF = 0.5), and (**c**) 8-story RCMRF peak story shear (SF = 0.5)

upper stories at a scale factor of 2.0. Story shears and overturning moments were generally underestimated by modal response spectrum analysis at a scale factor of 0.5, and generally are overestimated at scale factors of 1.0 and 2.0.

6.6 Comparing NSP and NDP Methods with Respect to Global Response Parameters

6.6.1 Comparison in the Framework of NSP

NRHA can be used for the seismic performance assessment of structures in the framework of Incremental Dynamic Analysis (IDA) (Vamvatsikos and Cornell 2002). IDA involves repeatedly running NRHAs using a suite of ground motions scaled to different factors such that the response to each ground motion is obtained at many different intensities. Specifically, for any Engineering Demand Parameter

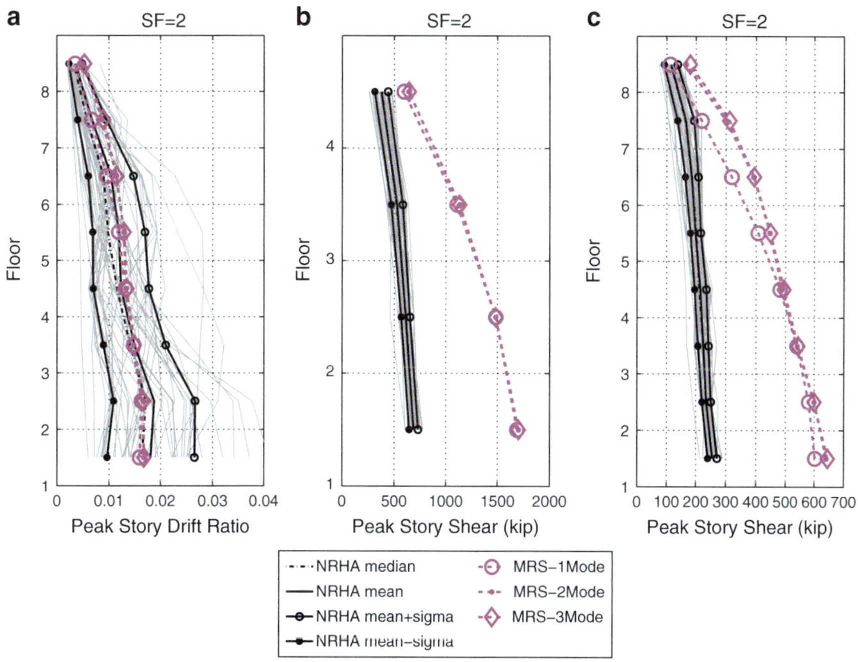

Fig. 6.21 Selected results for the Modal Response Spectrum Analysis: (**a**) 8-story RCMRF peak story drift (SF = 2.0), (**b**) 4-story RCMRF peak story shear (SF = 2.0) and (**c**) 8-story RCMRF peak story shear (SF = 2.0)

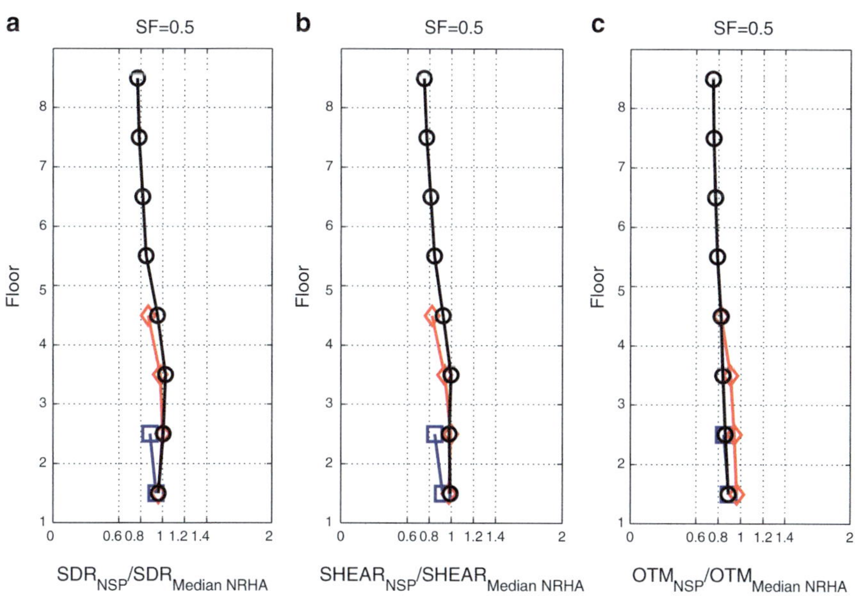

Fig. 6.22 Ratio of elastic MRSA results and NRHA: (**a**) peak story drifts (SF = 0.5), (**b**) peak story shears (SF = 0.5), and (**c**) peak overturning moments (SF = 0.5)

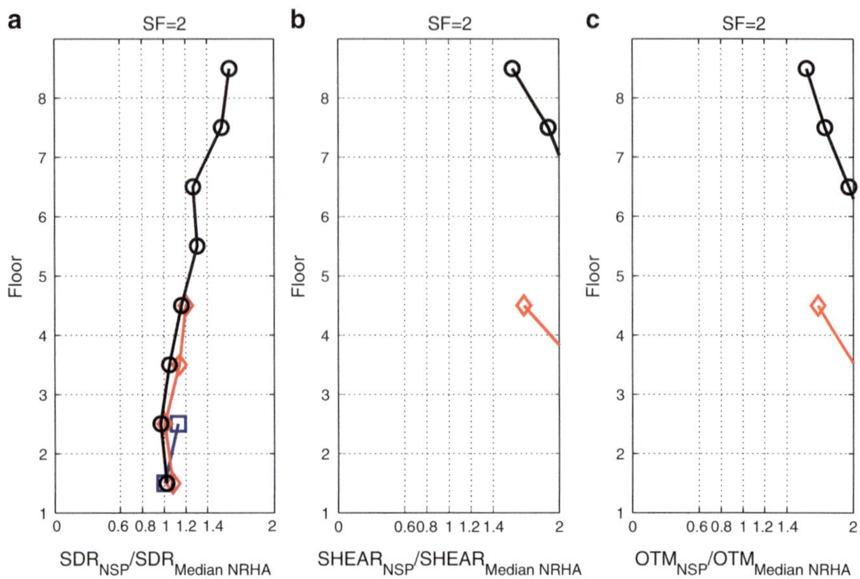

Fig. 6.23 Ratio of elastic MRSA and NRHA: (**a**) peak story drifts (SF = 2.0), (**b**) peak story shears (SF = 2.0), and (**c**) peak overturning moments (SF = 2.0)

(EDP) used to characterize structural response and an Intensity Measure (IM), e.g. the 5%-damped, first-mode spectral acceleration $S_a(T_1,5\%)$, we can generate IDA curves consisting of the EDP plotted as a function of the IM for each record (Fig. 6.24a). Conventionally, the response EDP (dependent parameter) is plotted on the abscissa, and the IM (independent variable) is plotted on the ordinate. Given these IDA curves, the statistical distribution of response as a function of input can be summarized by curves that represent the 16%, 50% and 84% fractiles.

Alternatively, the results of IDA can be plotted using the same coordinates as in the NSP, resulting in the so-called "dynamic capacity curve", calculated for each ground motion. The dynamic capacity curves can plot roof displacement, roof drift (θ_{roof}) (i.e. roof displacement normalized by the building's height) or any other EDP as a function of base shear. The effect of plotting on the ordinate the base shear (V_b) instead of $S_a(T_1,5\%)$ is shown in Fig. 6.24b, where the common EDP is θ_{max}, the maximum interstory drift over the height of the building. It should be noted that base shear is a structural response parameter, covariate with the drift, and therefore an EDP that should not be considered to be an IM. In order to consider base shear as the IM, it would be necessary to develop a building-specific ground motion prediction equation to derive appropriate seismic hazard curves for base shear, thereby eliminating the benefit of using a general-purpose IM. Thus, we will differentiate between the two different approaches by using the terms *IDA setting* and *IDA curves* when having an IM on the ordinate versus *pushover setting* and

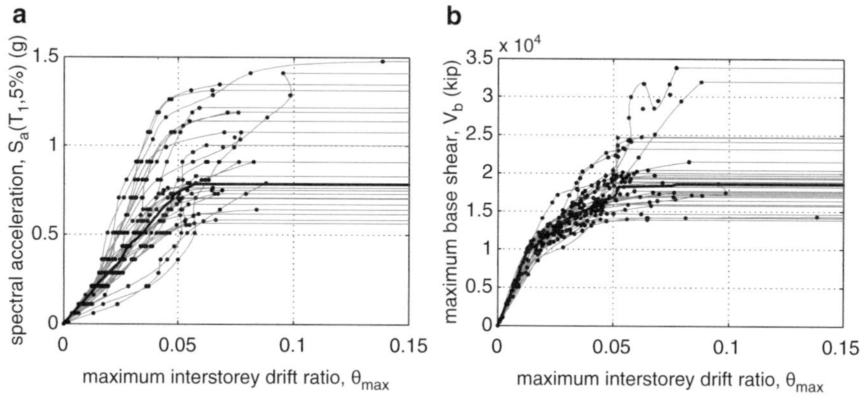

Fig. 6.24 IDA results for a nine-story steel building plotted as (**a**) IDA curves plotted as a function of $S_a(T_1,5\%)$, (**b**) dynamic pushover curves plotted as a function of the peak base shear V_b

dynamic pushover/capacity curves (or *DPO curves*) when plotting the base shear instead.

Due to the large variability observed in the single-record IDAs (Fig. 6.24), it is customary to summarize them with their median curve, plotted here with a solid dark line and the 16%, 84% fractiles that denote the dispersion. The median IDA can be calculated either as the median of the EDP given IM, or as the median IM given EDP. Both approaches will yield approximately the same results. The dispersion around the median, or in other words the variability of the IDAs, can be measured with beta, β, i.e. the standard deviation of the natural logarithms of the IM values for a given EDP value. If the data follow a lognormal distribution, β is equal to half the difference of the 16 and 84% fractile IM values. The dispersion can be used to obtain an estimate of the likelihood of a single-record IDA curve being close or away from the median curve.

The practice of plotting IDA results in a pushover setting can be found in several publications (Antoniou and Pinho 2004; Mwafy and Elnashai 2001) and has its roots in nonlinear mechanics where it is customary to visualize the nonlinear response of the structure with a force-displacement plot. Furthermore, when $S_a(T_1,5\%)$ is used in IDA, the independent variable is plotted on the ordinates although it is customary to have it on the abscissas. This practice is preferable, since it results to curves that are more familiar to engineers, having a clear initial elastic branch and terminating at a horizontal flatline that indicates the seismic intensity that the building collapses.

Looking at Fig. 6.24 it is clear that the dispersion around the median IDA is considerably larger than the dispersion around the median dynamic pushover curve. To further investigate the difference of the two plots of Fig. 6.24, we obtain similar results for a much simpler, single-degree-of-freedom (SDOF) system, shown in Fig. 6.25. Again significant dispersion exists in the IDA setting, where the strength

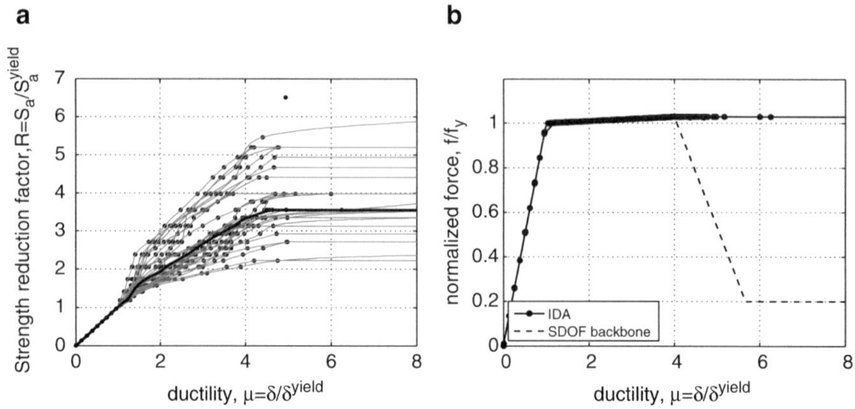

Fig. 6.25 IDA results of a nonlinear SDOF system with a backbone having in-cycle degradation (negative stiffness segment). Different intensity measures are plotted on the vertical axis: (**a**) strength reduction factor R, (**b**) peak oscillator's force f_s

reduction factor ($R = S_a(T_1,5\%\ )/S_a^{yield}(T_1,5\%)$) is used as the IM. On the other hand, in a pushover setting, no dispersion is evident. In this, case the dynamic curves all lie on the backbone of the oscillator (dashed line in Fig. 6.25b), as long as the backbone has a positive slope. The curves become horizontal where the backbone descends.

From the qualitative comparison of Figs. 6.24 and 6.25 useful conclusions can be drawn. When plotting either in IDA or pushover setting, the source that produces the dispersions observed in IDA analysis is related to the different damage patterns, or collapse mechanisms, activated by the ground motion record, which depend on the building's design and the ground motion characteristics. If we look at an advance stage of the inelastic response, e.g. θ_{max} values beyond 0.07, in the first case the dispersion that appears in the plots on the left (Figs. 6.24a and 6.25a) is caused by the variability in the $S_a(T_1,5\%)$ value of each ground motion. To investigate the dispersion of the plots on the right (Figs. 6.24b and 6.25b), the variability of the records is no longer revealed in the case of the SDOF oscillator (Fig. 6.25b), where only one damage pattern is possible. Therefore the IDAs show the usual record-to-record variability, while the dynamic pushover curves are all the same, coinciding with the backbone, i.e. with the force-deformation relationship that describes the capacity of the oscillator. Similarly, in the MDOF case, records that trigger similar damage mechanisms, or, in general, damage mechanisms that reach similar values of base shear, result in dynamic curves that are closer. Thus, we get the impression of a reduced dispersion around the median since much of the ground motion variability is now hidden. In the MDOF case, variability is also observed in the elastic domain, since for the same base shear value, the seismic forces result from the interaction of the various modes over time, with peak value resulting from different load patterns that vary with the ground motion record. No such elastic-level dispersion is observed for the SDOF oscillator, where only one

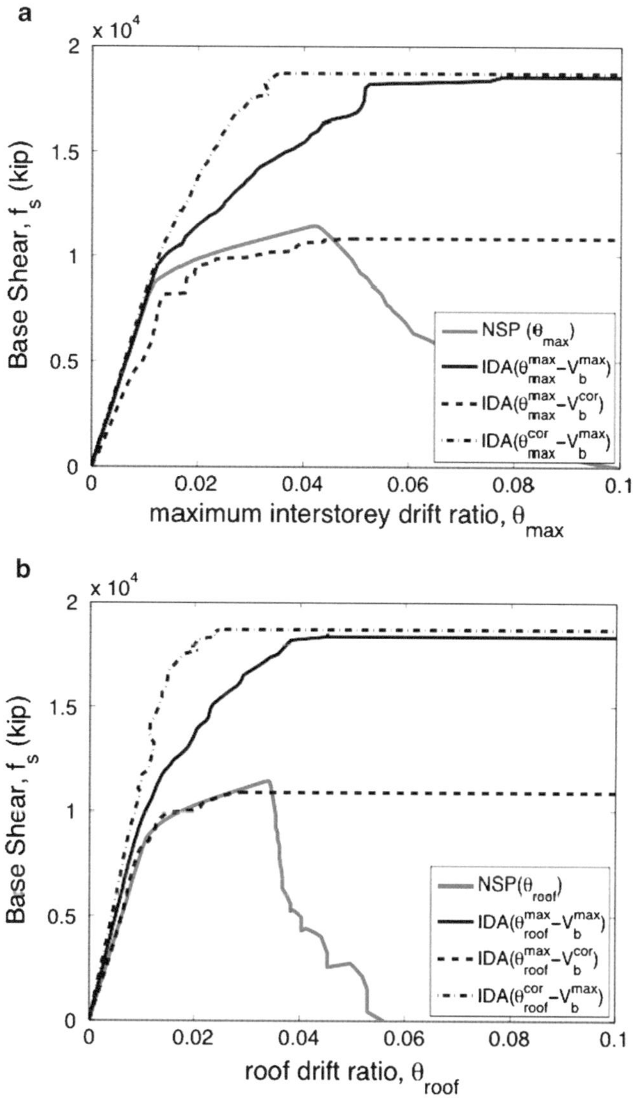

Fig. 6.26 Median dynamic pushover curves versus the corresponding static pushover curves of a nine-story steel building, (**a**) plotted against maximum interstory drift, θ_{max}, (**b**) plotted against roof drift, θ_{roof}. The superscripts "*max*" and "*cor*" denote that the quantity is the maximum over the entire timehistory, or it corresponds to the time instant that the other parameter is maximized, respectively. For example θ^{cor}_{max} is the θ_{max} value at the instant that the base shear is maximum and θ^{max}_{max} is the maximum θ_{max} value during the timehistory

degree-of-freedom exists and both IDA and dynamic pushover curves coincide before yielding (Fig. 6.25a, b).

As shown in Fig. 6.26, the dynamic pushover curves do not descend as the NSP curves do (gray lines). Moreover, since base shear and drift (θ_{max} or θ_{roof}) do not take their maximum values at the same instants in time, three variations for generating the dynamic capacity curves are investigated in Fig. 6.26. The first corresponds to the case where the maximum drift and maximum base shear (V_b) are plotted, while in the second case the V_b values considered are the values when the drift is maximized. In the third case, we use the drift values at the instant of maximum V_b. Another practice, not examined here, is considering the maximum drift and the peak base shear of a time window, e.g. ± 0.5 s, around the instant that the maximum drift occurs (Antoniou and Pinho 2004). Since θ_{max} is the EDP most commonly used in IDA while θ_{roof} is customarily plotted on the horizontal axis of the NSP curves, we perform the comparison for both EDPs. According to Fig. 6.26, median DPO and NSP curves will always have distinct differences, which may look smaller or larger when we show single records instead of the medians. For both EDPs the elastic slopes of the DPO and the NSP capacity curves (Fig. 6.26) practically coincide. However, significant differences are observed in the maximum base shear capacity for both EDPs, when we plot the maximum base shear versus the maximum or the corresponding drift value. Good prediction of the maximum base shear capacity is observed only when the base shear corresponds to the time instant that the drift is maximized. Furthermore, the DPO curves are not able to follow the negative slope of the NSP curves. This is a data-processing issue, since in DPO plotting we consider maximum force and/or displacement values over the time history.

6.6.2 Comparison in the Framework of Incremental Dynamic Analysis

Another approach to perform the comparison between NSP and IDA is to express the NSP curve in the IM and EDP coordinates chosen for the IDA. To facilitate direct comparison, we divide the base shear force by the building mass and adjust the "elastic stiffness" (or slope) of the NSP to that of the IDA, i.e. by matching their elastic segments. The results of such a procedure are shown in Fig. 6.27 where we plot the NSP curve, obtained using a first-mode lateral load pattern, against the median IDA for a 20- story steel moment-resisting frame having ductile connections. Qualitatively, we can make some general observations, which permit inference of the approximate shape of the median IDA simply from the characteristics of the NSP curve (Vamvatsikos and Cornell 2002). More specifically:

- By construction, the elastic region of the NSP curve matches well the IDA, including the first sign of non-linearity appearing at the same values of IM and EDP for both.
- A subsequent reduced, but still non-negative stiffness region of the NSP curve correlates on the IDA with the approximate 'equal-displacement' rule (for moderate-period structures), i.e. a near continuation of the elastic regime slope; in fact, this near-elastic part of the IDA is often preceded by a hardening portion.

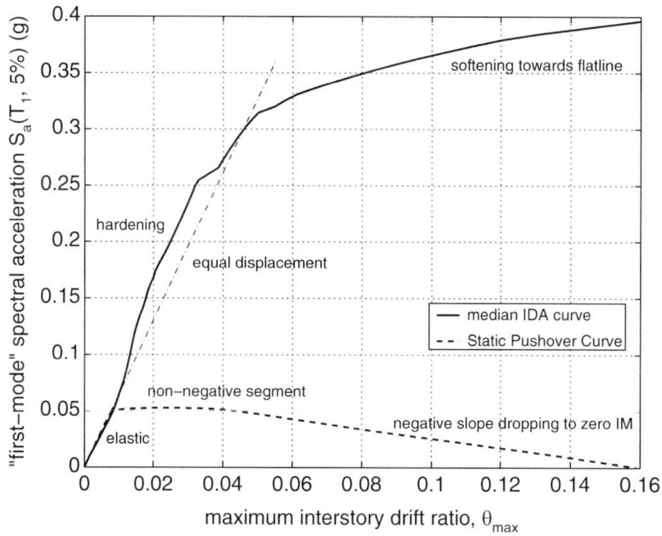

Fig. 6.27 Comparison of NSP and IDA curves for a 20-story steel frame (taken from (Vamvatsikos and Cornell 2002))

- A negative slope on the NSP curve translates to a "softening" region of the IDA, which can lead to collapse (indicated by flattening of the IDA to horizontal) unless it is arrested by a non-negative segment of the NSP curve before it reaches zero in IM terms.
- A non-negative region of the NSP curve that follows after a negative slope that has caused a significant IM drop, apparently presents itself in the IDA as a new, modified 'equal displacement' rule (i.e. an near-linear segment that lies on a secant) that has lower "stiffness" than the elastic.

6.6.3 IDA Capacity of SDOF Systems Using Approximate Methods

Quantitatively, it is possible to approximate the results of IDA using R-C_1-T (or R-μ-T) relationships available for SDOF systems in the literature. Among such relationships, the SPO2IDA tool and the IN2 method (Dolsek and Fajfar 2005) can approximate the median IDAs over the entire range of response for single- and multi-degree-of-freedom systems utilizing information from the force-deformation envelope (or backbone) of the static pushover. In the discussion that follows we adopt SPO2IDA for our calculations. Because the SPO2IDA set of equations (Vamvatsikos and Cornell 2006) incorporates fairly sophisticated routines to fit

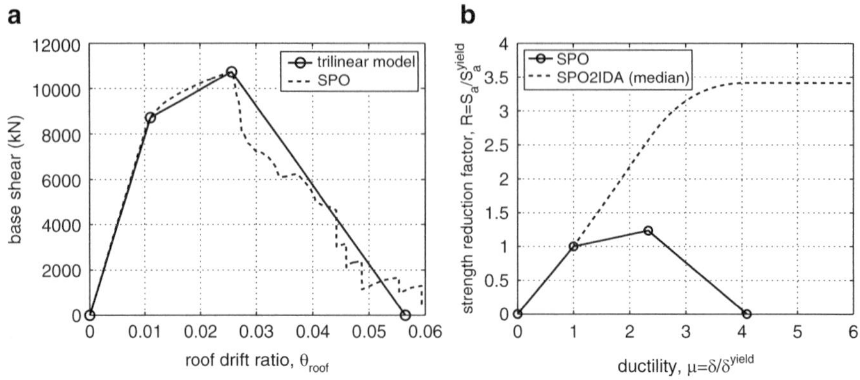

Fig. 6.28 (**a**) The static pushover curve for a nine-story steel structure and its trilinear approximation, and (**b**) the SPO2IDA prediction in normalized R-μ coordinates

response data for the particular oscillator parameters of interest, estimates have greater accuracy than the closed-form relationships used in R-C_I-T (or R-μ-T) relationships over a large range of oscillator parameter. The response data used for the SPO2IDA estimates are 5% damped SDOF systems featuring backbones that range from simple bilinear to complex quadrilinear. This data allows the SPO2IDA tool to provide estimates of response statistics (median and 16th and 84th percentile) considering record-to-record (aleatory) randomness. In the case of static pushover analysis, estimates of the global response (θ_{max} or θ_{roof}) on an ESDOF system can be obtained using a multilinear approximation of the static pushover curve.

For SDOF structures, IDA curves can be represented in normalized coordinates of the strength reduction factor, R and ductility μ. The strength reduction factor R is defined as the ratio $S_a(T_1,5\%)/S_a^{yield}(T_1,5\%)$, where $S_a^{yield}(T_1,5\%)$ is the $S_a(T_1,5\%)$ value to cause yield (equal to the base shear force at yield divided by the oscillator mass), while the ductility, μ, is the peak displacement of the oscillator, δ, normalized by the yield displacement, δ^y. Thus, once the period and the properties of the force-displacement relationship are known for the SDOF system, SPO2IDA directly provides estimates of the 16th, 50th, and 84th fractile demand and capacity in normalized R, μ coordinates. The application of the method requires a multilinear approximation of the static pushover curve to determine the properties of the backbone curve (Fig. 6.28). A trilinear or a quadrilinear approximation of the structure's pushover curve is recommended to allow characterization of the collapse, defined as the point that large increases in lateral displacement occur with diminishingly small changes in $S_a(T_1,5\%)$ or R.

Capacity of MDOF systems

Once the approximation of the IDA curve is available in R-μ coordinates, a set of algebraic calculations is carried out to characterize the IDA capacities of the corresponding MDOF structure. In the discussion that follows, we assume that

SPO2IDA is the R-C_I-T relationship, without any loss of generality. A thorough discussion on the procedure suggested can be also found in references (Vamvatsikos and Cornell 2005; Fragiadakis and Vamvatsikos 2010).

Since the capacities of SPO2IDA are in dimensionless R-μ coordinates, they need to be scaled to another pair of IM, EDP coordinates, more appropriate for MDOF systems, such as the $S_a(T_1,5\%)$ and the maximum inter story drift ratio θ_{max}. The scaling from R-μ to $S_a(T_1,5\%)$-θ_{max} is performed with simple algebraic calculations:

$$\begin{aligned} \boldsymbol{S_a}(T_1,5\%) &= \mathbf{R}\, S_a^{yield}(T_1,5\%) \\ \boldsymbol{\theta}_{\rm roof} &= \boldsymbol{\mu}\, \theta_{\rm roof}^{yield} \end{aligned} \tag{6.5}$$

The bold font denotes a vector quantity, thus $\mathbf{R}$ and $\boldsymbol{\mu}$ are the ordinates and abscissas of the IDAs in normalized coordinates, respectively. Once $\boldsymbol{\theta}_{\rm roof}$ is known, $\boldsymbol{\theta}_{\rm max}$ can be extracted from the results of the static pushover, since for every load increment the correspondence between the two EDPs is always available. Prior to applying Equations (5) we have to determine the values of $S_a(T_1,5\%)$ and $\theta_{\rm roof}$ at yield. This task is trivial for SDOF systems, but it is not straightforward for MDOF structures. Due to the effect of higher modes some records will force the structure to yield earlier and others later, thus yielding will always occur at different levels of $S_a(T_1,5\%)$ and $\theta_{\rm roof}$. Taking advantage of the approximation of the static pushover curve, we assume that the yield roof drift is approximately equal to the yield point of the multilinear approximation. This assumption is not strictly true for MDOF structures. It is precise only if the first mode is dominant, but it is sufficient for our purpose. Therefore, the accurate estimation of $S_a^{yield}(T_1,5\%)$comes down to approximating the elastic "slopes" of the median IDA curves plotted with $\theta_{\rm roof}$ as the EDP. The slope, denoted as $k_{\rm roof}$, is the median value obtained using elastic response history analysis with a few ground motion records, or simply by using standard response spectrum analysis. For first-mode dominated systems, a quick estimate can also be obtained by employing a first-mode approximation via the roof displacement participation factor, e.g. the C_0 factor defined in ASCE/SEI 41-06 (ASCE 2007).

For example, using the target displacement equation of ASCE/SEI 41-06 and given that in the elastic range the coefficients C_1, C_2, C_3 are equal to one, the roof drift and the IDA slope $k_{\rm roof}$ are obtained as:

$$\theta_{\rm roof} = \frac{\delta_{\rm roof}}{H} = C_0 S_a \frac{T_1^2}{4\pi^2 H} g \tag{6.6}$$

$$k_{\rm roof} = \frac{S_a(T_1,5\%)}{\theta_{\rm roof}} = \frac{4\pi^2 H}{C_0 T_1^2 g} \tag{6.7}$$

where H is the height of the building and g the surface gravity acceleration in appropriate units. Finally, we obtain:

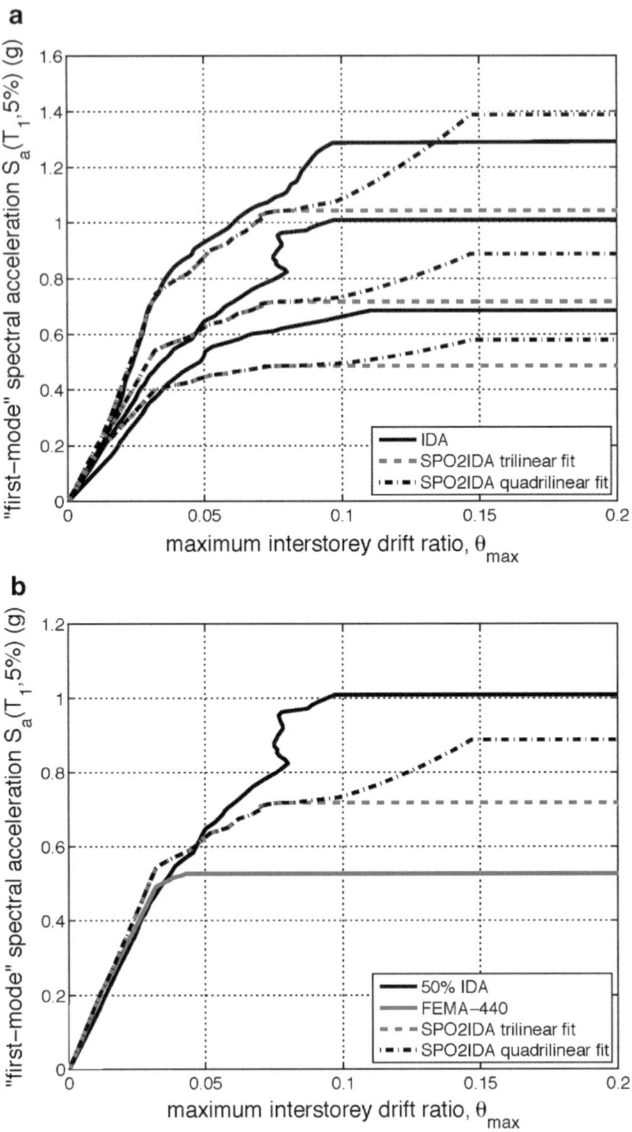

Fig. 6.29 (**a**) IDA median and 16, 84% capacity curves and its corresponding approximation using the proposed approach. The difference between the trilinear and the quadrilinear approximation is demonstrated. (**b**) Approximation of the median IDA when different R-C_l-T approximations are followed (taken from (Fragiadakis and Vamvatsikos 2010))

$$S_a^{yield}(T_1, 5\%) = k_{\mathrm{roof}}\theta_{\mathrm{roof}}^{yield} \tag{6.8}$$

For the SPO curve of Fig. 6.29 the median IDA obtained with SPO2IDA and the actual IDA curve using thirty ordinary ground motion records are shown in Fig. 6.29. For our model, the error in the conditional $S_a(T_1,5\%)$ capacities is typically 10–20%, while the computing time comes down from 2–3 h required for a single IDA to just a couple of minutes for SPO2IDA, approximately two orders of magnitude less. In addition, while a preliminary design must have been established prior to the NRHAs required for the IDAs, SPO2IDA can be used to establish constraints required to ensure the preliminary design will have acceptable seismic performance. It is worthwhile to note that compared to the quadrilinear pushover approximation, for the 9-story steel frame, the trilinear curve slightly biases our IDA results towards lower S_a-capacities. For comparison purposes we also show the capacities obtained with the FEMA-440 relationship. FEMA-440 is accurate enough only for low elastic and nearly-elastic $S_a(T_1,5\%)$ intensities, since its has been suggested for a different purpose and therefore is deliberately conservative.

6.7 Conclusions

A qualitative and quantitative comparison between NSP methods and nonlinear response history analysis has been presented. It is shown that the accuracy of NSP methods depends on the properties of the building, the EDP studied and the level of inelastic demand. Both the standard, 1st mode-based NSP, and more elaborate NSPs were included in the comparison. It is concluded that no simple method exists that is consistently reliable and generally applicable to multistory buildings, and therefore NRHA remains the most powerful approach for seismic performance evaluation. NSPs can be used to provide insight to the building's capacity, helping the engineer to understand how the system will respond from a global perspective, and can be used as in aid in preliminary design in performance-based earthquake engineering. To this cause we investigate the relationship between the global results of static pushover and Incremental Dynamic Analysis. The direct comparison is possible and can be either performed in the setting of IDA or NSP by plotting versus an intensity measure or versus base shear, respectively. Distinct similarities and differences appear in each setting, offering different insight into the structural behavior. All in all, while important conclusions can be derived from NSP results, appropriate care is advised in all applications of NSP methods when used for quantitative, rather than qualitative, estimation of seismic performance.

Acknowledgments This chapter relies, in part, on results obtained under Task Order 6 of the NEHRP Consultants Joint Venture (a partnership of the Applied Technology Council and Consortium of Universities for Research in Earthquake Engineering), under Contract SB134107CQ0019, Earthquake Structural and Engineering Research, issued by the National Institute of Standards and Technology. The views expressed do not necessarily represent those of the organizations represented above. The authors also want to acknowledge Dr. Curt Haselton for providing the OpenSees input files of the RC moment frames.

References

Antoniou S, Pinho R (2004) Development and verification of a displacement-based adaptive pushover procedure. J Earthquake Eng 8(5):643–661

ASCE (2007) Seismic rehabilitation of existing buildings, ASCE Standard ASCE/SEI 41-06. American Society of Civil Engineers, Reston

Aschheim M, Tjhin T, Comartin C, Hamburger R, Inel M (2007) The scaled nonlinear dynamic response. Eng Struct 29:1422–1441

Aydinoglou MN (2003) An incremental response spectrum analysis procedure based on inelastic spectral displacements for multi-mode performance evaluation. Bull Earthquake Eng 1:3–36

Chopra AK, Goel RK (2002) A modal pushover analysis procedure for estimating seismic demands for buildings. Earthquake Eng Struct Dyn 31(3):561–582

NEHRP Consultants Joint Venture (2010) ATC 76-6: Improved nonlinear static seismic analysis procedures –multiple-degree-of-freedom modeling, Applied Technology Council for the Federal Emergency Management Agency, Washington, DC

Dolsek M, Fajfar P (2005) Simplified non-linear seismic analysis of infilled reinforced concrete frames. Earthquake Eng Struct Dyn 34(1):49–66

Eurocode 8 (2004) Design of structures for earthquake resistance — Part 1: General rules, seismic actions and rules for buildings. CEN, European Committee for Standardization, Brussels

Fajfar P (1999) Capacity spectrum method based on inelastic demand spectra. Earthquake Eng Struct Dyn 28(9):979–993

Fajfar P, Fischinger M (1988) N2 – A method for non-linear seismic analysis of regular buildings. Proceedings of the 9th world conference on earthquake engineering, vol 5, Tokya, pp 111–116

FEMA (2009) Quantification of seismic performance factors. FEMA P-695 Report, prepared by the Applied Technology Council for the Federal emergency Management Agency, Washington, DC

Fragiadakis M, Vamvatsikos D (2010) Fast performance uncertainty estimation via pushover and approximate IDA. Earthquake Eng Struct Dyn 39(6):683–703

Fragiadakis M, Ioannidou D, Papadrakakis M (2007) Assessment of nonlinear static analysis procedures in the framework of performance-based design. 8th National Congress on Mechanics (HSTAM2007), Patras, Greece, 12–17 June 2007

Goel RK, Chopra AK (2005) Extension of modal pushover analysis to compute member forces. Earthquake Spectra 21(1):125–139

Haselton C, Deierlein G (2007) Assessing seismic collapse safety of modern reinforced concrete moment frame buildings. Report No. 156, John A. Blume Earthquake Engineering Center. Department of Civil and Environmental Engineering, Stanford University, Stanford

Haselton CB, Liel AB, Deierlein GG, Dean BS, Chou JH (2010) Seismic Collapse Safety of Reinforced Concrete Buildings: I. Assessment of Ductile Moment Frames. J Struct Eng. doi:10.1061/(ASCE)ST.1943-541X.0000318

International Code Council (2002) 2003 International Building Code (IBC). International Code Council, Country Club Hills

Kunnath S, Erduran E (2008) Pushover procedures for seismic assessment of buildings: issues, limitations and future needs. Nonlinear static methods for design/assessment of 3D structures. IST press, Lisbon

Lignos DG, Krawinkler H (2009) Sidesway collapse of deteriorating structural systems under seismic excitations, Report No. TB 172, The John A. Blume Earthquake Engineering Research Center, Stanford University, Stanford

McKenna F, Fenves GL, Scott MH, Jeremic B (2008) Open System for Earthquake Engineering Simulation, http://OpenSees.berkeley.edu

Mwafy AM, Elnashai AS (2001) Static pushover versus dynamic collapse analysis of RC buildings. Eng Struct 23:407–24

Poursha M, Khoshnoudian F, Moghadam AS (2009) A consecutive modal pushover procedure for estimating the seismic demands of tall buildings. Eng Struct 31(2):591–599

Vamvatsikos D, Cornell CA (2002) Incremental dynamic analysis. Earthquake Eng Struct Dyn 31(3):491–514

Vamvatsikos D, Cornell CA (2005) Direct estimation of seismic demand and capacity of multidegree-of-freedom systems through incremental dynamic analysis of single degree of freedom approximation. J Struct Eng 131(4):589–599

Vamvatsikos D, Cornell CA (2006) Direct estimation of the seismic demand and capacity of oscillators with multi-linear static pushovers through IDA. Earthquake Eng Struct Dyn 35(9):1097–1117

Chapter 7
Seismic Design Modification Factors for Steel SMRFs for Uniform Collapse Safety

Farzin Zareian, Dimitrios G. Lignos, and Helmut Krawinkler

Abstract This chapter summarizes a study utilizing the methodology for evaluation of design modification factors: R, Ω_0, and C_d, for Steel Moment-resisting Frames (SMFs). Archetypes comprising 3-bay special SMFs that serve as lateral load resisting system of buildings ranging from 1 to 20 stories are designed based on current US code provisions. Nonlinear models are developed using recent advances in structural component modeling and are analyzed to predict the collapse capacities of each design. The adjusted collapse margin ratios (ACMR) are evaluated and compared to acceptance criteria. The results indicate that presently employed R- factors along with current design provisions for SMFs can lead to unacceptable ACMR values for long period SMFs. An increase in Column-Beam moment ratio from the minimum code requirement of 1.0 to a larger value can significantly improve long period SMFs behavior leading to an acceptable ACMR values.

Keywords Steel Moment-resisting frames • Seismic Design • Collapse • Strength reduction factor • Column-beam moment strength ratio

---

F. Zareian (✉)
Department of Civil and Environmental Engineering, University of California, Irvine, CA 92697, USA
e mail: zareian@uci.edu

D.G. Lignos
Department of Civil Engineering and Applied Mechanics, McGill University, Montreal, QC H3A 2K6, Canada
e-mail: dimitrios.lignos@mcgill.ca

H. Krawinkler
Department of Civil and Environmental Engineering, Stanford University, Stanford, CA 94305, USA
e-mail: krawinkler@stanford.edu

M. Dolšek (ed.), *Protection of Built Environment Against Earthquakes*,
DOI 10.1007/978-94-007-1448-9_7, © Springer Science+Business Media B.V. 2011

7.1 Introduction

This chapter summarizes an evaluation of collapse performance of steel Special Moment-Resisting Frames (SMFs) designed according to current US seismic provisions. The methodology outlined in (FEMA P695 2009) is utilized for this purpose. An effective way for enhancing the collapse performance for those SMFs that fail to achieve the targeted collapse performance objective is proposed. In particular, we are interested to investigate whether current SMF design procedures provide an acceptable margin of safety against collapse, and show that by utilizing a larger Column-Beam moment ratio compared to the minimum (AISC ANSI/AISC 341–05 2005) requirement such an acceptable margin of safety against collapse can be achieved for those SMFs that are designed according to present US design requirements but do not pass the performance test.

7.2 FEMA P695 Collapse Safety Assessment Methodology

FEMA P695 provides a methodology for quantifying building system performance in the context of collapse safety. In this methodology, the safety margin against collapse of a lateral load resisting system designed with a specific response modification coefficient (R factor) is quantified and compared with an acceptable safety margin obtained based on a tolerable probability of collapse and level of variability in estimation of such probability.

In the FEMA P695 collapse performance assessment process design provisions are gathered, substantiated by component test information and professional design experience. Structures are designed that are representative of the current building stock and follow the aforementioned design provisions. These designs are denoted as "Archetypes". Analytical models of these archetypes are developed using state-of-the-art modeling techniques that take advantage of experimental data. By incorporating uncertainty from various sources including *Modeling Quality*, *Design Requirements*, and *Test Data*, and by utilizing nonlinear response history analysis for a prescribed set of ground motions that are scaled up until collapse is predicted, one can estimate the safety margin against collapse for each archetype model. The acceptance criteria, according to FEMA P695, for collapse safety are twofold: (1) a minimum required collapse margin for individual archetypes equivalent to less than 20% probability of collapse at the MCE (Maximum Considered Earthquake) hazard level, and (2) a minimum required collapse margin for a family of archetypes (denoted as Performance Groups that represent a group of archetypes with defined characteristics) equivalent to less than 10% average probability of collapse at the MCE hazard level for the performance group. If an individual system's collapse safety margin or the performance group collapse safety margin does not meet the required performance, seismic response factor R should be modified and archetypes should be re-designed. The iterative loop continues until the proposed seismic response factors can provide adequate collapse safety.

7.3 Application of FEMA P695 for Evaluation of Seismic Collapse Performance of Steel SMFs

7.3.1 Design Provisions

We have utilized AISC 341–05 design requirements and connection design criteria provided in (AISC ANSI/AISC 358–05 2005) for design of archetypes used for evaluation of seismic collapse performance of SMFs. Member sizes are determined based on strength, drift and P-Delta criteria. For SMFs a very important consideration is the selection of the seismic design procedure. Within limitations, both the Equivalent Lateral Force (ELF) and the Response Spectrum Analysis (RSA) procedures are used. The designs resulting from the two procedures are substantially different in stiffness and strength properties. Use of the RSA procedure will result often in a design base shear smaller than that based on the ELF procedure (although limited to a minimum of 85% of the ELF base shear). But more important for taller structures, the (ASCE 7–05 2006) minimum base shear requirements must be considered in drift design if the ELF procedure is used, whereas they do not have to be considered if the RSA procedure is employed. For taller structures this can result in very large differences in member sizes.

7.3.2 Archetype Configurations

All archetypes comprise 3-bay moment resisting frames that serve as the lateral load resisting system of buildings whose plan view is shown in Fig. 7.1a. We have varied the number of stories (e.g., 1, 2, 4, 8, 12, 20), with the height of the first story equal to 15 ft, and the height of all other stories equal to 13 ft. SMF archetype configurations used in this study cover the design space with variations in seismic design categories (e.g., D_{max}, and D_{min}), and design procedure (e.g., ELF, and RSA). Although it is rare to use the ELF as the design procedure for buildings taller than 4 stories, we have considered this variation to assess the collapse potential of a complete set of buildings designed according to ASCE 7–05. In accordance with FEMA P695 (and exception to ASCE 7–05) we assumed $C_d = R = 8$ (ASCE 7–05 uses $C_d = 5.5$). All connections are RBS (Reduced Beam Section) connections designed in accordance with AISC 358–05. The RBS parameters are specified as follows: $a = 0.625b_f$, $b = 0.75d_b$, and $c = 0.250b_f$. Column bases of the 1- and 2-story SMFs are hinged and the column bases of all other (taller) structures are fixed.

Properties of archetype designs used in this study are tabulated in the left part of Table 7.1. Seismic demands are represented by the maximum and minimum criteria of Seismic Design Category (SDC) D, in accordance with Sect. 5.2.1 of

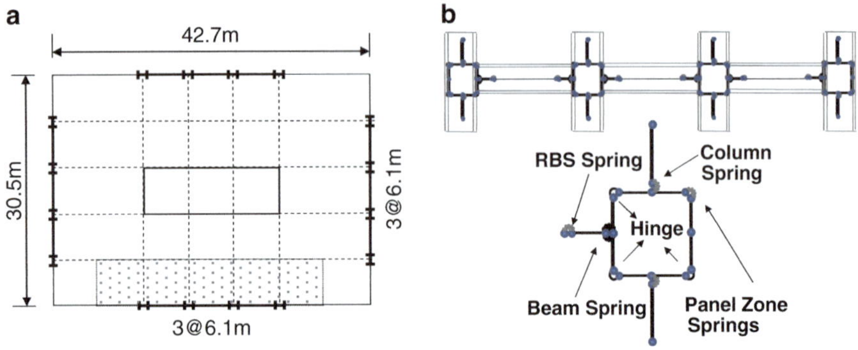

Fig. 7.1 Typical floor plan and corresponding modeling technique used for SMF archetypes: (**a**) Archetype plan view, (**b**) Typical beam-to-column connection model

FEMA P695: $S_{DS} = 1.0$ g and $S_{D1} = 0.60$ g for SDC D_{max}, and $S_{DS} = 0.50$ g and $S_{D1} = 0.20$ g for SDC D_{min}. T represents the code maximum period from equation $T = C_u T_a$ and T_1 represents the period from the computer analysis model. The value *V/W* is the base shear coefficient for code strength design, but this value did not control most of the member sizes because all structures are drift or P-Delta controlled.

Different code provisions control the final archetype designs. Exterior columns of lower stories in taller buildings are often controlled by M-P interaction or code overstrength requirements. Since the archetype designs are based on $C_d = R = 8$ rather than the code required 5.5, and because drift considerations control many of the member sizes, the archetypes designed as part of this study are often stiffer and stronger than required by ASCE 7–05 design requirements. The difference in member sizes between ELF and RSA designs are relatively small for low-rise SMFs, with the RSA member sizes being slightly smaller, and become larger as the number of stories increase. When the RSA procedure is employed, the minimum base shear requirements are explicitly excluded from consideration in drift calculation, but this is not the case when the ELF procedure is employed leading to significant differences in design of tall structures using ELF and RSA procedures. For the 20-story D_{max} archetypes the ratio of base shears for drift design is about 3.0 for ELF versus RSA designs. This is reflected in a difference in the elastic stiffness of the structures by a factor of 3.

7.3.3 Archetype Structural Models

We have developed 2-dimensional models for evaluation of collapse performance of archetypes using element models that incorporate stiffness and strength deterioration and permit modeling of response to collapse in a modified version of the

Table 7.1 SMF archetype design properties and collapse performance evaluation

Design configuration				Pushover and IDA results							Accept. check (Ind.)	Accept. check (Gen.)	
No. stories	T [sec.]	T_1 [sec.]	V/W [g]	Static Ω	S_{MT} [g]	S_{CT} [g]	CMR	μ_T	SSF	ACMR	Accept. ACMR	Avg. ACMR	Accept. ACMR
Performance group No. PG-1 ELF D_{max}													
1	0.34	0.71	0.125	4.63	1.50	3.81	2.54	3.8	1.21	3.08	1.52(P)	2.97	1.90(P)
2	0.56	0.87	0.125	3.69	1.50	3.42	2.28	4.7	1.26	2.86	1.52(P)		
Performance group No. PG-2 ELF D_{max}													
4	0.95	1.30	0.079	2.88	0.95	1.72	1.81	5.3	1.35	2.45	1.52(P)	1.94	1.75(P)
20	3.37	2.48	0.044	4.45	0.27	0.35	1.33	1.1	1.08	1.43	1.36(P)		
Performance group No. PG-3 ELF D_{min}													
1	0.37	1.62	0.062	2.55	0.75	1.63	2.18	2.5	1.16	2.53	1.47(P)	2.53	1.81(P)
Performance group No. PG-4 ELF D_{min}													
2	0.60	1.74	0.042	3.44	0.50	1.15	2.31	2.2	1.15	2.66	1.44(P)	3.13	1.77(P)
4	1.02	1.94	0.024	4.71	0.29	0.95	3.24	4.0	1.31	4.23	1.52(P)		
20	3.61	3.44	0.022	4.67	0.08	0.18	2.14	1.5	1.17	2.51	1.39(P)		
Performance group No. PG-1 RSA D_{max}													
1	0.34	0.71	0.106	5.46	1.50	3.31	2.54	3.8	1.21	3.08	1.52(P)	2.91	1.90(P)
2	0.56	0.91	0.106	3.94	1.50	3.33	2.22	4.0	1.23	2.73	1.52(P)		
Performance group No. PG-2 RSA D_{max}													
4	0.95	1.62	0.067	2.21	0.95	1.38	1.46	4.8	1.33	1.95	1.52(P)	1.76	1.81(F)
8	1.64	2.29	0.039	3.27	0.55	0.78	1.42	2.7	1.30	1.85	1.50(P)		
12	2.25	3.12	0.037	2.68	0.40	0.55	1.38	2.4	1.27	1.75	1.46(P)		
20	3.37	4.47	0.037	2.27	0.27	0.32	1.21	1.9	1.22	1.49	1.42(P)		
Performance group No. PG-3 RSA D_{min}													
1	0.37	1.66	0.053	2.84	0.75	1.77	2.36	2.3	1.15	2.72	1.46(P)	2.72	1.78(P)
Performance group No. PG-4 RSA D_{min}													
2	0.6	1.83	0.035	3.59	0.50	1.49	2.97	2.9	1.19	3.55	1.51(P)	3.28	1.84(P)
4	1.02	2.62	0.021	3.27	0.29	0.52	2.10	2.7	1.24	2.60	1.49(P)		
8	1.75	3.55	0.019	2.71	0.17	0.43	2.52	3.6	1.29	3.24	1.48(P)		
12	2.41	4.43	0.019	2.48	0.12	0.37	2.99	3.0	1.33	3.97	1.52(P)		
20	3.61	5.74	0.019	2.61	0.08	0.20	2.40	2.3	1.27	3.04	1.46(P)		

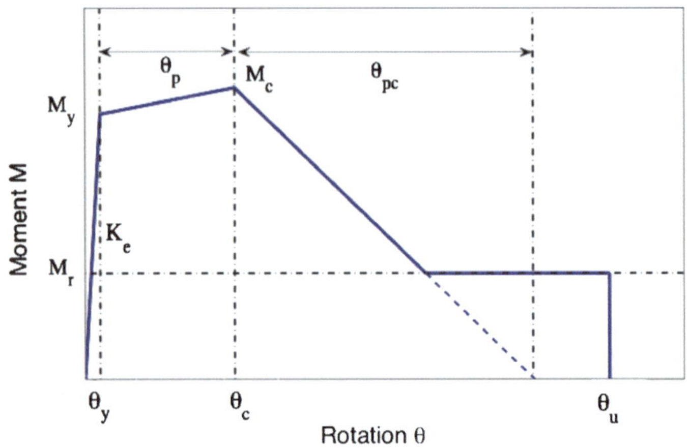

Fig. 7.2 Parameters of the monotonic backbone curve of the modified Ibarra-Krawinkler model. Effective yield strength and rotation (M_y and θ_y); Effective elastic stiffness $K_e = M_y / \theta_y$; Strength cap and associated rot. for monotonic loading (M_c and θ_c); Pre-capping plastic rotation for monotonic loading θ_p; Post-capping rotation θ_{pc}Residual strength M_r; Ultimate rotation θ_u

program Drain-2DX (Prakash et al. 1993). Each beam and column is modeled as an elastic element, with all inelastic behavior concentrated in plastic hinge regions at the member ends. A typical floor model is shown in Fig. 7.1b. It consists of three elastic beam elements spanning between points of RBS connections, six elastic beam elements between the RBS connections and the column faces, four parallelograms representing the joint panel zones, and elastic column elements framing into the parallelograms. Details of the spring models can be found in (PEER/ATC 72–1 2010). Salient features of the analytical model are:

- P-Delta effects are modeled with a leaning column with zero flexural stiffness placed in parallel to the frame. This leaning column is loaded with a vertical load at each floor level that represents 1.05D + 0.25 L of half of the structure where D represents design dead load and L represents unreduced live load.
- The panel zone is modeled with eight rigid elements connected with hinges at three corners and with two bilinear rotational springs in parallel at the fourth corner (shown with a rotational spring on the upper right corner of the panel zone model). The rotational springs represent the panel zone shear force – shear deformation behavior with a trilinear hysteretic model.
- The inelastic behavior at plastic hinge regions in beams (at RBS sections) is represented by rotational springs with appropriate strength, stiffness, and deterioration properties (see Fig. 7.2). A monotonic backbone curve that is capable of capturing monotonic deterioration in stiffness and strength is used in the analytical model. A bilinear hysteresis model is used to simulate the basic cyclic characteristics of plastic hinges in steel beams and columns. Cyclic deterioration rules developed by (Ibarra et al. 2005) and modified by

(Lignos and Krawinkler 2007, 2009) are used to model cyclic deterioration in steel components. The rate of cyclic deterioration depends on the deterioration parameter Λ, which defines a reference energy dissipation capacity for the component expressed as $E_t = \Lambda M_y$, with $\Lambda = \lambda\theta_p$ denoting the cumulative plastic rotation capacity. Utilization of these cyclic deterioration rules permits modeling of basic and post-capping strength deterioration as well as unloading stiffness deterioration. Parameters of this deterioration model have been calibrated based on moment rotation relationships deduced from a database of about 300 steel components (Lignos and Krawinkler 2007, 2009). For this specific study we used effective yield strength $M_y = 1.1{\times}M_p$, with M_p based on expected yield strength of 379.2 MPa, $M_c/M_y = 1.1$, $M_r/M_y = 0.4$, $\theta_u = 0.2$, θ_p, θ_{pc}, and Λ as obtained for RBS connections from regression equations derived by (Lignos and Krawinkler 2009) and presented in (PEER/ATC 72–1 2010). The effects of a composite floor slab on modeling parameters is not considered.

- In order to account, approximately, for the effect of axial force on column bending strength, axial force is estimated from the pushover analysis as $P_{grav} + 0.5P_{E,max}$, where $P_{E,max}$ is the maximum axial force due to the first mode lateral load pattern. The reduced bending strength is determined from this axial force using the AISC P-M interaction equation, and this reduced bending strength is used in the response history analysis. It is recognized that the bending strength will vary as a function of axial force, but this compromise had to be made because the presently employed deterioration models cannot account for the effect of a variable axial force on bending strength.

7.3.4 *Treatment of Uncertainty in Estimation of Collapse Capacity*

Considering the criteria described in FEMA P695 we have categorized the *Test Data* as "B-Good," *Modeling Quality* as "B-Good," and *Design Requirements* as "A-Superior" to calculate the total system collapse uncertainty β_{TOT}. Details justification on this issue can be found in (NIST 2010). The rating of "B-Good" for *Test Data* is in light of shortage of available test data on the inelastic behavior of deep columns subjected to high axial forces and cyclic bending moments, and shortage of beam-to-column subassembly tests that quantify the effects of a composite slab on component strength and stiffness. The structural model (*Modeling Quality*) is given the "B-Good" rating because there is still room for improvement in the model, particularly in modeling plastic hinging in columns. Finally, *Design Requirements* for SMFs are categorized as "A-Superior" since they represent many years of development and include lessons learned from major earthquakes.

7.3.5 Quantification of SMFs Margin of Safety Against Collapse

Nonlinear response history analysis is performed to estimate the collapse margin ratio (*CMR*). *CMR* is obtained as the ratio of the median *Sa* of the "scaled" 44 far filed ground motion set provided by FEMA P695 at the code period C_uT_a (denoted as $\hat{S}_{CT}$) to the MCE level ground motion spectral demand (S_{MT}) at the same period. The ground motion set is "scaled" such that 22 out of 44 motions cause the structural system to collapse. It is assumed that Rayleigh damping of 2.5% is assigned at the first mode period of the analytical model T_1 and at $T = 0.2T_1$.

Performance evaluation is accomplished by comparing the adjusted collapse margin ratio, *ACMR*, of the structure with an acceptable *ACMR*. The adjusted collapse margin ratio accounts for the effect of the spectral shape on the median collapse capacity using a spectral shape factor (*SSF*). *SSF* depends on the fundamental period (*T*), period-based ductility (μ_T), and seismic design category. The adjusted collapse margin ratio (*ACMR*) is computed by multiplying *SSF* and *CMR* values. Nonlinear static (pushover), in accordance with FEMA P695, is used to compute μ_T and Ω (system overstrength factor). The *SSFs*, together with the μ_T values on which they are based, Ω values, and the resulting *ACMR* ratios are listed in Table 7.1.

In order to assess acceptability of performance, the composite uncertainty (β_{TOT}) in collapse capacity is needed. In accordance with FEMA P695, this composite uncertainty is computed as $\beta_{TOT} = \sqrt{\beta_{RTR}^2 + \beta_{DR}^2 + \beta_{TD}^2 + \beta_{MDL}^2}$ where $\beta_{DR} = 0.1$ (superior *Design Requirements*), $\beta_{TD} = 0.2$ (good *Test Data*), $\beta_{MDL} = 0.2$ (good *Modeling Quality*), and $\beta_{RTR} = 0.40$ for systems with $\mu_T \geq 3$, and $\beta_{RTR} = 0.1 + 0.1\mu_T$ for $\mu_T < 3.0$.

Table 7.1 shows the estimates of *ACMR* values for SMF archetypes along with acceptable *ACMR* values for the each collapse safety criterion. The *ACMRs* do not follow regular patterns. For archetypes in the D_{max} design category they usually decrease with the number of stories, but for archetypes in the D_{min} design region they often increase with the number of stories. The reasons for irregular patterns are many, but include dominance of different design criteria for different structures. Member sizes for archetypes in the D_{max} design region they are drift controlled, whereas lower story member sizes in the D_{min} design region are mostly P-Delta controlled. These variations in controlling design conditions have a dominant effect on the collapse capacity of individual archetypes. All individual archetypes and almost all performance groups pass the acceptability check. The exception is PG-2 RSA (long-period RSA designs in the D_{max} region), whose relatively poor performance is dominated by low *ACMRs* of the 12 and 20-story archetypes. In the next section we will discuss the source of this relatively poor performance and recommend a remedy for performance enhancement.

7.4 Significance of Column-Beam Moment Ratio in Enhancement of Collapse Performance of Long Period SMFs

The long-period RSA designs in the D_{max} region (PG-2 RSA) did not pass the acceptability check. For these long period structures, the amplification of story drifts in the lower stories, which is caused by P-Delta effects, dominates response in the highly inelastic range and leads to development of a partial mechanism that involves the lower few stories of the structural system. We expect that the collapse performance of these structures (PG-2 RSA) can be enhanced by increasing the minimum Column-Beam Moment (CBM) strength ratio from the suggested code value of unity to a larger value (e.g., 1.2 and 2.4). CBM strength ratio for a joint is defined as the ratio of the sum of the nominal flexural strength in the column above and below the joint to the sum of expected flexural strength in the beams at the joint. Zareian and Krawinkler (2009) have shown that an increase in design CBM ratio can change the failure mechanism to a mode in which column hinging is postponed and therefore the structure is more capable of redistributing the nonlinear deformations to beams in several stories.

In order to investigate the effect of an increase in design minimum CBM strength ratio on ACMR estimates of steel SMFs in the PG-2 RSA performance group, we have redesigned these structures using design minimum CBM strength ratios of 1.2 and 2.4. Figure 7.3 shows the deformation profile and distribution of plastic hinges along the height of three 12-Story SMFs designed for seismic design category D_{max} using RSA method subjected to first model shape lateral load pattern. The three designs are different in column sizes, from left to right in which design minimum CBM strength ratio is increased from 1.0 to 1.2, and 2.4. The gray color circles show the location of plastic hinges and the radius of each circle shows the relative magnitude of the plastic rotations. Figure 7.3 clearly shows the benefit in increasing the minimum CBM strength ratio used for design of SMFs by changing the failure mechanism and postponing column hinging. The enhancement in the deformation capacity has been to the extent that for 12-story SMF designed with CBM > 2.4, we are able to push the structure up to roof drift ratio of 0.040 compared to the smaller value of 0.023 and 0.020 for 12-story SMF designed with CBM > 1.2 and CBM > 1.0, respectively. Similar plots can be generated for other SMFs in the PG-2 RSA performance group, showing the same trend in enhancement of overall structural performance.

The global pushover curve tells much about the effects of the employed design minimum CBM strength ratio and about anticipated behavior under nonlinear dynamic analysis and the collapse potential of the archetypes. Figure 7.4a shows the global pushover curves for the three variations of the 12-Story archetype in PG-2 RSA. The impact of using a larger value of minimum CBM strength ratio in the design process can be evaluated from the differences in maximum strength and post-elastic deformation capacity. Figure 7.4a in combination with Fig. 7.3 shows that the P-Delta generated amplification of story drifts in the lower stories becomes

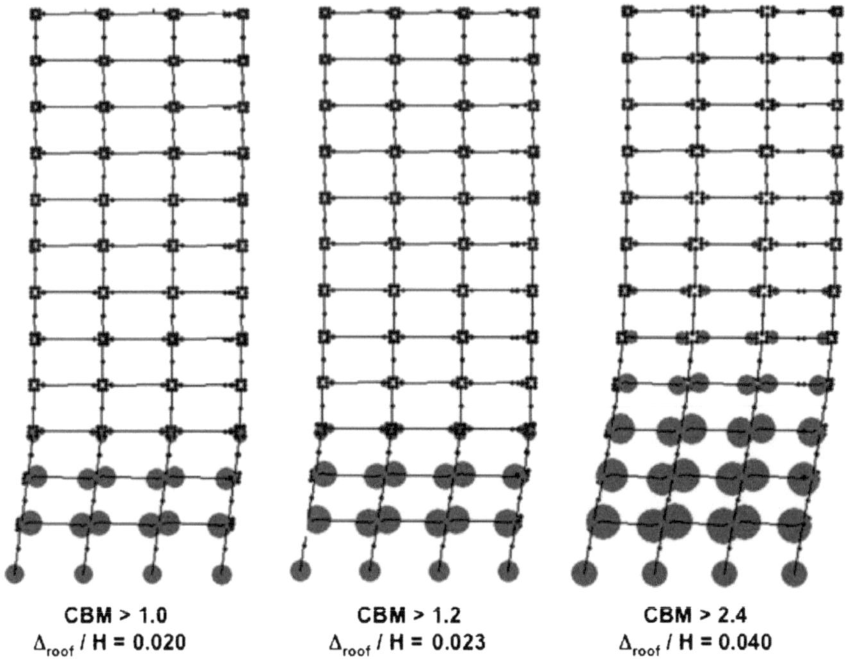

Fig. 7.3 Deformation profile and distribution of plastic hinges along the height of 12-story SMF designed for seismic design category D_{max} using RSA method subjected to first model shape lateral load pattern

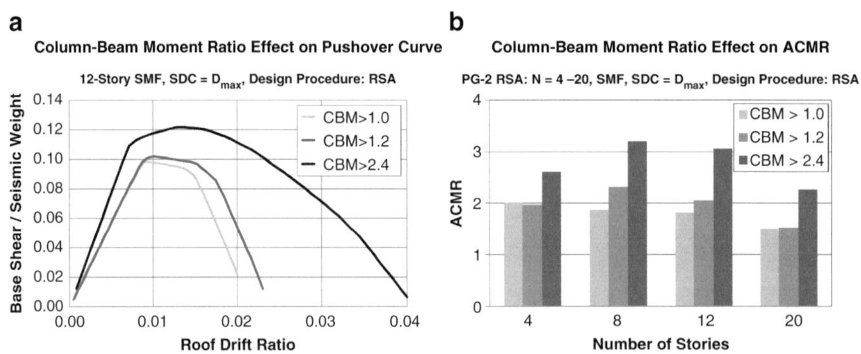

Fig. 7.4 Relative performance of 12-story SMF designed for seismic design category D_{max} using RSA method as function of design minimum CBM ratio: (**a**) Pushover curve, (**b**) ACMR

less dominant by increasing design minimum CBM strength ratio, which leads to development of a mechanism that involves more of the lower stories of the structural system. In the context of FEMA P695, maximum strength and post-elastic deformation capacity are described in terms of Ω, and μ_T, respectively.

The value of (Ω, μ_T) is increased from (2.68, 2.35) for CBM > 1.0, to (2.75, 2.61) and (3.28, 3.97) for CBM > 1.2 and CBM > 2.4, respectively, quantifying the enhancement in overall behavior of a SMF system when inelastic action is distributed among more stories and more beams.

The computed ACMRs, obtained from nonlinear response history analysis, are presented graphically in Fig. 7.4b for 4–20 story SFM archetypes designed for seismic design category D_{max} using RSA method with design minimum CBM equal to 1.0, 1.2, and 2.4. The performance of these structures is enhanced once a larger minimum CBM strength ratio is used in the design process, leading to an increase in the average ACMR values from 1.80 for CBM > 1.0, to 1.96 and 2.79 for CBM > 1.2 and CBM > 2.4, respectively. When these ACMR estimates are compared to the Acceptable ACMR values of 1.82, 1.84, and 1.90 for CBM > 1.0, CBM > 1.2, and CBM > 2.4, it is seen that the PG-2 RSA passes the acceptability check if a CBM > 1.2 is adopted in the design process.

7.5 Conclusions

This chapter summarizes a study focused on evaluating the collapse performance of steel Special Moment-resisting Frames (SMFs) by application of the FEMA P695 methodology. The collapse performance of steel SMFs is enhanced by increasing the minimum requirement for Column-Beam Moment strength ratio from unity to a larger value. Our findings and conclusions are based on evaluating 3-bay SMF archetypes that serve as lateral load resisting system of buildings ranging up to 20 stories designed using Equivalent Lateral Force (ELF) procedure and Response Spectrum Analysis (RSA) procedure for seismic design categories $D_{\max}$ and $D_{\min}$. The caveat to any general conclusion on collapse safety and adequacy of presently employed *R* factors is that most member sizes are controlled by stiffness requirements (drift limitations and P-Delta considerations) and not by strength requirements directly related to an *R* factor. All individual archetypes did pass the FEMA P695 acceptability check (conditional collapse probability $\leq$ 20% at the MCE level), but taller RSA designed archetypes in category D_{max} did pass the check with only a small margin. The performance group of "long period" archetypes designed according to the RSA method in design category D_{max} (PG-2 RSA) did not pass the acceptability check of a conditional collapse probability $\leq$ 10% at the MCE level.

Increasing the minimum design requirement for Column-Beam Moment (CBM) strength ratio from unity to a larger value is the solution proposed in this chapter to enhance the collapse performance of steel SMFs. It is shown that an increase in the design minimum CBM strength ratio from 1.0 to 1.2 (and even better to a larger ratio) can increase the collapse margin ratio such that the performance group PG-2 RSA passes the collapse safety acceptability check. The increase in ACMR is due to the change in the collapse mechanism of the structural system, whereby an increase in the design minimum CBM strength ratio reduces the concentration of story drift demand in the first few stories and inelastic deformation is distributed among the beams in a larger number of stories.

Acknowledgments The work presented here was conducted as part of the ATC-76-1 Project "Quantification of Building System Performance and Response Parameters," funded by the NEHRP Consultants Joint Venture (a partnership of the Applied Technology Council and Consortium of Universities for Research in Earthquake Engineering), under Contract SB134107CQ0019, Earthquake Structural and Engineering Research, issued by the National Institute of Standards and Technology. Any opinions, findings, and conclusions or recommendations expressed in this chapter are those of the authors and do not necessarily reflect the views of the sponsors.

References

AISC ANSI/AISC 341–05 (2005) Seismic provisions for structural steel buildings. American Institute of Steel Construction Inc., Chicago

AISC ANSI/AISC 358–05 (2005) Prequalified connections for special and intermediate steel moment frames for seismic applications. American Institute of Steel Construction Inc., Chicago

ASCE 7–05 (2006) Minimum design loads for buildings and other structures. American Society of Civil Engineers

FEMA P695 (2009) Quantification of building seismic performance factors. Federal Emergency Management Agency, Washington, DC

Ibarra LF, Medina RA, Krawinkler H (2005) Hysteretic models that incorporate strength and stiffness deterioration. Earthq Eng Struct Dyn 34(12):1489–1511

Lignos DG, Krawinkler H (2007) A database in support of modeling of component deterioration for collapse prediction of steel frame structures. ASCE Structures Congress, Long Beach

Lignos DG, Krawinkler H (2009) Sidesway collapse of deteriorating structural systems under seismic excitations. Rep. No. TB 172. The John A. Blume Earthquake Engineering Research Center, Stanford

NIST (2010) Evaluation of the FEMA P695 methodology for quantification of building seismic performance factors. Rep NIST GCR 10- 10-917-8, prepared by the NEHRP Consultants Joint Venture for the National Institute of Standards and Technology, Gaithersburg

PEER/ATC 72–1 (2010) Modeling and acceptance criteria for seismic design and analysis of tall buildings. Applied Technology Council, Redwood City

Prakash V, Powell GH, Campbell S (1993) DRAIN-2DX: basic program description and user guide. UCB/SEMM-1993/17, University of California, Berkeley

Zareian F, Krawinkler H (2009) Simplified performance based earthquake engineering. Rep. No. TB 169. The John A. Blume Earthquake Engineering Research Center, Stanford

Chapter 8
Tools and Strategies for the Performance-Based Seismic Assessment of Masonry Buildings

Andrea Penna

Abstract Performance-based earthquake engineering became popular in the last decades for both assessment and design of structures. This new trend in research determined the innovation of design codes for all structural typologies with the incorporation of seismic assessment procedures based on pushover analysis and the identification of damage states in terms of displacement thresholds. Within such a framework, the application of performance-based engineering to masonry structures requires the solution of specific problems and the development of suitable methods and dedicated computational tools. Indeed, in existing masonry buildings the lack of proper connections between orthogonal walls and between walls and floors is rather common and can facilitate the activation of local failure modes, mainly related to the out-of-plane response of walls. Early local damage modes may prevent the development of a global building response governed by the in-plane behaviour of masonry walls and the floor in-plane stiffness. On the other hand, the presence of very flexible diaphragms (i.e. timber floors and roofs) makes the adoption of nonlinear static analysis procedures more complicated and requires to take into account specific issues which can be normally neglected for the global capacity assessment of buildings belonging to other structural typologies. All these issues, together with some lack of experimental information on the capacity limits of different masonry typologies, make dealing with the extension of performance-based seismic assessment approach to masonry buildings a more complex subject than its application to other structures. Recent research advances and the availability of computational tools based on frame-type macro-element modelling suggest possible strategies for a consistent evaluation of the seismic performance of masonry buildings. The need for further experimental, numerical and theoretical research on this topic is still evident.

A. Penna (✉)
European Centre for Training and Research in Earthquake Engineering, Via ferrata 1, Pavia, Italy
e-mail: andrea.penna@eucentre.it

M. Dolšek (ed.), *Protection of Built Environment Against Earthquakes*,
DOI 10.1007/978-94-007-1448-9_8,

Keywords Masonry structures • Performance-based assessment • Local failure modes • Macro-element modeling • Building knowledge • Experimental testing

8.1 Introduction

In newly designed masonry buildings, the architectural conception typically governs the wall distribution and few free degrees of freedom are left to the structural designer. Apart from some code constrains (maximum wall slenderness, minimum distance between one opening and the end of the wall, etc.), the architect plans the geometry of the construction based on different factors (e.g. use, thermal insulation, acoustic insulation, other) and even the choice of the type of masonry blocks can be governed by non-structural reasons (energy efficiency, zero carbon, zero emissions, etc.).

In some rare cases the result of the architectural design is a compact, reasonably symmetric and regular building, with a number of storeys and a percentage of masonry walls in the two perpendicular directions satisfying the code requirements for seismic design of "simple buildings". More likely it is a "complex" structure, often mixed masonry and RC frame, sometimes with flexible roof and floors.

In all these cases the task of the structural engineer is to estimate the seismic performance of the conceived building, rather than a free design of the structure. This is particularly true for unreinforced masonry buildings for which the distinction between structural and non-structural elements is not often trivial.

Therefore, a common approach to seismic performance assessment can be applied both to new and existing masonry buildings with some specific additional cares to be used in case of existing structures.

The main difference between newly designed and existing masonry buildings for what concerns structural performance assessment is the incomplete knowledge of material properties and construction details in existing structures which are instead assumed as completely known for new ones.

Although masonry buildings are often regarded as non-engineered structures and their seismic behavior is generally much more complex that the one of other structural types, the assessment of their seismic performances can still be carried out according an approach consistent with those suggested for frame structures. It certainly requires specific assessment strategies to account for some important issues, in particular for existing buildings. Among others, the following points can be regarded as critical ones:

- Consideration of local failure modes, mainly related to out-of-plane response of poorly connected external walls;
- Removal of the commonly adopted rigid floor hypothesis for three-dimensional analysis of global building response;
- Accounting for uncertainties in the material properties, as well as their member to member variability.

Specific analysis tools developed for the nonlinear modeling of masonry buildings are now available both at the research and professional level.

Some considerations on available solutions for this issues are reported in the following together with a few lines of application to be developed in the future.

8.2 Performance-Based Seismic Assessment of Masonry Buildings

The paradigm of modern performance-based earthquake engineering (PBEE) was essentially developed for frame structures (e.g. steel structures in the SAC project, (Hamburger et al. 2003)), starting from the identification of different performance levels for different return period events (SEAOC 1995). It requires, in particular:

- The definition of objective indicators of structural limit states corresponding to performance limit conditions related to the building use, to the reparability of structural/non structural damage, to the economic and social impact;
- The availability of reliable analysis tools and methods correlating seismic action and structural demand for the considered structural system.

In case of the seismic performance assessment of masonry structures, the general PBEE philosophy can be followed, with some specific adjustments and particular issues to be accounted for. The following steps can be hence envisaged in the seismic assessment of a masonry building:

- Knowledge of the structure, including construction phases and history, past events and structural accidents (fire, earthquakes, etc.), building geometry, structural details, material properties;
- Definition of the seismic action for the considered return periods. In many countries (e.g. Italy) this can be simply done using available national hazard models providing spectral accelerations and displacements and tools for the appropriate selection of spectrum-compatible time-histories (e.g. Corigliano et al. 2011; Iervolino et al. 2010);
- Identification of structural capacity by means of appropriate models for the analysis of the global response (mainly associated to the in-plane response of walls and floors) and local collapse modes (essentially due to the out-of-plane walls response);
- Identification of structural performance levels based on damage and/or displacement/deformation indicators;
- Computation of demand, generally in terms of displacements/deformations to be compared to limit state thresholds in order to assess the expected performance.

The limited interest in seismic performance of masonry structures in several countries around the world, either due to limited presence of masonry constructions or to low seismic hazard, left the investigation on the issues specific for masonry

Fig. 8.1 Examples of local out-of-plane failure mode (*left*) and global behavior, involving wall in-plane response (*right*), of masonry buildings during the 2009 Abruzzi earthquake in Italy

outside from the main international research on PBEE. Advances on these topics, based on recently developed Italian research works, are summarized in the following sections.

8.2.1 Analysis of Local Failure Modes

Geometry constrains (e.g. maximum out-of-plane wall slenderness ratio) and structural detailing requirements (e.g. presence of continuous RC ring beams at the intersection of walls and floors) which are typically enforced by seismic design codes permit to prevent the occurrence of local (mostly out-of-plane) failure modes in new masonry buildings. Hence, in seismic design of masonry structures, although codes (e.g. EN 1998a) provide verification methods and formulas for the assessment of wall out-of-plane stability, safety checks are easily satisfied.

In existing masonry buildings the presence of seismic devices such as steel tie-rods may allow contrasting the activation of local failure modes, but frequently the lack of any appropriate structural detail with poor connections between orthogonal walls and between walls and floors induces local collapses, typically associated to the wall out-of-plane response. Hence the assessment procedure must include the potential activation of early local mechanisms which may do not allow the development of a global response of the structure up to its ultimate conditions.

Figure 8.1 presents two examples of masonry buildings damaged by the 2009 Abruzzi earthquake. In the left picture a clear example of local out-of-plane collapse of a wall is reported, while in the right picture it is possible to observe the in-plane damage of the walls in a building with good connections which exhibited a global response.

Figure 8.2 shows typical failures which can be observed in multiple leaf stone masonry walls with crumbling of the external veneer. They are frequently observed collapse modes, common in undressed stone masonry walls. They can develop if

Fig. 8.2 Similar cases of out-of-plane collapse of the external veneer in double-leaf stone masonry in Pakistan (*left*, Kashmir earthquake 2005) and in Italy (*right*, Umbria-Marche earthquake 1997)

Fig. 8.3 Observed collapse mechanism during the shaking table test of a full-scale stone masonry building with flexible floors, without tie-rods (Magenes et al., 2010)

the different leaves are not connected through the walls thickness and are subject to different levels of vertical compression.

If transversal connections are effective by means of the presence of through stones, even for multiple leaf masonry the local response can be approximated by rocking of rigid bodies. This was also directly observed in the shaking table tests performed at EUCENTRE (Magenes et al. 2010) on full-scale stone masonry building prototypes (Fig. 8.3).

These observations allowed the development of a method for the seismic assessment of local failure modes based on limit analysis. Such an approach was already introduced in the analysis of historical masonry structures, first for the estimation ultimate static capacity of masonry systems under vertical loads (Heyman 1966) and later for horizontal loads simulating seismic forces (Como and Grimaldi 1985; Giuffrè 1993). The basic idea that single structural portions can develop independent seismic behaviors was suggested by the observation of earthquake damage (Doglioni et al. 1994; D'Ayala and Speranza 2003). Further refinement of these

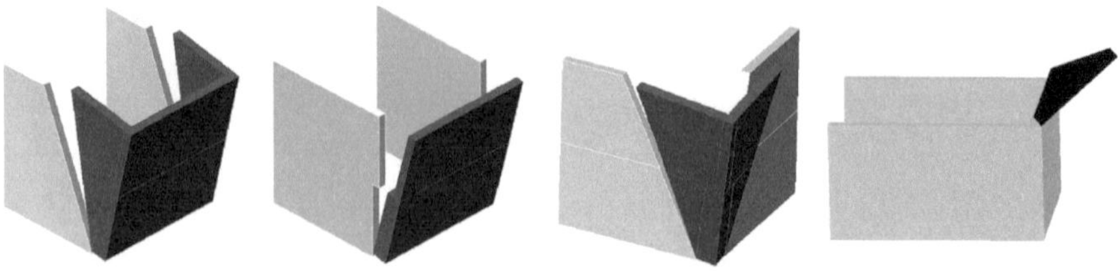

Fig. 8.4 Schemes of basic local collapse mechanisms commonly observed in masonry buildings

methods (de Felice and Giannini 2001; Doherty et al. 2002; Griffith et al. 2003; Sorrentino et al. 2008) allowed the introduction in the Italian code (NTC 2008) of a new method for the seismic analysis of local collapse mechanisms based on equilibrium limit analysis.

Such a methodology is composed by the following steps:

- Transformation of one part of the structure into a mechanism (kinematic chain) by means of identification of rigid bodies capable of rotating or sliding. Some examples of common simple mechanisms are reported in Figure 8.4;
- Estimation of the static horizontal load multiplier that causes the activation of the mechanism by the application of equilibrium equations (e.g. via the Principle of Virtual Works or via direct equilibrium in the simplest cases);
- Estimation of the evolution of the static horizontal load multiplier with increasing displacements of a control point of the kinematic chain until vanishing of the horizontal seismic force (nonlinear capacity curve) and identification of displacement limit states;
- Idealization of the equivalent single degree of freedom structure and conversion of the capacity curve into acceleration-displacement spectral coordinates;

  Determination of the maximum acceleration (force-based linear kinematic analysis) or displacement (displacement-based nonlinear kinematic analysis) demand for the local system by means of proper dynamic amplification of the design ground motion considering the position of the portion involved in the mechanism along the building height and the interaction between the dynamic properties of the local and global systems. The Italian code recommendations (Circ. NTC08 2009) also suggest a simplified formula for the estimation of the maximum displacement demand for the equivalent SDOF system representative of the mechanism located in the building at a certain height within the building, considering the interaction between the first period of vibration of the building and the effective period of vibration of the local mechanism;
- Assessment of the expected performance associated with each considered local collapse mechanism by comparing capacity and demand parameters for the considered damage states.

The proposed mechanical approach allows the identification of peak values of ground motion parameters corresponding to the attainment of the considered damage states for the local systems considered. The minimum of them represents the building vulnerability to local collapse modes and can be directly compared with the corresponding parameter computed for the global response, allowing the assessment of the capacity of exploiting the global structural resources prior to the activation of early local failures.

It is worth recalling that the presence of proper connections between walls and floors and out-of-plane slenderness limitations in seismic design of new buildings normally prevent the occurrence of local failure modes.

The repetition of the shaking table test after the insertion of perimeter steel tie-rods and roof cable bracings on the nearly collapsed building (Magenes et al. 2010) showed that even low-cost traditional connection interventions can be capable of contrasting local failure modes and hence allowing the development of a global structural behavior governed by the in-plane response of walls. This was also observed in previous shaking table tests on reduced scale building specimens (e.g. Tomaževič and Lutman 1996; Benedetti et al. 1998; Mazzon et al. 2009).

8.2.2 Analysis of Global Building Response

The need for nonlinear analysis tools for complete masonry buildings arose in the late 1970s in Italy and in Slovenia when simplified modeling techniques and analysis methods were developed and adopted in practice (Tomaževič 1978; Braga and Dolce 1982; Tomaževič 1987). In the following decades several other nonlinear models were developed and some of them make now possible rather reliable nonlinear pushover analysis of masonry structures (Magenes and Della Fontana 1998; Brencich and Lagomarsino 1998; Galasco et al. 2004; Chen et al. 2008). These methods, generally based on the equivalent frame approach (Magenes 2000; Roca et al. 2005; Kappos et al. 2002; Belmouden and Lestuzzi 2009) and the macro-element discretization (single 2-node elements modeling structural members such as piers and spandrel beams), require a limited computational burden since the number of degrees of freedom and elements in the structural model is limited. Computational tools for nonlinear pushover analyses of masonry buildings based on equivalent frame modeling are already available to practitioners (e.g. Magenes et al. 2006; Lagomarsino et al. 2006).

An efficient equivalent-frame formulation which allows the dynamic global analysis of whole buildings, when only in-plane response of walls is considered, is available in TREMURI model (Lagomarsino et al. 2007).

The macro-element model implemented in TREMURI is based on the original formulation proposed by Gambarotta and Lagomarsino (Gambarotta and Lagomarsino 1996). The algorithms embedded in the program are described in detail in several literature works (Lagomarsino et al. 2007; Penna 2002; Gambarotta and Lagomarsino 1997). The nonlinear macro-element model allows to represent the two main in-plane masonry failure modes (i.e. bending-rocking and shear-sliding

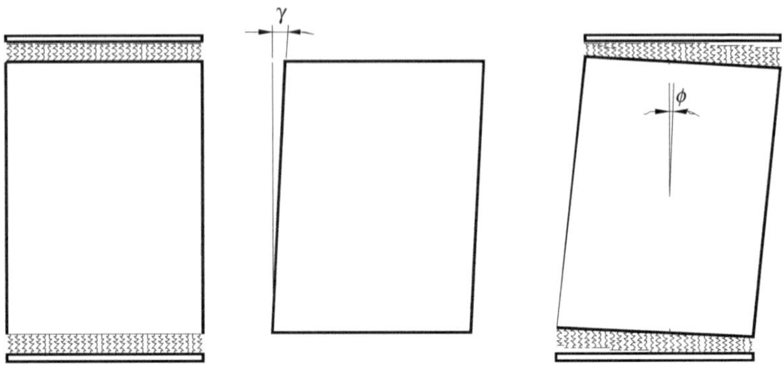

Fig. 8.5 Macro-element model implemented in the TREMURI program: shear deformable central body and end interfaces where coupled axial and flexural response are concentrated

with friction), with a limited number of degrees of freedom (8), on the basis of only mechanical assumptions. This model considers, by means of internal variables, the shear-sliding damage evolution, which controls the strength deterioration (softening) and the stiffness degradation.

A macro-element is subdivided in three sub-structures, as shown in Fig. 8.5: two layers, one at the top and the other at the bottom of the element, in which the bending and axial effects are concentrated, and a central part that can have only shear deformations and presents no evidence of axial or bending deformation. It is assumed that the ends of the element have an infinitesimal thickness. A complete 2D kinematic model should take into account three degrees of freedom for each node "i" and "j" at the extremities: axial displacement, horizontal displacement and rotation. Moreover, there should be two degrees of freedom for the central part: axial displacement and rotation. Thus, the kinematics is described by an eight degrees of freedom vector, for each macro-element. The bending-rocking mechanism, occurring due to the absence of tensile strength in masonry, is modelled by a unilateral elastic contact between the interfaces of the two layers. The constitutive equations are uncoupled until the limit condition, corresponding to partialisation of the section, occurs.

The panel shear response is expressed considering a uniform shear deformation distribution in the central part and imposing a relationship between the kinematic quantities (horizontal displacements at the two nodes and rotation) and the shear stress. The cracking damage is usually located on the diagonal, while the displacement takes place along the joints and is represented by an inelastic deformation component, which is activated when the Coulomb's limit friction condition is reached. The damage effects on the panel mechanical characteristics are described by a damage variable governing the evolution of the failure criterion. The resistance is hence a function of this damage parameter, increasing up to the critical value (corresponding to the damage parameter equal to one) and then decreasing. In this way, the model can represent stiffness degradation, strength deterioration and pinching effects.

The macro-element shear model is a macroscopic representation of a continuum model (Gambarotta and Lagomarsino 1997), in which the parameters are directly correlated to the mechanical properties of the masonry elements. The macro-element parameters should be considered as representative of an average behavior. In addition to its geometrical characteristics, the macro-element is defined by six parameters: shear modulus, axial stiffness, masonry shear strength, a non-dimensional coefficient controlling the inelastic deformation, global friction coefficient and a factor controlling the softening phase. The macro-element used in the program takes also into account the effect (especially in bending-rocking mechanisms) of the limited compressive strength of masonry (Penna 2002).

The TREMURI program allows to perform nonlinear seismic analyses of unreinforced masonry buildings, using a set of analysis procedures: incremental static analysis (modified Newton-Raphson method) both with force and displacement control, 3D pushover analysis with fixed and adaptive load patterns and 3D time-history dynamic analysis, using Newmark integration method and Rayleigh viscous damping.

8.2.3 *Identification of Global Performance Limit States*

As recognized by many authors (e.g. Priestley 1997) displacement/deformation thresholds are the best indicators for identifying limit states corresponding to damage levels, either structural or non-structural, and hence performances.

For what concerns the in-plane response of masonry structures, since a moderately ductile behavior can be expected both in flexural/rocking and in shear failure modes, different deformation limits at the structural element level can be obtained for the two damage modes. The main source of information on this values is certainly represented by cyclic in-plane tests on masonry piers (e.g. Tomaževič et al. 1993; Anthoine et al. 1995; Magenes and Calvi 1997; Tanner et al. 2005; Magenes et al. 2008; Galasco et al. 2010). Obviously such drift thresholds for the different damage states are significantly influenced by the masonry typology, the level of axial load, the effective boundary conditions and other construction details (such as presence of filled/unfilled headjoints, thickness of bed-joints). Although in the recent years several testing campaigns have been carried out the need of experimental information is still high since some limitations reported in codes and recommendations do not perfectly apply to all structural typologies.

In order to identify global displacement thresholds representative of structural limit states from a pushover analysis different approaches are possible:

- Identification of significant displacements on the global pushover curve (e.g. maximum base shear, significant stiffness reduction, relevant drop of the

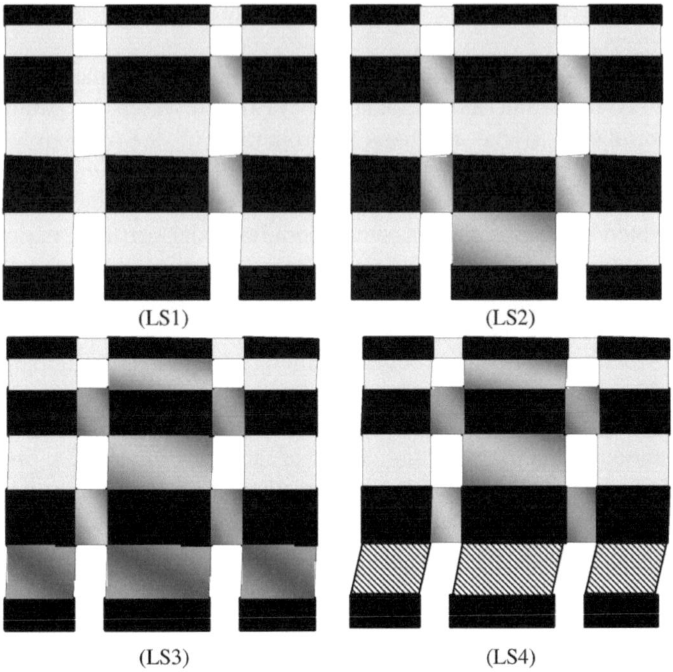

Fig. 8.6 Identification of limit states based on the shear damage parameter in the TREMURI: LS1 corresponds to shear cracking of spandrel beams, LS2 to first shear cracking of a pier, LS3 to the attainment of the shear failure of the piers in one. LS4 can be only identified by the element shear drift limit

resistance after the peak force based on progressive elimination of the contribution of structural elements reaching element drift limits);

- Identification of global displacement thresholds corresponding to the attainment of inter-storey drift limits;
- Detection of local damage and diffusion based on model parameters.

While the first and, to some extents, the second approaches are already included in codes (e.g. NTC 2008; EN 1998b), the third is possible either using damage indicators developed on purpose or a nonlinear model providing damage evolution parameters. This is the case of the TREMURI model, which provides, for each masonry member, a shear damage parameter and other indicators related to toe crushing associated to flexural response. Figure 8.6 shows a possible identification of three damage state thresholds (slight, LS1, moderate, LS2, and extensive, LS3, damage limit states) based on the evolution and diffusion of shear damage in piers and spandrels in a single wall model. The near collapse condition can not be predicted using this approach, but it can only identified based on drift limits (complete damage, LS4, in Fig. 8.6).

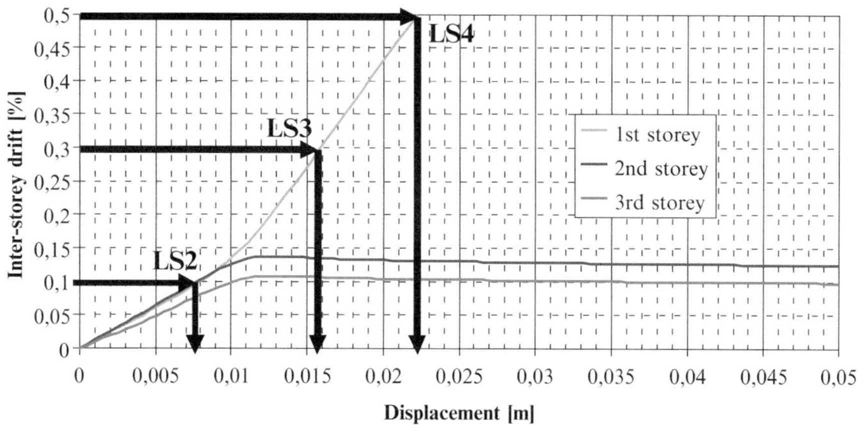

Fig. 8.7 Identification of displacement limit state thresholds based on inter-storey drift (as proposed by Calvi (Calvi 1999))

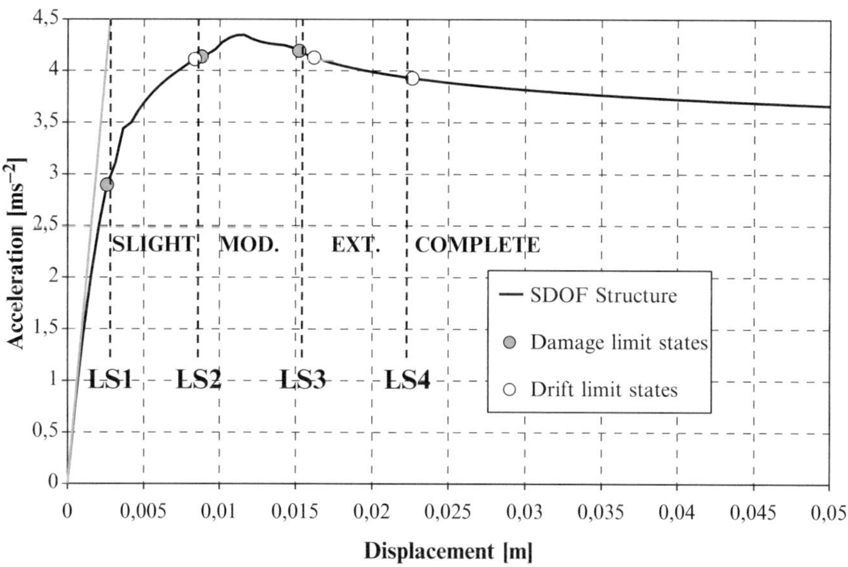

Fig. 8.8 Comparison of damage state thresholds identified on the pushover curve based on the macro-element shear damage parameter (*grey dots*) and inter-storey drift limits (*white dots*)

The comparison of the global displacements representative of the different limit conditions obtained by this approach and the displacement thresholds identified based on inter-storey drift limits (Calvi 1999, Fig. 8.7) shows in this case a rather good fit for the two corresponding limit states (LS2 and LS3), as reported in Fig. 8.8.

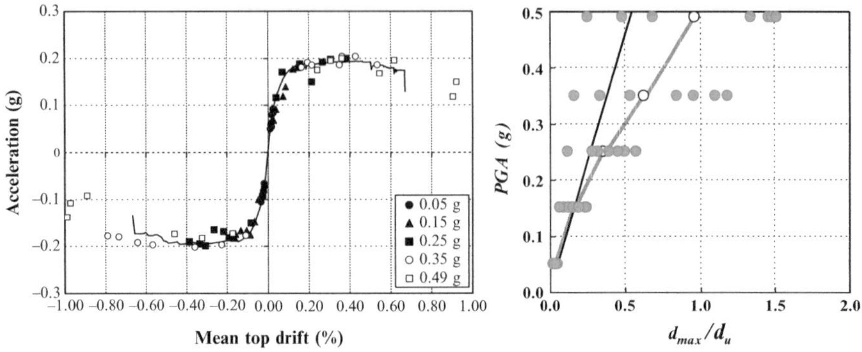

Fig. 8.9 Comparison of pushover and time-history analysis results (*left*) and comparison of displacement demand prediction by N2 method (*black line*) and time-history analyses (*right*) (Costa et al. 2010)

8.2.4 Computation of Displacement Demand

Several procedures are available for the computation of displacement demand based on pushover analysis results (e.g. Freeman 1998; Fajfar 1999; ATC 40 1996). They are commonly applied in codes and recommendations (e.g. EN 1998a; NTC 2008), usually regardless to the structure hysteretic behavior, presence of stiffness and strength degradation, period of vibration.

In particular, the applicability of such methods to masonry buildings, which are low period structures with hysteretic behavior characterized by strength deterioration (softening) and stiffness degradation has still to be confirmed.

Indeed, as shown in Fig. 8.9, the comparison of the results obtained by nonlinear static procedures (N2 method) and time-history analysis for some masonry building prototypes yielded the following conclusions (Costa et al. 2010):

- the simplified pushover-based prediction of maximum displacement is not generally capable of reproducing the variation of displacement demand for increasing levels of ground motion, as obtained from nonlinear dynamic analysis;
- from the incremental dynamic analysis procedure a very large scatter in the displacement demand has been observed, increasing with PGA.

The latter may be certainly due to the higher level of structural non linearity involved in the structural models, making the response more sensitive to the characteristics of the ground motion records. On the other hand, the natural accelerograms used for the analysis, which have been selected based on the compatibility criterion with the acceleration response spectrum, showed a high scatter in terms of displacement response spectrum.

8.3 Building and Material Knowledge

Typical problem with existing structures is the incomplete knowledge of the structure, including construction details and material properties. In particular, the in-situ direct experimental measurement of material parameters in existing masonry buildings is often not feasible or not reliable.

The in-situ measure of shear strength of stone masonry is presently possible only with destructive testing of panels of significant dimensions, rarely smaller than 1.0 × 1.0 m, through a self-equilibrated diagonal compression testing or more complex shear-compression test procedures. The conditions for the feasibility of such tests may not be always satisfied, depending on the quality and texture of masonry, on the thickness of the walls, on the number of storeys of the building, on the availability of adequate experimental equipment and, in case of ancient constructions, on the issues related to the conservation of cultural heritage.

In the new Italian norms it was therefore felt essential to define specific criteria for masonry regarding the different knowledge levels. First, full geometric survey is always required, and information regarding structural details should specifically address: quality of connections between vertical walls, quality of connections between floor/roof and walls, and presence of ring beams or other tying devices, presence of structurally efficient architraves/lintels above openings, presence of elements which can equilibrate horizontal thrusts, presence of structural or non structural elements of high vulnerability, typology of masonry (stone or brick, regular or irregular units, single-leaf or multi-leaf , etc.).

Regarding the quantification of material parameters, *limited in situ investigations* are required and the mechanical properties of the material are estimated after visual inspections. Plaster is typically removed in selected areas to assess the texture and the connection between orthogonal walls. Other visual inspections through the thickness allow the determination of the internal level of connection of the leaves and the ability of the wall to behave monolithically through the thickness. Qualitative assays can be also performed to assess the consistency of mortar. Such recognition of the typology and quality of the material is then used to associate it with the mechanical parameters reported in a reference table (an extract is given in Table 8.1), which was compiled on the basis of the experimental data available on the most common typologies.

The next level is defined as *extensive in-situ investigations*. At such level, the visual inspections described in the previous level are carried out extensively and systematically with superficial and internal samples for every type of masonry present, while tests with double flat jacks and tests for characterization of the mortar (type of binding agent, type of aggregate, binding agent/aggregate ratio, etc.) and eventually on stone and/or bricks (physical and mechanical characteristics) are required to verify the correspondence of the masonry to the typology defined in the reference table. A test for every type of masonry present in the building is required. Non-destructive testing procedures (sonic tests, sclerometer tests), penetrometer test for mortar, etc. may be utilized as complementary to the required tests.

Table 8.1 Reference values of the mechanical parameters and average specific weights for selected types of masonry (extract from Table C8A.2.2. of Circ. NTC08 (Circ. NTC08 2009))

Masonry typology	f_m (MPa)	τ_o (MPa)	E (MPa)	G (MPa)	W (kN/m^3)
Irregular stone masonry (pebbles, erratic, irregular stone)	1.0–1.8	0.020–0.032	690–1,050	230–350	19
Uncut stone masonry with facing walls of limited thickness and infill core	2.0–3.0	0.035–0.051	1020–1440	340–480	20
Cut stone with good bonding	2.6–3.8	0.056–0.074	1,500–1,980	500–660	21
Soft stone masonry (tuff, limestone, etc.)	1.4–2.4	0.028–0.042	900–1,260	300–420	16
Dressed rectangular (ashlar) stone masonry	6.0–8.0	0.090–0.120	2,400–3,200	780–940	22
Solid brick masonry with lime mortar	2.4–4.0	0.060–0.090	1,200–1,800	400–600	18

Finally, *exhaustive in-situ investigations* serve to obtain direct quantitative information on the material strength. Apart from the visual inspections of the internal samples and the tests mentioned in the previous levels, a further series of experimental tests have to be carried out, both in quantity and quality, in order to be able to estimate the mechanical characteristics of the masonry. The measurements of the mechanical characteristics of the masonry are obtained by means of in-situ and laboratory tests (on undisturbed elements extracted from the structure). The tests can generally include diagonal compression tests on panels or combined tests of vertical compression and shear. Non-destructive testing methods can be used in combination, but not as a substitute, of the aforementioned tests.

The results of the tests have to be examined and considered within a general typological frame of reference which takes into account the results of the experimental tests available in the literature up to that time for the masonry typology under investigation, and that allows the estimation of an effective representative of the values found, even in statistical terms. The results of the tests have to be utilized with reference to the values reported in the reference table.

8.4 Seismic Assessment Procedure Including Uncertainties

The procedure proposed in Rota et al. (Rota et al. 2010), based on the capabilities of the TREMURI program, aims at rigorously including all sources of uncertainties in the seismic performance assessment of partly known masonry buildings.

The procedure is composed by the following steps:

- Identification of the building geometry (deterministic);
- Definition of mechanical damage state (DS*i*) thresholds and their probabilistic distribution (probability density functions, Pdf(DS*i*)) based on available experimental data on structural members made with the same construction technology;

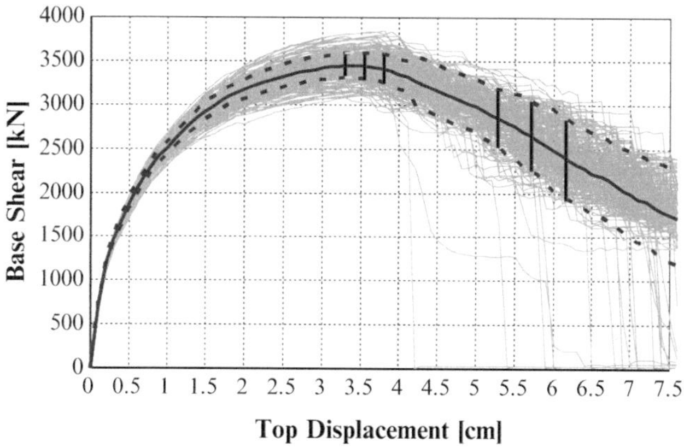

Fig. 8.10 Examples of stochastic pushover analysis based on Monte Carlo simulation of masonry mechanical properties and deformation capacities. Mean (*continuous thick line*) and mean plus and minus standard deviation (*dashed lines*) curves are superimposed to the single analysis curves. The vertical lines identify mean and mean plus and minus standard deviation of global damage states (DS3 and DS4)

- Monte Carlo pushover analysis for the stochastic characterization of structural capacity and global limit state thresholds (Fig. 8.10). The analyses are based on randomly defined material properties, which are then randomly assigned to the different structural elements to account for the fact that material properties of masonry buildings can vary significantly from element to element, even within the same structure;
- Selection of a set of recorded ground motions compatible with response spectrum at the site;
- Incremental dynamic analysis for increasing intensity measures (typically PGA since spectrum compatibility is enforced);
- Convolution of the probability that displacement demand exceeds global displacement, 1-Cdf(PGA), values with the probability, Pdf(DS*i*), that those values represent damage state thresholds obtaining the probability of exceeding each damage state, Pf(DS*i*) (Fig. 8.11);
- Derivation of fragility curves by fitting the fragility points, obtained from the convolution of demand and capacity distributions for discrete values of reference PGA, with standard lognormal distributions (Fig. 8.12).

8.5 Seismic Performance Assessment for the Acceptance of Innovative Masonry Systems

As envisaged in the FEMA 349 Action plan (EERI 2000), one of the aims of performance-based earthquake engineering is to develop consistent criteria for the acceptance of innovative structural systems.

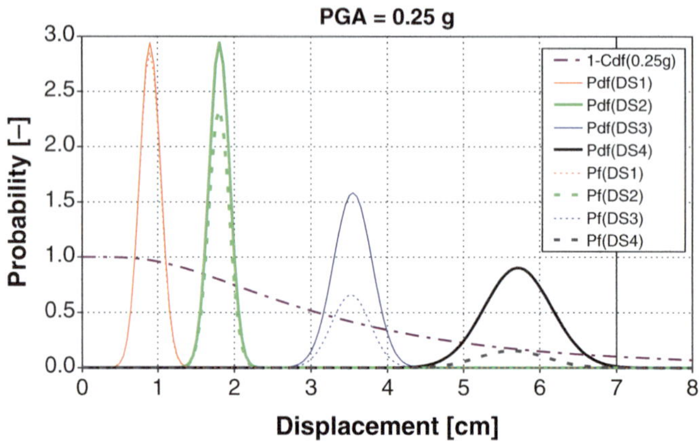

Fig. 8.11 Convolution of probabilities of displacement demand for a specific value of PGA and capacities for the considered 4 damage states

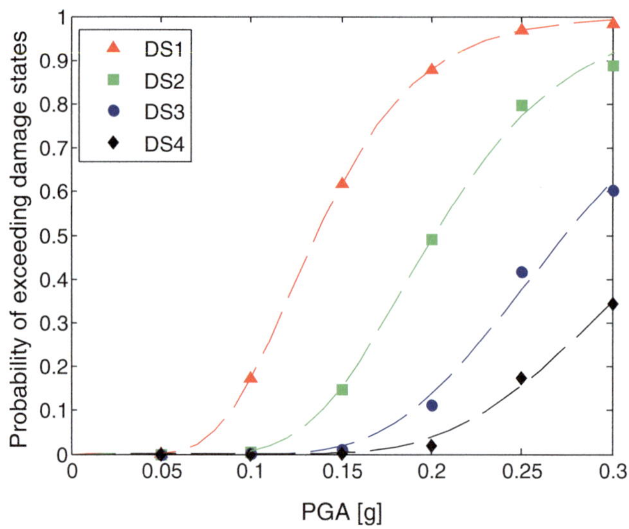

Fig. 8.12 Lognormal fragility curves obtained by fitting the fragility points obtained from the convolution of capacity and demand

The general performance requirement for a masonry structural system is to guarantee adequate resistance and safety against excessive damage and structural collapse under the combined action of gravity and seismic loads.

The following procedure for the assessment of the structural response under the effects of combined static and dynamic (seismic) loading is proposed:

1. Execution of monotonic tests on small size specimens to identify mechanical properties such as compressive strength and shear strength. An adequate number of specimens shall be tested for the following types of test:

   (a) vertical compression test;
   (b) test for determining the initial shear strength.

2. Cyclic shear-compression tests on full scale components (walls with size and aspect ratio normally employed in the constructions) shall be performed on a minimum number of four specimens, in order to assess the influence of in-plane slenderness and axial load, and provide information on both shear and flexural failure modes. Several examples of these tests are reported in the literature (e.g. Anthoine et al. 1995; Magenes et al. 2008; Ötes A and Löring 2003; Fehling and Stürz 2008), but only a few codes incorporate rules for performing cyclic tests for the qualification of masonry typologies (e.g. the Mexican code NTCM-2004 (NTCM-2004 2004));
3. Numerical simulations shall be carried out to evaluate the seismic performance of a number of typical building configurations. They can be based on frame-type nonlinear modeling of the in-plane response of walls and pushover analyses. Analyses shall also provide an appropriate estimate of the reliability of the results, based on realistic variations of the mechanical properties.

In modern masonry structures, indeed, the enforcement of constructive details, such as minimum wall thickness and maximum out-of-plane slenderness ratio, together with proper connections between walls and floors, allows to prevent out-of-plane failure modes. Therefore the global seismic response is governed by the in-plane behavior of masonry walls. In-plane cyclic testing on masonry piers, performed under controlled static and kinematic conditions, is hence one of the main sources of information on the structural performance of modern masonry buildings. This kind of tests provides important information on the possible failure modes, the associated lateral strength, the displacement capacity and the cyclic response.

Once the experimental characterization of the behavior of such structural elements is completed, a model of the building can be created in order to analyze its global seismic behavior, directly assessing its limit states and expected seismic performance for increasing values of ground motion.

The assessment procedure may consist in the following steps:

- Identification from experimental results of the values of mechanical properties and drift limits to be adopted in the model;
- Selection of representative building prototypes;
- Nonlinear numerical modeling of the selected buildings;
- Nonlinear static assessment of the building seismic performance, with reference to significant limit states and corresponding safety checks.

This strategy was applied to the seismic performance assessment of Autoclaved Aerated Concrete (AAC) masonry buildings. The need to assess the seismic performance of AAC masonry arose in different countries in the last years and it is still the subject of ongoing research (e.g. Tanner et al. 2005; Costa et al. 2010). The use of AAC for load-bearing walls is quite common in low seismicity areas in Central and Northern Europe, where its thermal insulation properties are particularly appreciated. An increasing attention to energy efficient buildings is now supporting the adoption of a material with such characteristics also in higher seismicity regions. The light weight of this material reduces the seismic inertial forces with respect to other materials. However, the thermal insulation capacity of AAC decreases with increasing material density which, in turn, favourably affects compressive strength. The limited compressive strength of AAC units typically adopted in non seismic areas could be seen as a hindrance towards its use in seismic areas. However, since material density (and hence compressive strength) can be controlled during the production process, it is interesting to evaluate the seismic performance of different types of unit, in order to identify a satisfactory behaviour both in terms of structural safety and thermal insulation properties, with reference to the level of seismic hazard expected for the site.

In the case of AAC masonry (Costa et al. 2010; Penna et al. 2010), seismic performance assessment based on nonlinear analysis of entire building models was carried out. Among other results, including maximum expected PGA leading to the attainment to ultimate limit states based on different building characteristics (number of storeys, masonry wall density with respect to the floor area, building regularity and compactness in plan), a value of the behavior factor for linear design was proposed. Both the approaches, one (Penna et al. 2010) performed using procedure based on pushover analysis and the N2 method (Fajfar 1999) as implemented in the Eurocode 8 (EN 1998a) and the other one (Costa et al. 2010) based on incremental dynamic analysis (Vamvatsikos and Cornell 2002) allowed to justify the adoption of a minimum basic factor equal to 2 to be multiplied by an overstrength factor (Magenes 2006; Magenes 2010) equal to 1.5, leading to a minimum behavior factor equal to 3.

8.6 Conclusions

Performance-based earthquake engineering represented in the last years a significant step forward in the direction of a rationale orientation of design and assessment procedures, with the updating of design codes and an effective communication of the design objectives. In fact, although the concept of limit states in structural design dates back to several decades ago and it was already introduced in design codes in its semi-probabilistic formulation since more than 30 years, the adoption of this concept in a new framework which explicitly accounts for nonlinear analysis becomes much more effective and understandable for earthquake engineering in particular.

For masonry structures, the need for seismic performance assessment based on (simplified) nonlinear analyses was already recognized at the end of the 1970s, but at the moment masonry buildings are not treated completely according to PBEE philosophy world, due to formerly discussed issues.

Even though the performance-based assessment framework needs specific adaptations to be applied to masonry buildings, what is reported in the previous sections shows that relevant work has been done in Italy and several other masonry-sensitive world countries and that practical tools and methods are already available. As previously highlighted, further analytical and experimental research is still needed on several open issues related to the peculiar characteristics of masonry structures and their seismic response.

References

Anthoine A, Magenes G, Magonette G (1995) Shear-compression testing and analysis of brick masonry walls. In: Proceedings of the 10th European conference on earthquake engineering, Balkema, Rotterdam, pp 1657–1662

ATC 40 (1996) Seismic evaluation and retrofit of existing concrete buildings. Applied Technology Council, Redwood City

Belmouden Y, Lestuzzi P (2009) An equivalent frame model for seismic analysis of masonry and reinforced concrete buildings. Constr Build Mater 23(1):40–53

Benedetti D, Carydis G, Pezzoli P (1998) Shaking table tests on 24 simple masonry buildings. Earthquake Eng Struct Dyn 27(1):67–90

Braga F, Dolce M (1982) A method for the analysis of antiseismic multi-storey masonry buildings. In: Proceedings of the 6th I.B.Ma.C., Roma, 1982, pp 1088–1099, in Italian

Brencich A, Lagomarsino S (1998) A macro-elements dynamic model for masonry shear walls, A. Proc. STRUMAS IV 4th international symposium on computer methods in structural masonry, E&FN Spon, London, 1998, pp 67–75

Calvi GM (1999) A displacement-based approach for vulnerability evaluation of classes of buildings. J Earthquake Eng 3(3):411–438

Chen S-Y, Moon FL, Yi T (2008) A macroelement for the nonlinear analysis of in-plane unreinforced masonry piers. Eng Struct 30(8):2242–2252

Circ. NTC08 (2009) Circolare del Ministero delle Infrastrutture e dei Trasporti, n. 617 del 2 febbraio 2009, Suppl. Ord. n. 27 alla G.U. n. 47 del 26 febbraio 2009, Istruzioni per l'applicazione delle nuove norme tecniche per le costruzioni di cui al decreto ministeriale 14 gennaio 2008, in Italian

Como M, Grimaldi A (1985) A unilateral model for the limit analysis of masonry walls. In: Del Piero e G, Maceri F (eds) Unilateral problems in structural analysis: CISM Courses and Lectures, vol 288. Springer-Verlag, Wien New York, pp 25–45

Corigliano M, Lai CG, Rota M, Strobbia C (2011) ASCONA: automated selection of compatible natural accelerograms. Earthquake Spectra (submitted)

Costa AA, Penna A, Magenes G (2010) Seismic performance of Autoclaved Aerated Concrete (AAC) masonry: from experimental testing of the in-plane capacity of walls to building response simulation. J Earthquake Eng 15(1):1–31

D'Ayala D, Speranza E (2003) Definition of collapse mechanisms and seismic vulnerability of masonry structures. Earthquake Spectra 19(3):479–509

de Felice G, Giannini R (2001) Out-of-plane seismic resistance of masonry walls. J Earthquake Eng 5(2):253–271

Doglioni F, Moretti A, Petrini V (1994) Le chiese e il terremoto. LINT, Trieste, in Italian

Doherty K, Griffith MC, Lam N, Wilson L (2002) Displacement-based seismic analysis for out-of-plane bending of unreinforced masonry walls. Earthquake Eng Struct Dyn 31 (4):833–850

EERI (2000) Action plan for performance-based seismic design. FEMA 349, Report, prepared by the Earthquake Engineering Research Institute for the Federal Emergency Management Agency, Washington, DC

EN 1998–1 (2005) Eurocode 8: Design of structures for earthquake resistance, Part 1: General rules, seismic actions and rules for buildings, CEN

EN 1998–3 (2005) Eurocode 8: Design of structures for earthquake resistance, Part 3: Assessment and retrofitting of buildings, CEN

Fajfar P (1999) Capacity spectrum method based on inelastic spectra. Earthquake Eng Struct Dyn 28(9):979–993

Fehling E, Stürz J (2008) Experimentelle Untersuchungen zum Schubtragverhalten von Porenbetonwandscheiben, Report University of Kassel, Germany, in German

Freeman SA (1998) The capacity spectrum method as a tool for seismic design. In: Proceedings of the 11th European conference of earthquake engineering, Paris, Balkema, Rotterdam

Galasco A, Lagomarsino S, Penna A, Resemini S (2004) Non-Linear Seismic Analysis of Masonry Structures. In: Proceedings of the 13th World Conference on Earthquake Engineering, Vancouver

Galasco A, Magenes G, Penna A, Da Paré M (2010) In-plane cyclic shear tests of undressed double leaf stone masonry panels. In: Proceedings of the 14th European conference on earthquake engineering, paper no 1435, Ohrid

Gambarotta L, Lagomarsino S (1996) On the dynamic response of masonry panels. In: Proceedings of the national conference "Masonry Mechanics Between Theory and Practice", Messina, in Italian

Gambarotta L, Lagomarsino S (1997) Damage models for the seismic response of brick masonry shear walls. Part II: the continuum model and its applications. Earthquake Eng Struct Dyn 26:441–463

Giuffré A (ed) (1993) Safety and conservation in historical centres the case of Ortigia. Laterza, Italy, in Italian

Griffith MC, Magenes G, Melis G, Picchi L (2003) Evaluation of out-of-plane stability of unreinforced masonry walls subjected to seismic excitation. J Earthquake Eng 7(1):141–169

Hamburger RO, Foutch DA, Cornell CA (2003) Translating research to practice: FEMA/SAC performance-based design procedures. Earthquake Spectra 19(2):255–267

Heyman J (1966) The stone skeleton. Int J Solids Struct 2:249–279

Iervolino I, Galasso C, Cosenza E (2010) REXEL: computer aided record selection for code-based seismic structural analysis. Bull Earthquake Eng 8:339–362. doi:10.1007/s10518-009-9146-1

Kappos A, Penelis GG, Drakopoulos CG (2002) Evaluation of simplified models for lateral load analysis of unreinforced masonry buildings. J Struct Eng ASCE 128(7):890–897

Lagomarsino S, Penna A, Galasco A (2006), TREMURI program: seismic analysis program for 3D masonry buildings, University of Genoa, Genoa

Lagomarsino S, Galasco A, Penna A (2007) Non linear macro-element dynamic analysis of masonry buildings. In: Proceedings of the ECCOMAS thematic conference on computational methods in structural dynamics and earthquake engineering, Rethymno, Crete

Magenes G (2000) A Method for Pushover Analysis in Seismic assessment of Masonry Buildings. In: Proceedings of the 12the world conference on earthquake engineering, CD-ROM, Auckland

Magenes G (2006) Masonry building design in seismic areas: recent experiences and prospects from a European point of view. In: Proceedings 1st European conference on earthquake engineering and seismology, Geneva, Keynote Address K9, CD-ROM

Magenes G (2010) Earthquake resistant design of masonry structures: rules, backgrounds, latest findings, Keynote Speech. In: Proceedings of the 8th international masonry conference, Dresden

Magenes G, Calvi GM (1997) In-plane seismic response of brick masonry walls. Earthquake Eng Struct Dyn 26:1091–1112

Magenes G, Della Fontana A (1998) Simplified non-linear seismic analysis of masonry buildings. In: Proceedings of the British Masonry Society, No. 8, Oct 1998, pp 190–195

Magenes G, Remino M, Manzini C, Morandi P, Bolognini D (2006) SAM II, Software for the simplified seismic analysis of masonry buildings, University of Pavia and EUCENTRE

Magenes G, Morandi P, Penna A (2008) Test results on the behaviour of masonry under static cyclic in plane lateral loads, ESECMaSE project, report RS-01/08, Department of Structural Mechanics, University of Pavia

Magenes G, Penna A, Galasco A (2010) A full-scale shaking table test on a two-storey stone masonry building. In: Proceedings of the 14th European conference on earthquake engineering, paper no 1432, Ohrid

Mazzon N, Valluzzi MR, Aoki T, Garbin E, De Canio G, Ranieri N, Modena C (2009) Shaking table tests on two multi-leaf stone masonry buildings. In: Proceedings of the 11th Canadian Masonry symposium, Toronto

NTC (2008) Norme Tecniche per le Costruzioni D.M. 14 gennaio 2008, Suppl. ord. no 30 alla G.U. n. 29 del 4/02/2008

NTCM-2004 (2004) Normas tecnicas complementarias para diseño y construccion de estructuras de mamposteria, Reglamento de Construcciones para el Distrito Federal, Gaceta Oficial del Departamento del Distrito Federal, in Spanish

Ötes A, Löring S (2003) Tastversuche zur Identifizierung des Verhaltensfaktors von Mauerwerksbauten für den Erdbebennachweis, Research Report, University of Dortmund, in German

Penna A (2002) A macro-element procedure for the nonlinear dynamic analysis of masonry buildings, Ph.D. thesis, Politecnico di Milano, Milan, in Italian

Penna A, Rota M, Magenes G, Frumento S (2010) Seismic performance assessment of AAC masonry. In: Proceedings of the 14th European conference on earthquake engineering, paper no 1431, Ohrid

Priestley MJN (1997) Displacement-based seismic assessment of reinforced concrete buildings. J Earthquake Eng 1(1):157–192

Roca P, Molins C, Marì AR (2005) Strength capacity of masonry wall structures by the equivalent frame method. J Struct Eng ASCE 131(10):1601–1610

Rota M, Penna A, Magenes G (2010) A methodology for deriving analytical fragility curves for masonry buildings based on stochastic nonlinear analyses. Eng Struct 32(5):1312–1323

SEAOC (1995) Performance based seismic engineering of buildings, vols. I and II: conceptual framework. Structural Engineers Association of California, Sacramento

Sorrentino L, Kunnath S, Monti G, Scalora G (2008) Seismically induced one-sided rocking response of unreinforced masonry façades. Eng Struct 30(8):2140–2153

Tanner JE, Varela JL, Klingner RE, Brightman MT, Cancino U (2005) Seismic testing of autoclaved aerated concrete shearwalls: a comprehensive review. ACI Struct J 2005:374–382

Tomaževič M (1978) The computer program POR, Report ZRMK, Ljubljana, in Slovenian

Tomaževič M (1987) Dynamic modelling of masonry buildings: storey mechanism model as a simple alternative. Earthquake Eng Struct Dyn 15(6):731–749

Tomaževič M, Lutman M (1996) Seismic upgrading of behaviour of old brick-masonry urban houses: tying of walls with steel ties. Earthquake Spectra 12(3):599–622

Tomaževič M, Lutman M, Petković M (1993) Seismic behaviour of masonry walls: experimental simulation. J Struct Enrg, ASCE 122(9):1040–1047

Vamvatsikos D, Cornell CA (2002) Incremental dynamic analysis. Earthquake Eng Struct Dyn 31(3):491–514

Chapter 9
Explicit Probabilistic Seismic Design of RC Structures Through an Elastic Proxy

Paolo Franchin and Paolo Emilio Pinto

Abstract A method is proposed for seismic design of reinforced concrete structures to meet multiple structural performance requirements expressed in terms of exceedance probabilities. The method is approximate in nature and rests on two main results: the closed form solution for the mean annual rate of exceedance of a limit state due to Cornell et al. (J Struct Eng 128:526–533, 2002), and the so-called equal displacement rule. Compliance with the design objectives is obtained through a gradient-based search algorithm in the space of the design variables with reference to a linear elastic proxy of the structure. To this purpose analytical gradients for the Cornell's formula are derived. Two applications illustrate the method and its validation through inelastic time-history analysis. From the limited investigation carried out the method appears to offer satisfactory accuracy.

Keywords Seismic • Inelastic analysis • Mean annual rate • Risk • Sensitivity

9.1 Introduction

In a recent paper by the authors (Franchin and Pinto 2010), advancing a proposal for explicitly probabilistic performance-based design, a brief survey is given of the state of progress regarding seismic performance-based design, as contrasted to assessment.

Most (of the relatively few) approaches employ the concepts and tools of optimisation theory to arrive at a design solution that is optimal in some sense (e.g., amongst others, Beck et al. 1999; Vamvatsikos and Papadimitriou 2005;

P. Franchin (✉) • P.E. Pinto
Department of Structural Engineering & Geotechnics, University of Rome "La Sapienza", Via Gramisci, 53-00197 Rome, Rome, Italy
e-mail: paolo.franchin@uniroma1.it

M. Dolšek (ed.), *Protection of Built Environment Against Earthquakes*,
DOI 10.1007/978-94-007-1448-9_9, © Springer Science+Business Media B.V. 2011

Lagaros and Papadrakakis 2007; Lagaros et al. 2006; Zou and Chan 2005a,b). Review of the available optimisation-based approaches shows that, though conceptually appealing, they are still associated with a prohibitive computational burden, preventing their application in most real-sized situations.

To the knowledge of the authors only two, non optimisation-based approaches are available: namely, that in Krawinkler et al. (2006) and in the already mentioned proposal in Franchin and Pinto (2010). The first design procedure iteratively enforces satisfaction of two performance targets in terms of cost, associated with 50/50 and 2/50 hazard levels, respectively. The procedure makes use of *median* incremental dynamic analysis (IDA) curves to relate the hazard levels with the corresponding demand parameters, as well as of average loss curves, for both structural and non structural damage, to relate response with damage/cost.

The design variables are the fundamental period T_1 and the base shear ratio γ. The procedure requires a prior production of so-called *design aids* in the form of alternative median IDA curves for different values of the design variables.

The second proposal takes as a starting point the approach in Krawinkler et al. (2006), but: (a) employs constraints formulated directly in terms of mean annual frequency (MAF) of exceedance of chosen performance-levels/limit-states; (b) eliminates the need for design-aids by making use of structure-specific analysis. In line with Krawinkler et al. (2006) it works in terms of global design variables, like T_1 (period) and γ (strength) of the building, resulting in a procedure that though it is not completely automatic, it gives guidance towards choosing target stiffness and strength in each iteration, while leaving control and flexibility to the designer on the means to achieve them. The fact that the constraints are expressed probabilistically in terms of their respective MAFs is regarded as a step towards a more rational design approach, since all possible seismic intensities leading to exceedance of a given performance-level are accounted for.

The procedure, which is based on plain, single-mode static pushover analysis, hits its limits when extension is attempted to the design of structures whose dynamic response is significantly contributed by more than one mode. A further difficulty may arise in the case of large structures, due to the indirect control that the designer has on the period and base shear ratio through the basic design variables (members cross-section dimensions and reinforcement).

To solve the latter problem this paper presents a method for performance-based seismic design of RC structures which is still based on explicitly probabilistic design constraints in terms of MAFs, but works directly in terms of basic variables. The method, which is an approximate one, rests on the validity of two basic results of earthquake engineering: the closed-form expression for the MAF of exceedance of a limit-state from Cornell et al. (2002) and the well-known, so-called (empirical) "equal-displacement" rule. Limits of validity of the above results are clearly recognised and are inherited by this proposal.

The next section illustrates the method, whose approximation is then explored in the following one with reference to both a 5-storey and a 15-storey RC plane frame structure.

9.2 Methodology

The method iteratively modifies a design solution until it satisfies multiple probabilistic constraints, i.e. constraints on the MAFs of multiple performance-levels (e.g. light damage, collapse, etc), employing the gradients of the MAFs with respect to the design variables. By virtue of the assumed validity of the "equal-displacement" rule, the iteration process is carried out on a (cracked) elastic model of the structure whose deformed shape, as obtained from multi-modal response spectrum analysis, is taken as a proxy for the true inelastic shape. The assumption of elasticity allows explicit analytical evaluation of the gradients of the MAFs, needed in the search algorithm employed to find a design that satisfies the constraints (see later), a fact that increases manifold the computational effectiveness of the procedure. Flexural reinforcement is only designed when the iteration process on the cross-section dimensions has ended. Shear reinforcement is capacity-designed as the last step.

9.2.1 *Gradients of the MAF*

The "SAC/FEMA" closed-form expression for the mean annual frequency of exceedance λ of a structural limit-state reads (Cornell et al. 2002):

$$\lambda = \lambda_{IM}\left(IM_{\hat{D}=\hat{C}}\right)\exp\left[\frac{1}{2}\frac{k^2}{b^2}\left(\beta_D^2+\beta_C^2\right)\right] = k_0\left(\frac{\hat{C}}{a}\right)^{-\frac{k}{b}}\exp\left[\frac{1}{2}\frac{k^2}{b^2}\left(\beta_D^2+\beta_C^2\right)\right] \tag{9.1}$$

where $IM_{\hat{D}=\hat{C}} = \sqrt[b]{\hat{C}/a}$ is the intensity measure (IM) value that induces a median demand $\hat{D}$ equal to the median limit-state capacity $\hat{C}$, while $\beta_D = \sigma_{\ln D}$ and $\beta_C = \sigma_{\ln C}$ are the demand and capacity dispersions (log-standard deviation). The above expression holds under the assumption that demand and capacity distribute lognormally, that the median demand varying with the IM according to the power law $\hat{D} = aIM^b$, and that the MAF of the IM (hazard) approximates as $\lambda_{IM}(x) = k_0\,x^{-k}$. The parameters a and b of Eq. 9.1 are established by linear regression, usually on the results of inelastic time-history analyses carried out with unscaled recorded motions selected to cover the IM range of interest (i.e. a so-called "cloud" analysis).

Under the assumed validity of the equal-displacement rule, the response of the structure depends only on the elastic properties of its elements, i.e. the section axial and flexural stiffnesses, EA and EI. For RC structures the effect of cracking on flexural stiffness is taken care of by means of reduction factors. Cross-section dimensions are grouped into a vector **d**.

The gradient of the MAF in Eq. 9.1 with respect to the design variables **d** can be expanded as:

$$\nabla_{\mathbf{d}}\lambda = \frac{\partial \lambda}{\partial k_0}\frac{\partial k_0}{\partial \mathbf{d}} + \frac{\partial \lambda}{\partial k}\frac{\partial k}{\partial \mathbf{d}} + \frac{\partial \lambda}{\partial a}\frac{\partial a}{\partial \mathbf{d}} + \frac{\partial \lambda}{\partial b}\frac{\partial b}{\partial \mathbf{d}} \tag{9.2}$$

where:

$$\begin{aligned} \frac{\partial \lambda}{\partial k_0} &= \frac{\lambda}{k_0} \\ \frac{\partial \lambda}{\partial k} &= \frac{\lambda}{b}\left[\frac{k}{b}\left(\beta_D^2 + \beta_C^2\right) - \ln\left(\frac{\hat{C}}{a}\right)\right] \\ \frac{\partial \lambda}{\partial a} &= \lambda \frac{k}{ab} \\ \frac{\partial \lambda}{\partial b} &= \lambda \frac{k}{b^2}\left[\ln\left(\frac{\hat{C}}{a}\right) - \frac{k}{b}\left(\beta_D^2 + \beta_C^2\right)\right] \end{aligned} \tag{9.3}$$

Eq. 9.2 does not contain terms with the derivatives of λ with respect to the dispersions β_D and β_C since the dependence of the latter on the design through response is generally minor and thus they are *assumed* to remain constant throughout iteration.

The derivatives of the hazard parameters k_0 and k are given by:

$$\begin{aligned} \frac{\partial k_0}{\partial \mathbf{d}} &= \frac{\partial k_0}{\partial T_1}\frac{\partial T_1}{\partial \mathbf{d}} \\ \frac{\partial k}{\partial \mathbf{d}} &= \frac{\partial k}{\partial T_1}\frac{\partial T_1}{\partial \mathbf{d}} \end{aligned} \tag{9.4}$$

where the $\partial k_0/\partial T_1$ and $\partial k/\partial T_1$ can be readily obtained for every site of interest from a polynomial interpolation of k_0, k values numerically obtained for a finite number of structural periods T_1. For a linear elastic system the gradients of the eigenvalues $\eta_k = \omega_k^2 = (2\pi/T_k)^2$ with respect to the design variables (in terms of stiffness **K** and mass matrices **M**) is available in analytical form, together with the gradients of the corresponding mode shapes $\boldsymbol{\varphi}_k$ (Lin et al. 1996):

$$\frac{\partial T_k}{\partial d_j} = -\frac{\pi}{\eta_k^{3/2}}\left[\boldsymbol{\varphi}_k^T \frac{\partial \mathbf{K}}{\partial d_j}\boldsymbol{\varphi}_k - \eta_k \boldsymbol{\varphi}_k^T \frac{\partial \mathbf{M}}{\partial d_j}\boldsymbol{\varphi}_k\right] \tag{9.5}$$

$$\frac{\partial \boldsymbol{\varphi}_k}{\partial d_j} = \sum_{\substack{s=1 \\ s\neq k}}^{N} \frac{1}{\eta_k - \eta_s}\boldsymbol{\varphi}_s^T\left(\frac{\partial \mathbf{K}}{\partial d_j} - \eta_k \frac{\partial \mathbf{M}}{\partial d_j}\right)\boldsymbol{\varphi}_k\boldsymbol{\varphi}_s - \frac{1}{2}\boldsymbol{\varphi}_k^T \frac{\partial \mathbf{M}}{\partial d_j}\boldsymbol{\varphi}_k\boldsymbol{\varphi}_k \tag{9.6}$$

Equation 9.6 provides the derivative of the k-th mode shape as a function of all other mode shapes, which is not efficient for large structures where usually only a reduced number of modes need to be evaluated for response determination. Refined methods allow the determination of $\partial \boldsymbol{\varphi}_k / \partial d_j$ employing only the k-th mode shape and associated eigenvalue (Lin et al. 1996).

The derivatives of the demand parameters a and b with respect to the design variables can be obtained by differentiating the least square expression that relates them to the actual intensity-demand points that are used to fit the $D = aIM^b\varepsilon$ power-law:

$$\ln D = \ln a + b \ln IM + \ln \varepsilon \rightarrow \begin{Bmatrix} \ln a \\ b \end{Bmatrix} = \left(\mathbf{Z}^T\mathbf{Z}\right)^{-1}\mathbf{Z}^T \ln \mathbf{D} = \mathbf{Z}^* \ln \mathbf{D} \tag{9.7}$$

where $\mathbf{Z} = [\mathbf{1} \quad \ln \mathbf{IM}]$ is the "design" matrix. It follows that:

$$\begin{aligned} \frac{\partial a}{\partial \mathbf{d}} &= \frac{\partial e^{\ln a}}{\partial \ln a}\left(\frac{\partial \ln a}{\partial \mathbf{z}_1^*}\frac{\partial \mathbf{z}_1^*}{\partial \mathbf{IM}}\frac{\partial \mathbf{IM}}{\partial T_1}\frac{\partial T_1}{\partial \mathbf{d}} + \frac{\partial \ln a}{\partial \ln \mathbf{D}}\frac{\partial \ln \mathbf{D}}{\partial \mathbf{u}}\frac{\partial \mathbf{u}}{\partial \mathbf{d}}\right) \\ &= a\left(\ln \mathbf{D}\frac{\partial \mathbf{z}_1^*}{\partial \mathbf{IM}}\frac{\partial \mathbf{IM}}{\partial T_1}\frac{\partial T_1}{\partial \mathbf{d}} + \mathbf{z}_1^*\frac{\partial \ln \mathbf{D}}{\partial \mathbf{u}}\frac{\partial \mathbf{u}}{\partial \mathbf{d}}\right) \end{aligned} \tag{9.8}$$

and:

$$\begin{aligned} \frac{\partial b}{\partial \mathbf{d}} &= \left(\frac{\partial b}{\partial \mathbf{z}_2^*}\frac{\partial \mathbf{z}_2^*}{\partial \mathbf{IM}}\frac{\partial \mathbf{IM}}{\partial T_1}\frac{\partial T_1}{\partial \mathbf{d}} + \frac{\partial b}{\partial \ln \mathbf{D}}\frac{\partial \ln \mathbf{D}}{\partial \mathbf{u}}\frac{\partial \mathbf{u}}{\partial \mathbf{d}}\right) \\ &= \ln \mathbf{D}\frac{\partial \mathbf{z}_2^*}{\partial \mathbf{IM}}\frac{\partial \mathbf{IM}}{\partial T_1}\frac{\partial T_1}{\partial \mathbf{d}} + \mathbf{z}_2^*\frac{\partial \ln \mathbf{D}}{\partial \mathbf{u}}\frac{\partial \mathbf{u}}{\partial \mathbf{d}} \end{aligned} \tag{9.9}$$

in which $\mathbf{z}_1^*$ and $\mathbf{z}_2^*$ are the first and second row of the $\mathbf{Z}^*$ matrix, respectively. The remaining terms in Eqs. 9.8 and 9.9 are:

- $\partial \mathbf{z}_1^* / \partial \mathbf{IM}, \partial \mathbf{z}_2^* / \partial \mathbf{IM}$ the derivatives of the row vectors $\mathbf{z}_1^*$ and $\mathbf{z}_2^*$, which are found from $\mathbf{Z}^* = \left(\left[\mathbf{1}_{n\times1} \quad \ln \mathbf{IM}_{n\times1}\right]^T\left[\mathbf{1}_{n\times1} \quad \ln \mathbf{IM}_{n\times1}\right]\right) \cdot \left[\mathbf{1}_{n\times1} \quad \ln \mathbf{IM}_{n\times1}\right]^T$
- $\partial \mathbf{IM} / \partial T_1$ the derivative of the IM values with respect to the fundamental period, and are known given the spectral shape of the records
- $\partial \ln \mathbf{D} / \partial \mathbf{u}$ the derivative of the chosen demand parameter (e.g. peak interstorey drift ratio $\theta_{\max}$) with respect to the full response vector $\mathbf{u}$,
- $\partial \mathbf{u} / \partial \mathbf{d}$ the derivative of the response vector with respect to the design parameters.

The $\partial \mathbf{u} / \partial \mathbf{d}$ derivative depends on the particular analysis method employed for response determination. In the presented design method the response is obtained by means of modal analysis with unreduced elastic response spectrum.

The SRSS combination of modal contributions to the i-th displacement component reads:

$$u_i = \sqrt{\sum_{k=1}^{n} u_{ik}^2} \tag{9.10}$$

where the contribution of the k-th mode is given by:

$$u_{ik} = p_k S_d(T_k)\varphi_{ik} \tag{9.11}$$

in which the $S_d(T_k)$ and φ_{ik} are the displacement spectral ordinate at the period of the k-th mode and the i-th component of the k-th mode shape, respectively, while the modal participation factor is given by:

$$p_k = \frac{\boldsymbol{\varphi}_k^T \mathbf{Mt}}{\boldsymbol{\varphi}_k^T \mathbf{M} \boldsymbol{\varphi}_k} \tag{9.12}$$

Differentiation of Eq. 9.10 with respect to the j-th design variable yields:

$$\frac{\partial u_i}{\partial d_j} = \frac{1}{2u_i} \sum_{k=1}^{n} 2u_{ik} \frac{\partial u_{ik}}{\partial d_j} \tag{9.13}$$

which is a function of the corresponding derivative of Eq. 9.11

$$\frac{\partial u_{ik}}{\partial d_j} = \frac{\partial p_k}{\partial d_j} S_d(T_k)\varphi_{ik} + p_k \frac{\partial S_d(T_k)}{\partial T_k} \frac{\partial T_k}{\partial d_j} \varphi_{ik} + p_k S_d(T_k) \frac{\partial \varphi_{ik}}{\partial d_j} \tag{9.14}$$

where $\partial S_d(T_k)/\partial T_k$ is known from the response spectrum used in the response analysis, and the derivative of the modal participation factor reads:

$$\begin{aligned} \frac{\partial p_k}{\partial d_j} = & \left[\left(\frac{\partial \boldsymbol{\varphi}_k^T}{\partial d_j} \mathbf{Mt} + \boldsymbol{\varphi}_k^T \frac{\partial \mathbf{M}}{\partial d_j} \mathbf{t} \right) \boldsymbol{\varphi}_k^T \mathbf{M} \boldsymbol{\varphi}_k \right. \\ & \left. - \boldsymbol{\varphi}_k^T \mathbf{Mt} \left(2\boldsymbol{\varphi}_k^T \mathbf{M} \frac{\partial \boldsymbol{\varphi}_k}{\partial d_j} + \boldsymbol{\varphi}_k^T \frac{\partial \mathbf{M}}{\partial d_j} \boldsymbol{\varphi}_k \right) \right] \left(\boldsymbol{\varphi}_k^T \mathbf{M} \boldsymbol{\varphi}_k \right)^{-2} \end{aligned} \tag{9.15}$$

Consistently with the equal displacement rule, the demand exponent b is set constant and equal to 1. As a consequence the derivative in Eq. 9.9 is identically zero and only one response spectrum analysis for an arbitrary intensity level suffices to establish the value of the demand coefficient a.

9.2.2 *Search Algorithm*

The performance constraints that the structure must comply with, as anticipated, are expressed in terms of λ of the limit-state violations for a number of limit-states of interest, e.g. a *serviceability* limit-state, such as *light damage* (LD), and a *safety*-related one, such as *collapse prevention* (CP). For example (the limits on the frequencies being arbitrary):

$$\begin{aligned} \lambda_{LD} &\le \lambda^*_{LD} = 1/100 \text{ years} \\ \lambda_{CP} &\le \lambda^*_{CP} = 1/2500 \text{ years} \end{aligned} \tag{9.16}$$

When working with multiple constraints it is useful to normalize each of them according to the expression: $\tilde{\lambda} = \lambda/\lambda^* - 1$. This allows to define the governing constraint at each iteration, as the one having the largest value of $\tilde{\lambda}$. At the end of the process only one of the constraints is satisfied in equality, while the remaining ones are satisfied with more or less wide margins.

For simple cases with few design variables the search for the design solution can be performed with a steepest-descent (Newton) algorithm. In larger size applications, however, the latter method is not acceptably reliable/accurate.

The search for a feasible design solution, i.e. the problem of finding a zero for the $\tilde{\lambda} = \tilde{\lambda}(\mathbf{d})$ function, is carried out by means of a quasi-Newton method, transforming it into the problem of finding a minimum for $\tilde{\lambda}^2$, where the gradient $\nabla_{\mathbf{d}}\tilde{\lambda}^2 = \mathbf{0}$. In practice, since the feasible design must also satisfy a number of other practical constraints related, e.g., to construction, the problem is cast in the form of a constrained optimization:

$$\begin{cases} \min\limits_{\mathbf{d}} s = \tilde{\lambda}^2 \\ \text{subject to } \mathbf{c} \le 0 \end{cases} \tag{9.17}$$

where the vector $\mathbf{c}$ collects the n constraints $c_i(\mathbf{d})$ which are formulated to take upon positive values whenever the corresponding constraint is violated. Typical constraints employed in practice are of the form: (a) $\gamma d_j \le d_{j+1} \le d_j$ regulating column tapering (with $\gamma < 1$ and the ordering of column members increases upward), (b) $d_j \le d_{j,\max}$ limiting from above the cross-section dimension, or (c) $d_j \ge d_{j,\min}$ limiting to a minimum (slenderness, axial load, etc) the cross-section dimension. These constraints can all collectively be put in the form $\mathbf{c}(\mathbf{d}) = \mathbf{Ad} + \mathbf{b} \le 0$.

The gradient of the function s in Eq. 9.17 is given by:

$$\nabla_{\mathbf{d}} s = 2\tilde{\lambda}\nabla_{\mathbf{d}}\tilde{\lambda} = \frac{2}{\lambda^*}\left(\frac{\lambda}{\lambda^*} - 1\right)\nabla_{\mathbf{d}}\lambda \tag{9.18}$$

The quasi-Newton method provides the variation of the design variables at the i-th step as a function of the gradient $\nabla_{\mathbf{d}} s$ and the inverse Hessian $\mathbf{H}$ (α being the step size):

$$\Delta\mathbf{d}_i = \mathbf{d}_{i+1} - \mathbf{d}_i = -\alpha_i \mathbf{H}_i \nabla_{\mathbf{d}} s(\mathbf{d}_i) \tag{9.19}$$

Several efficient algorithms are available for the determination of the Hessian matrix at step $i + 1$, or, most efficiently, directly of its inverse, as a function of the previous (inverse) Hessian and of the gradients difference $\mathbf{y}_i = \nabla_{\mathbf{d}} s_{i+1} - \nabla_{\mathbf{d}} s_i$. Herein the well-known Broyden–Fletcher–Goldfarb–Shanno (BFGS) algorithm (Luenberger and Ye 2008) is used:

$$\mathbf{H}_{i+1} = \left(I - \frac{\mathbf{y}_i \Delta \mathbf{d}_i^T}{\mathbf{y}_i^T \Delta \mathbf{d}_i} \right)^T \mathbf{H}_i \left(I - \frac{\mathbf{y}_i \Delta \mathbf{d}_i^T}{\mathbf{y}_i^T \Delta \mathbf{d}_i} \right) + \frac{\Delta \mathbf{d}_i \Delta \mathbf{d}_i^T}{\mathbf{y}_i^T \Delta \mathbf{d}_i} \tag{9.20}$$

9.2.3 *Reinforcement Design*

Design of longitudinal reinforcement is carried out for a "seismic" combination of gravity loads and a seismic action characterized by a given average return period. This latter is chosen to limit structural damage (yielding) for frequent earthquakes, therefore design of longitudinal reinforcement is carried out for a seismic action with an average return period related to the λ_{LD}^* limit on the light damage performance-level. Since the frequency of exceedance of the response according to the Cornell's formula is the product of the MAF of the seismic action inducing a median demand equal to the median limit state capacity $\lambda_{IM_{\hat{D}=\hat{C}}}$, times an exponential amplification factor commonly between 1.2 and 2.2 (depending essentially on the hazard slope k, for usual values of β_D and β_C), one can conclude that the order of magnitude of the return period to be used for reinforcement design is in the order of $1.5/\lambda_{LD}^*$ (say, 150 years, according to Eq. 9.16).

Capacity design procedure is not strictly required for the relative flexural strength of beams and columns, since the drift, considered the critical control variable, is explicitly limited in the method. Adoption of capacity design, however, is a good practice that would certainly reduce the uncertainty associated with the drift prediction and increase the reliability of the proposed methods. Capacity design is clearly employed for determining the shear strength of members and joints. The distribution of flexural reinforcement along the height is carried out aiming at a "as uniform as possible" ratio of moment demand to capacity (see also Fig. 9.1).

9.3 Applications

The method has been implemented in a MATLAB© code that employs the built-in optimization functions (in particular the constrained minimization algorithms in `fmincon`) and performs response spectrum analysis of linear elastic frames, computing both response and its sensitivity. In the two applications to follow the method is illustrated and validation of the obtained design is carried out by means of inelastic time-history analysis within the finite-element code OpenSEES (McKenna and Fenves 2001).

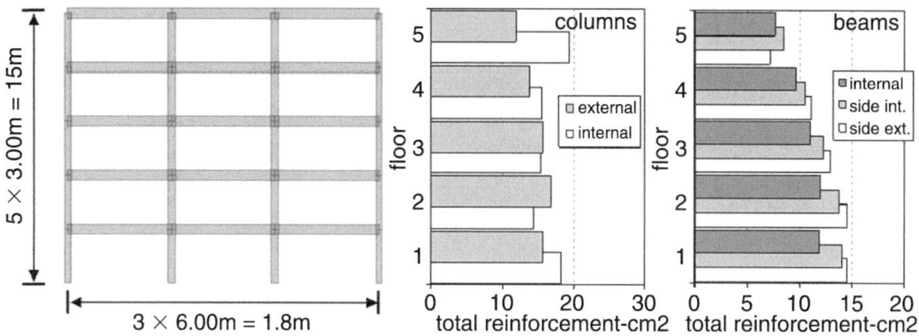

Fig. 9.1 Example 1: geometry and reinforcement of the five-storeys RC plane frame

Table 9.1 Example 1: Design variables, all dimensions in meters

Var.	Min.	Max.	Initial	Final
t_{ext}	0.30	0.50	0.30	0.47
t_{int}	0.30	0.50	0.30	0.42

9.3.1 *Five-Storey RC Frame*

The first test of the method is carried out on the simple five storeys RC plane frame shown in Fig. 9.1. , which also reports the designed reinforcement in cm^2 for the final iteration member dimensions. For this example two design variables are considered: the in-plane dimension of the external and internal columns (constant over height, out-of-plane dimension is 50, 40 and 30 cm for external columns, internal columns and beams, respectively). Two constraints are imposed on the design, namely:

$$\lambda_{LD} = \lambda(\theta_{\max} \leq 0.004) \leq \lambda^*_{LD} = 1/100 \text{ years}$$

$$\lambda_{CP} = \lambda(\theta_{\max} \leq 0.015) \leq \lambda^*_{CP} = 1/950 \text{ years.}$$

The variables are constrained between a minimum and a maximum value, as shown in Table 9.1, which reports also their initial and final values.

The value of the demand dispersion term is set equal to $\beta_D = 0.30$. Capacity terms are set to $\beta_C = 0.30$ and 0.0, for the CP and LD performance levels, respectively. Table 9.2 reports the evolution with the iterations of the fundamental period T_1, the hazard coefficients k_0 and k, the slope a of the demand-intensity relation, as well as the non normalized and normalized MAF for both limit-states. Inspection of the absolute values of $\tilde{\lambda}_{LD}$ and $\tilde{\lambda}_{CP}$ shows how the governing constraint is changing during the iterations. The light damage condition is the constraint satisfied "in equality" (tolerance set to 10^{-4} on $s = \tilde{\lambda}^2$) at the end of the iteration process.

Table 9.2 Example 1: Iterations

Iter	T_1 (s)	k_0	k	a	λ_{LD}	$\tilde{\lambda}_{LD}$	λ_{CP}	$\tilde{\lambda}_{CP}$
1	1.836	0.0027	1.880	0.0100	0.0175	0.7649	0.0017	0.6084
2	1.201	0.0059	1.880	0.0044	0.0083	−0.1655	0.0008	−0.2387
3	1.212	0.0058	1.880	0.0045	0.0084	−0.1503	0.0008	−0.2244
4	1.358	0.0047	1.880	0.0056	0.0104	0.0529	0.0010	−0.0389
5	1.293	0.0052	1.880	0.0051	0.0095	−0.0419	0.0009	−0.1253
6	1.316	0.0050	1.880	0.0053	0.0098	−0.0091	0.0010	−0.0955
7	1.323	0.0049	1.880	0.0053	0.0099	0.0019	0.0010	−0.0854

Table 9.3 Example 1: Modes for the initial and final designs

	Initial		Final	
Mode	T (s)	PMR (%)	T (s)	PMR (%)
1	1.836	85	1.323	83
2	0.611	10	0.425	10
3	0.368	3	0.240	4
4	0.272	1	0.165	2

Table 9.3 reports the modal periods and participating mass ratios for the initial and final iterations. Reinforcement is designed according to the previous section, for a seismic action with a 150 years return period.

In order to validate the design solution provided by the method, the structure resulting from the final iteration is subjected to nonlinear time-history analysis for a suite of 35 ground motion records. The motions are spectrum-compatible artificial records generated in groups of seven to match five uniform-hazard spectra of increasing intensity (mean return period ranging from 60 years to 2,000 years), in order to span a sufficiently large range of spectral accelerations. The acceleration response spectra of the records are shown in Fig. 9.2. (left).

It is recalled that the method relies on the validity of two basic approximations: the so-called, empirical "equal displacement rule" and the closed-form solution of the MAF by Cornell and co-workers. The predictive power of the elastic deformed shape as a proxy of the inelastic one is first checked by comparing the peak inter-storey drift profiles as obtained from SRSS of modal responses for the five target spectra versus the average profiles from each of the five groups of artificial records matching the same spectra. The profiles are shown in Fig. 9.2 (right), where it is apparent that the match is quite satisfactory for this simple example. A good prediction of the $\theta_{\max}$ profiles is a pre-requisite for the closeness of the risk λ to the target one λ^*. It can be observed how the average $\theta_{\max}$ value for the records with $1/\lambda^*_{CP} = 950$ years intensity is lower than the 1.5% limit, since the procedure enforces a constraint on the probability of exceedance, which accounts also for the dispersion in this limit (β_C) as well as in the demand (β_D).

The second possible source of error in the method is in the approximation of the Cornell's formula for risk. For this reason the risk value obtained by means of linear analysis at the end of the design iteration process needs to be compared with the

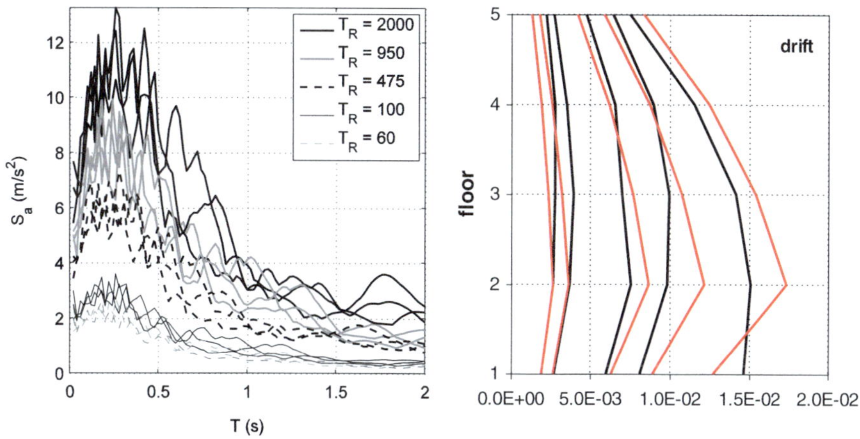

Fig. 9.2 (*Left*) Response spectra of the 5 × 7=35 records suite chosen for validation through nonlinear time-history analysis; (*right*) Example 1: Peak inter-storey drift profiles for the five return periods as obtained from linear (SRSS of modal responses, *red*) and nonlinear (average over seven records each, *black*) analyses

"true value" of the risk, based on "exact" convolution of hazard and fragility (Jalayer et al. 2007):

$$\begin{aligned}\lambda_{CP} &= \int_0^{\infty} P(\theta_{\max} > 0.015 | IM = x) d\lambda_{IM}(x) \\ &= \int_0^{\infty} P\left(S_{a,CP} < x\right) d\lambda_{IM}(x) = \int_0^{\infty} F_{S_{aCP}}(x) d\lambda_{IM}(x)\end{aligned} \tag{9.21}$$

The fragility is evaluated based on inelastic time-history analysis. For this purpose, the value $S_{a,CP}$ of spectral acceleration inducing collapse, i.e. the capacity of the structure in terms of seismic intensity, for each of the 35 records is obtained. The fragility is then evaluated assuming a lognormal shape and computing the log-mean and log-standard deviation from the sample of $S_{a,CP}$.

A meaningful comparison of the risk value obtained by means of linear analysis at the end of the design iteration process with the "exact" value must take into account the inter-analyst dispersion of "exact" values computed by different analysts employing different sets of records. To obtain an estimate of the distribution of "exact" risk values the bootstrap (Efron and Tibshirani 1993) technique has been used. The fragility has been computed for 900 sub-sets of size 20 taken randomly (without replacement) from the 35 values of $S_{a,CP}$. Fig. 9.3 shows two histograms of risk values. The first histogram (light grey) comes from convolution carried out according to Eq. 9.21. The second histogram (dark grey) is obtained with the Cornell's formula based on the fit of the demand-intensity power-law for the same 900 sub-sets as above. The two distributions are very close, confirming the good approximation of the closed-form expression for this case.

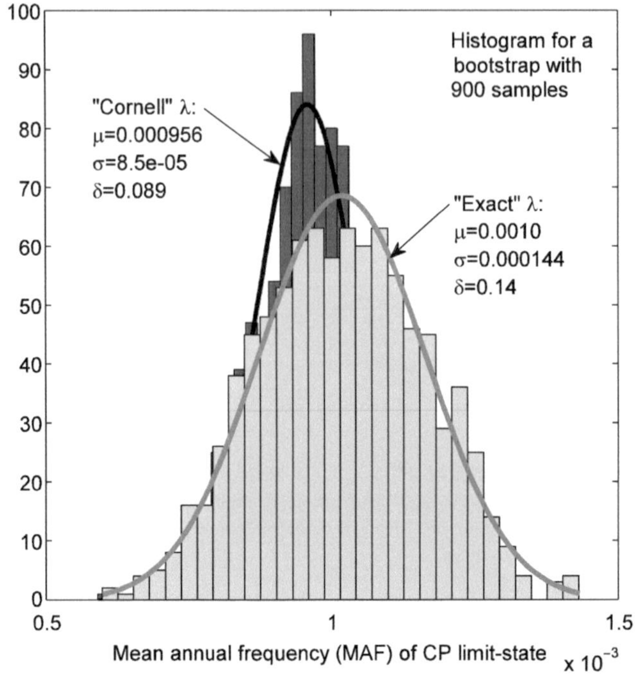

Fig. 9.3 Example 1: Histograms of MAF values obtained from "exact" integration of fragility and hazard, and from Cornell's formula on nonlinear time-history results

The $\lambda_{CP} = 1.1 \times 10^{-3}$ value obtained at the end of the iteration falls within both distributions. Not unexpectedly, given the good match of drift profiles in Fig. 9.2 (right), it is closer to the central value of the Cornell-based risk distribution, while it falls in the upper tail of the convolution-based risk distribution. Overall, it can be concluded that the actual value of the mean annual frequency λ_{CP} of the produced design is quite reasonably close to the target one.

9.3.2 Fifteen-Storey RC Frame

As a second step the fifteen storeys RC plane frame shown in Fig. 9.4 is analyzed. The figure shows the overall dimensions and reports the reinforcement, in cm^2, designed for the configuration obtained at the end of the iteration (reinforcement does not enter in the iteration process).

For this example seven design variables are considered: three variables for the in-plane dimension of the three orders of external columns (each order corresponding to five floors), three variables for the internal ones, and the seventh variable for beam height, constant for all floors. The out-of-plane dimensions for all

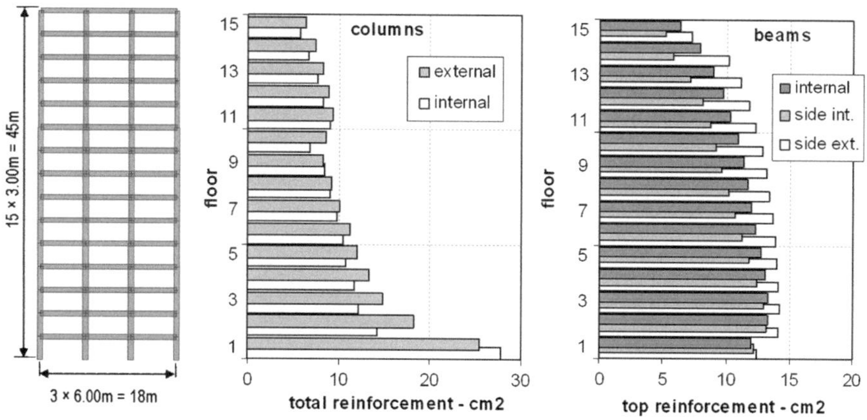

Fig. 9.4 Example 2: geometry and reinforcement of the fifteen-storeys RC plane frame

Table 9.4 Example 2: Design variables, all dimensions in meters

Var.	Min.	Max.	Initial	Final
$t_{1,ext}$	0.70	1.20	0.70	0.73
$t_{2,ext}$	0.50	1.00	0.50	0.73
$t_{3,ext}$	0.30	0.80	0.50	0.63
$t_{1,int}$	0.70	1.40	0.70	0.76
$t_{2,int}$	0.50	1.20	0.50	0.73
$t_{3,int}$	0.30	1.00	0.50	0.62
t_{beam}	0.55	0.65	0.55	0.65

members is kept constant and equal to 50, 40 and 30 cm for external columns, internal columns and beams, respectively, as for the previous example.

Two constraints as for the previous example are imposed on the design, namely $\lambda_{LD} = \lambda(\theta_{\max} \leq 0.004) \leq \lambda^*_{LD} = 1/100y$ and $\lambda_{CP} = \lambda(\theta_{\max} \leq 0.015) \leq \lambda^*_{CP} = 1/1950y$.

The variables are constrained between a minimum and a maximum value, as shown in Table 9.4, which reports also the initial and final values. Further, column dimensions have been constrained with the additional eight constraints:

$$\begin{cases} t_{i+1,ext} \leq t_{i,ext} \\ t_{i+1,ext} \geq 0.85 t_{i,ext} \\ t_{i+1,\text{int}} \leq t_{i,\text{int}} \\ t_{i+1,\text{int}} \geq 0.85 t_{i,\text{int}} \end{cases} \quad i = 1, 2 \tag{9.22}$$

Once again, the value of the demand dispersion term is set equal to $\beta_D = 0.30$, while capacity terms are set to $\beta_C = 0.30$ and 0.0, for the CP and LD performance levels, respectively. Table 9.5 reports the evolution with the iterations of the fundamental period, the hazard coefficients, the slope a of the demand-intensity relation, as well as the non normalized and normalized MAF for both limit-states.

Table 9.5 Example 2: Iterations

Iter	T_1 (s)	k_0	k	a	λ_{LD}	$\tilde{\lambda}_{LD}$	λ_{CP}	$\tilde{\lambda}_{CP}$
1	3.526	0.001	1.467	0.0146	0.0046	−0.5347	0.0007	0.3309
2	2.938	0.001	1.532	0.0097	0.0043	−0.5687	0.0006	0.1372
3	2.636	0.001	1.611	0.0076	0.0039	−0.6016	0.0005	−0.0211
4	2.703	0.001	1.591	0.008	0.004	−0.5961	0.0005	0.0103
5	2.680	0.0012	1.598	0.0079	0.004	−0.5979	0.0005	−0.0005

Table 9.6 Example 2: Modal properties for the initial and final iteration

	Initial		Final	
Mode	T (s)	PMR (%)	T (s)	PMR (%)
1	3.527	76	2.680	78
2	1.174	11	0.891	11
3	0.683	4	0.506	4
4	0.467	2	0.346	2

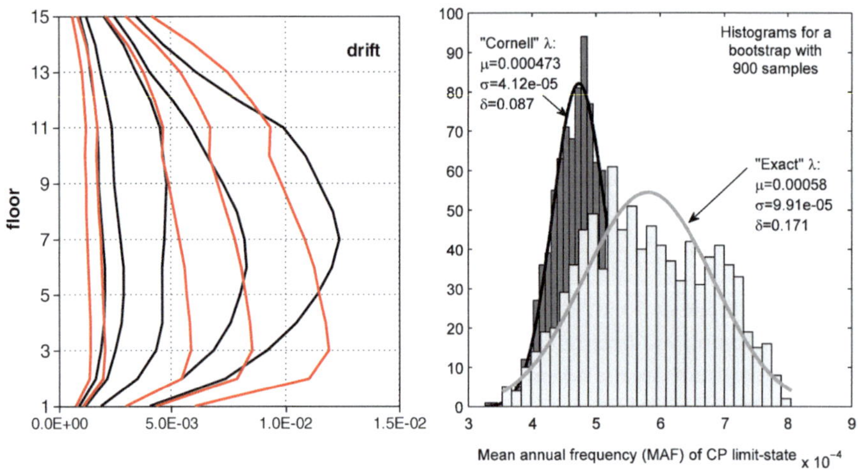

Fig. 9.5 Example 2: (*left*) Peak interstorey drift profiles for the five return periods as obtained from linear (SRSS of modal responses, *red*) and nonlinear (average over seven records each, *black*) analyses; (*right*) Histograms of MAF values obtained from "exact" integration of fragility and hazard (*light grey*), and from Cornell's formula on nonlinear time-history results (*dark grey*)

In this example the governing constraint is the collapse prevention one. Actually, the light damage limit state is already satisfied for the initial design. Table 9.6 reports the modal periods and participating mass ratios for the initial and final iterations. The frame exhibits a moderate second-mode contribution to response.

The design from the final iteration is then subjected to nonlinear time-history analysis for the suite of 35 ground motion records introduced in the first example. The drift profiles averaged over the seven records of each of the five groups are compared with corresponding profiles from SRSS of modal contributions in Fig. 9.5 (left).

The match between elastic and inelastic drift profiles is less close than for the previous case, but still satisfactory. The elastic profiles consistently overestimate the inelastic one at the lower floors indicating that the reduction factor at the lower levels should be somewhat higher. It is believed that the difference is mostly explained by the uniform reduction factor adopted for the elastic stiffness, which, for a better approximation, might be made member-dependent and function of axial load ratio and response level in a single global iteration.

As for the previous example the bootstrap technique (again, 900 sub-sets of size 20 taken randomly without replacement from the 35 values of $S_{a,CP}$) is employed to obtain two distributions, one of the risk values from convolution carried out according to Eq. 9.21, the other of risk values obtained with the Cornell's formula. The corresponding histograms are shown in Fig. 9.5 (right), where "exact" risk is in light grey and Cornell's risk is dark grey. The two distributions are not as close than for the first example. The $\lambda_{CP} = 5 \times 10^{-4}$ value obtained at the end of the iteration falls within both distributions, however, and quite close to their central values.

9.4 Conclusions

The method illustrated shows, as far as it is allowed by the limited validation that has been carried out, that it is feasible to design in an explicitly probabilistic manner a RC frame structure on the basis of two basic results: the so-called "Cornell" closed-form for seismic risk and the so-called "equal-displacement" rule. This latter rule appears to maintain its approximate validity also for MDOF structures, on the condition the reinforcement design is carried out to provide the structure with a "uniform" distribution of the actual to required strength ratio, in order to maintain similarity of the deformation pattern between the elastic and inelastic response regimes.

As a final comment, though the design iteration procedure employs a constrained minimization algorithm, it is emphasized that this is only a means to enforce the constraints on the MAF. In this sense the proposed method is not producing an "optimal" design, since there is no objective function, related e.g. to weight/cost. Minimization of the structural materials cost is not regarded as a primary goal, considering its minor contribution to the total initial cost. Inclusion of this objective, however, is not feasible in the present procedure since iteration is carried out on the concrete section dimensions, and reinforcement is designed only afterwards.

Acknowledgment The substantial contribution to this work given by Dr. Fabrizio Noto is gratefully acknowledged.

References

Beck JL, Chan E, Irfanoglu A, Papadimitriou C (1999) Multi-criteria optimal structural design under uncertainty. Earthquake Eng Struct Dyn 28:741–761

Cornell CA, Jalayer F, Hamburger RO, Foutch DA (2002) The probabilistic basis for the 2000 SAC/FEMA steel moment frame guidelines. J Struct Eng 128:526–533

Efron B, Tibshirani RJ (1993) An introduction to the bootstrap. Chapman & Hall, New York
Franchin P, Pinto PE (2010) Direct probability-based seismic design of RC structures. In: Proceedings ACES workshop, Springer Corfu, 5th–8th July 2009
Jalayer F, Franchin P, Pinto PE (2007) A scalar damage measure for seismic reliability analsysis of RC structures. Earthquake Eng Struct Dyn 36:2059–2079
Krawinkler H, Zareian F, Medina RA, Ibarra L (2006) Decision support for conceptual performance-based design. Earthquake Eng Struct Dyn 35:115–133
Lagaros ND, Papadrakakis M (2007) Seismic design of RC structures: a critical assessment in the framework of multi-objective optimization. Earthquake Eng Struct Dyn 36:1623–1639
Lagaros ND, Fotis AD, Stilianos AK (2006) Assessment of seismic design procedures based on the total cost. Earthquake Eng Struct Dyn 35:1381–1401
Lin RM, Wang Z, Lim MK (1996) A practical algorithm for the efficient computation of eigenvector sensitivities. Comput Methods Appl Mech Eng 130:355–367
Luenberger DG, Ye Y (2008) Linear and nonlinear programming, vol 116, 3rd edn, International series in operations research & management science. Springer, New York
McKenna F, Fenves GL (2001) The openSees command language manual, version1.2, Pacific Earthquake Engineering Research Center, University of California, Berkeley
Vamvatsikos D, Papadimitriou C (2005) Optimal multi-objective design of a highway bridge under seismic loading through Incremental Dynamic Analysis. In: Proceedings ICOSSAR'05, Rome
Zou XK, Chan CM (2005a) An optimal resizing technique for seismic drift design of concrete buildings subjected to response spectrum and time history loadings. Comput Struct 83:1689–1704
Zou XK, Chan CM (2005b) Optimal seismic performance-based design of reinforced concrete buildings using nonlinear pushover analysis. Eng Struct 27:1289–1302

Chapter 10
Structural Optimization: An Assessment Approach of Design Procedures Against Earthquake Hazard

Chara Ch. Mitropoulou, Nikos D. Lagaros, and Manolis Papadrakakis

Abstract The objective of this work is to assess seismic design procedures with reference to optimally design of 3D reinforced concrete structures. For this purpose four optimization problems are formulated corresponding to different values of the behaviour factor q. The optimization problems are formulated as sizing optimization problems where the size and steel reinforcement of the beams and the columns are considered as design variables, while the initial cost is the objective function to be minimized. Furthermore, life-cycle cost analysis is an assessment tool for the performance of systems; and has been used in many fields of science and engineering. In this chapter the influence of the analysis procedure, the number of seismic records imposed and the performance criterion used on the life-cycle cost analysis procedure is also investigated. In particular the nonlinear static and the multiple stripe incremental dynamic analysis are applied while the maximum inter-storey drift and the maximum floor acceleration are adopted as the performance criteria calculated in multiple hazard levels that are required for the life-cycle cost analysis. In the test example considered, the life-cycle cost was calculated taking into consideration the damage repair cost, the loss of contents cost due to structural damage that is quantified by the maximum inter-storey drift and the floor acceleration, the loss of rental cost, the income loss cost, the cost of injuries and the cost of human fatality.

Keywords Optimum design • Life-cycle cost assessment • Nonlinear static analysis • Incremental dynamic analysis • Reinforced concrete buildings

---

C.Ch. Mitropoulou • N.D. Lagaros (✉) • M. Papadrakakis
Institute of Structural Analysis & Antiseismic Research, National Technical University of Athens, 9, Iroon Polytechniou Street, Zografou Campus, Athens 15780, Greece
e-mail: chmitrop@central.ntua.gr; nlagaros@central.ntua.gr; mpapadra@central.ntua.gr

M. Dolšek (ed.), *Protection of Built Environment Against Earthquakes*,
DOI 10.1007/978-94-007-1448-9_10,

10.1 Introduction

Structural optimization problems are characterized by various objective and constraint functions, which are generally non-linear functions of the design variables. These functions are usually implicit, discontinuous and non-convex. The mathematical formulation of structural optimization problems with respect to the design variables, the objective and constraint functions depend on the type of the application. During the last three decades there has been a growing interest in problem solving systems based on algorithms which rely on analogies to natural processes. These systems have some selection process based on fitness of individuals and some recombination operators and they are used in this work for assessing design approaches against the earthquake hazard. The life-cycle cost analysis (LCCA) principles are based on economic theories, and have been used as decision-support tools in industrial and commercial projects. LCCA is mainly implemented to energy and water conservation projects as well as transportation projects, including highways, bridges, and pavements. When it comes to building structures, the application of LCCA is considered particularly important in the case of retrofitted/deteriorating structures. Especially in the case of steel and reinforced concrete building structures in seismic regions, LCCA is applied as a structural performance criterion for taking into account future damages due to earthquakes. Furthermore, it is also used as a decision making tool for the most cost-effective solution related to the construction of building structures in seismic regions.

In early 1960s LCCA was applied in the commercial area in the design of products considering the total cost of developing, producing, using and retiring. The introduction of LCCA was made in the field of infrastructures as an absolute investment assessment tool. In particular, in early 1980s it was used in USA as an appraisal tool for the total cost of ownership over the lifespan of an asset (Arditi and Messiha 1996; Asiedu and Gu 1998). Later, in view of large losses due to extreme hazards, like earthquakes and hurricanes, there was a need for new design procedures of facilities that could lead to life protection and reduction of damage and economical impact of such hazards to an acceptable level (Wen and Kang 2001a). In this context LCCA was introduced in the field of constructions as a complex investment appraisal tool incorporating a structural performance criterion (Sanchez-Silva and Rackwitz 2004).

A considerable amount of work has been done in estimating losses due to earthquakes. In particular in the work by Beck et al. (2003) a measure, to be incorporated into the seismic risk assessment framework for economic decision-making of buildings, was introduced, denoted as the probable frequent loss, which is defined as the mean loss resulting from shaking with 10% exceedance probability in 5 years. Liu et al. (2003) presented a two-objective optimization procedure for designing steel moment resisting frame buildings within a performance-based seismic design framework, where initial material and lifetime seismic damage costs are treated as two separate objectives. In the work by Sanchez-Silva and Rackwitz (2004) it is concluded that structures should be optimal with respect to economic investment, benefits derived from their use, expected consequences in case of failure and the degree of protection to human life. Takahashi et al. (2004) presented a decision methodology for the management of seismic risk where the decision criterion aims at minimizing the expected life-cycle cost, including the initial cost of the design and the expected cost of damage due to future earthquakes. Frangopol and Liu (2007) reviewed the recent development of life-cycle maintenance and management planning for deteriorating civil infrastructure with emphasis on bridges. Kappos and Dimitrakopoulos (2008) implemented decision making tools, namely cost-benefit and life-cycle cost analyses, in order to examine the feasibility of strengthening reinforced concrete buildings. A probabilistic framework to estimate long-term earthquake-induced economical loss for woodframe structures was proposed and demonstrated in the work by Pei and Van De Lindt (2009).

In order to take into account damage and other earthquake losses into the LCCA procedure a reliable tool for estimating the capacity of any structural system in multiple earthquake hazard levels is required. Among others (Fajfar 2000; Chopra and Goel 2002), incremental dynamic analysis (IDA) (Vamvatsikos and Cornell 2002) is proven to be an analysis procedure for obtaining good estimates of the structural performance in the case of earthquake hazards and it is considered an appropriate method to be incorporated into the LCCA procedure. In view of the complexity and the computational effort required by the 3D structural analysis models, that are employed to represent real buildings, simplified 2D structural simulations are used during the design procedure. This is mainly justified in plan-symmetric buildings and mostly in the case of steel framed buildings composed by 2D moment resisting frames. In 3D reinforce concrete (RC) buildings, however, the columns belong to two or more intersecting lateral-force-resisting systems, therefore it is not possible to implement a 2D simulation since the bidirectional orthogonal shaking effects are significant and should be taken into account. Moreover, 3D models should also be considered in the case of nonsymmetric in plan steel or RC buildings.

In this study the assessment of the European seismic design codes and in particular of EC2 and EC8 with respect to the recommended behaviour factor q is performed. The assessment is performed on a reinforced concrete multi-storey building, having symmetrical plan view, which was optimally designed under four different values of the behaviour factor. In the mathematical formulation of the optimization problem the initial construction cost is considered as the objective function to be minimized while the cross sections and steel reinforcement of the beams and the columns constitute the design variables. The provisions of Eurocodes 2 and 8 are imposed as constraints to the optimization problem. Life-cycle cost analysis, in conjunction with structural optimization, is believed to be a reliable procedure for assessing the performance of structures during their life time. Furthermore, the sensitivity of life-cycle cost with reference to the analysis procedure (static or dynamic), the number of seismic records imposed and the performance criterion used. In particular, nonlinear static analysis and multiple stripe analysis, which is a variation of IDA, were applied with respect to the maximum inter-storey drift and the maximum floor acceleration. In the test example considered, the life-cycle cost was calculated taking into consideration the damage repair cost, the cost of loss of contents due to structural damage that is quantified by the maximum inter-storey drift and floor acceleration, the loss of rental cost, the income loss cost, the cost of injuries and the cost of human fatalities.

10.2 Multiple-Stripe Dynamic Analysis

In the seismic assessment of structures a wide range of seismic records and more than one performance levels should be considered in order to take into account the uncertainties that the seismic hazard introduces into a performance-based seismic assessment problem. The methods used for the performance-based assessment implementing non-linear dynamic analyses are classified as single and multiple hazard level methods. Multiple-stripe dynamic analysis (MSDA) and incremental dynamic analysis (IDA) are the two most applicable methods, both considering multiple hazard levels.

The main objective of a MSDA or IDA study is to define a curve through a relation between the seismic intensity level and the corresponding maximum response of the structural system. The intensity level and the structural response are described through an intensity measure (IM) and an engineering demand parameter (EDP), respectively. The IDA study is implemented through the following steps: (i) define the nonlinear FE model required for performing nonlinear dynamic analyses; (ii) select a suit of natural or artificial records; (iii) select a proper intensity measure and an engineering demand parameter; (iv) employ an appropriate algorithm for selecting the record scaling factor in order to obtain the IDA curve performing the least required nonlinear dynamic analyses and (v) employ a summarization technique for exploiting the multiple records results.

Selecting IM and EDP is one of the most important steps of the MSDA study. In the work by Giovenale et al. (2004) the significance of selecting an efficient IM is discussed while an originally adopted IM is compared with a new one. The IM should be a monotonically scalable ground motion intensity measure like the peak ground acceleration (PGA), peak ground velocity (PGV), the $\xi = 5\%$ damped spectral acceleration at the structure's first-mode period ($S_A(T_1,5\%)$) and many others. In the current study the $S_A(T_1,5\%)$ is selected, since it is the most commonly used intensity measure in practice today for the analysis of buildings. On the other hand, the damage may be quantified by using any of the EDPs whose values can be related to particular structural damage states. A number of available response-based EDPs were discussed and critically evaluated in the past for their applicability in seismic damage evaluation (Ghobarah et al. 1999). In the work by Ghobarah et al. (1999) the EDPs are classified into four categories: engineering demand parameters based on maximum deformation, engineering demand parameters based on cumulative damage, engineering demand parameters accounting for maximum deformation and cumulative damage, global engineering demand parameters. In the current study the maximum interstorey drift θ_{max} and maximum floor acceleration are chosen. The reason for the two EDPs is because there is an established relation between interstorey drift and maximum floor acceleration values and performance-oriented descriptions such as immediate occupancy, life safety and collapse prevention (FEMA 273 1997). Furthermore, there is also a defined relation between both EDPs and damage-state (Ghobarah (2004); Elenas and Meskouris 2001) that is required for LCCA.

Multiple-stripe analysis is a non-linear dynamic analysis method that can be used for performance-based seismic assessment of structures for a wide range of ground motions and more than one performance levels. It refers to many groups of analyses (*stripes*) performed at multiple spectral acceleration levels, where at each stripe analysis a number of structural analyses are performed for a group of ground motion records that are scaled to a single value of spectral acceleration. The suite of ground motion records used for performing each stripe analysis should ideally be representative of the seismic threat at the corresponding spectral acceleration; however, it is common, but not necessarily always justified (e.g. Jalayer and Cornell 2009), to use the same suite of records for all the spectral acceleration levels. For a MSDA the intensity measure is usually the first mode spectral acceleration $S_A(T_1,5\%)$ for damping equal to 5%. The maximum inter-storey drift recommended by FEMA-350 (2000) as the most suitable performance criterion for frame structures is used in this study. Depending on the problem and the performance that is needed to be calculated different intensity measures and performance factors can also be used.

10.3 The Capacity Spectrum Method Adopted by ATC-40

The capacity spectrum method (CSM) was initially proposed by Freeman (1998). The method compares the capacity of a structure to resist lateral forces according to the demands of earthquake response spectra. This is depicted in a graphical representation allowing a visual evaluation of how the structure will perform when subjected to the ground motion represented by the corresponding response spectrum. Both curves are converted and plotted against an acceleration-displacement (A-D) graph making easy the evaluation of the point of equal demand and capacity, also known as performance point. The procedure of ATC-40 employed in this study for the calculation of the maximum inter-storey drifts (θ_i) for three hazard levels consists of the following steps:

1. Perform non-linear static analysis and determine the capacity curve expressed by the base shear V_b against top displacement of the MDOF structural system.
2. Convert the capacity curve of the MDOF system to that of an equivalent SDOF system using the transformation factor Γ and determine the idealized elastic perfectly plastic force-displacement relationship.
3. Plot both the force-displacement diagram and the 5%-damped elastic response spectrum diagram in an A-D format.
4. Estimate the peak deformation demand d_i where the response spectrum diagram intersects the capacity diagram and determine the corresponding

pseudo-acceleration A_i. Initially, assume $d_i = d(\mathrm{T_n},\zeta = 5\%)$, determined for period T_n from the elastic demand diagram.

5. Compute the ductility factor $\mu = d_i/u_y$.
6. Compute the equivalent damping ratio

$$\hat{\zeta}_{eq} = \zeta + \kappa \cdot \zeta_{eq} \tag{10.1}$$

7. Plot the elastic demand diagram for $\hat{\zeta}_{eq}$ and estimate the new peak deformation demand d_j.
8. Check for convergence

$$\text{if } \frac{(d_j - d_i)}{d_j} \begin{cases} \leq \textit{tolerance} \text{ then Stop} \\ > \textit{tolerance} \text{ then go to step 5} \end{cases} \tag{10.2}$$

If the convergence check is satisfied the target displacement for the SDOF system $d_t^* = d_j$ and the corresponding displacement for the MDOF system is equal to $d_t = \Gamma \cdot d_t^*$.

10.4 Life-Cycle Cost Assessment

The total cost C_{TOT} of a structure, may refer either to the design-life period of a new structure or to the remaining life period of an existing or retrofitted structure. This cost can be expressed as a function of time and the design vector $\boldsymbol{s}$ as follows (Wen and Kang 2001b)

$$C_{TOT}(t, \boldsymbol{s}) = C_{IN}(\boldsymbol{s}) + C_{LS}(t, \boldsymbol{s}) \tag{10.3}$$

where C_{IN} is the initial cost of a new or retrofitted structure, C_{LS} is the present value of the limit state cost; $\boldsymbol{s}$ is the design vector corresponding to the design loads, resistance and material properties that influence the performance of the structural system, while t is the time period. The term "initial cost" of a new structure refers to the cost required for construction. The initial cost is related to the material and the labour cost for the construction of the building which includes concrete, steel reinforcement, labour cost for placement as well as the non-structural component cost, in the case of a RC building. The term "limit state cost" refers to the potential damage cost from earthquakes that may occur during the life of the structure. It accounts for the cost of repair, the cost of loss of contents, the cost of injury recovery or human fatality and other direct or indirect economic losses related to loss of contents, rental and income, after an earthquake. The quantification of these losses in economical terms depends on several socio-economic parameters. It should be mentioned that in the calculation of C_{LS} a regularization factor is used that transforms the costs in present values. The most difficult cost to quantify

Table 10.1 Drift ratio and floor acceleration limits for bare moment resisting frames

Limit state	Interstorey drift (%) (Ghobarah 2004)	Floor acceleration (g) (Elenas and Meskouris 2001)
(I) – None	$\theta \leq 0.1$	$a_{\text{floor}} \leq 0.05$
(II) – Slight	$0.1 < \theta \leq 0.2$	$0.05 < a_{\text{floor}} \leq 0.10$
(III) – Light	$0.2 < \theta \leq 0.4$	$0.10 < a_{\text{floor}} \leq 0.20$
(IV) – Moderate	$0.4 < \theta \leq 1.0$	$0.20 < a_{\text{floor}} \leq 0.80$
(V) – Heavy	$1.0 < \theta \leq 1.8$	$0.80 < a_{\text{floor}} \leq 0.98$
(VI) – Major	$1.8 < \theta \leq 3.0$	$0.98 < a_{\text{floor}} \leq 1.25$
(VII) – Collapsed	$\theta > 3.0$	$a_{\text{floor}} > 1.25$

is the cost corresponding to the loss of a human life. There are a number of approaches for its estimation, ranging from purely economic reasoning to more sensitive that consider the loss of a human being irreplaceable. Therefore, the estimation of the cost of exceedance of the collapse prevention damage state will vary considerably according to which approach is adopted.

Damage may be quantified by using several damage indices (DIs) whose values can be related to particular structural damage states. The idea of describing the state of damage of the structure by a specific quantity, on a defined scale in the form of a damage index, is attractive because of its simplicity. So far a significant number of researchers have studied various damage indices for reinforced concrete or steel structures, a detailed survey can be found in the work by Ghobarah et al. (1999). Damage, in the context of life-cycle cost assessment (LCCA), refers not only to structural damage but also to non-structural damage. The latter including the case of architectural damage, mechanical, electrical and plumbing damage and also the damage of furniture, equipment and other contents. The maximum inter-story drift (θ) has been considered as the response parameter which best characterises the structural damage, associated with all types of losses. It is generally accepted that inter-storey drift can be used as a reliable limit state criterion to determine the expected damage. The relation between the drift ratio limits with the limit state, employed in this study (Table 10.1), is based on the work of Ghobarah (2004) for ductile RC moment resisting frames. On the other hand, the intensity measure which has been associated with the loss of contents, like furniture and equipment, is the maximum response floor acceleration. The relation of the limit state with the values of the floor acceleration (Table 10.1) is based on the work of Elenas and Meskouris (2001). Although in recent years there have been important improvements in the estimation of seismic demand on structures, there still exists a great deal of uncertainty in the estimation of the seismic capacity of structures, thus uncertainty on the limit state bounds can also be considered. Furthermore, it should be noted that simplified expressions are used to define relationship between limit states and interstorey drift. In general such relationship can be defined based on empirical or semi-empirical equations, which takes into account the quality of material amount of reinforcement, which is not taken into account in this chapter.

Table 10.2 Limit state cost – calculation formulas (Sanchez-Silva and Rackwitz 2004; Wen and Kang 2001b)

Cost category	Calculation formula	Basic cost
Damage/repair (C_{dam})	Replacement cost × floor area × mean damage index	1,500 MU/m^2
Loss of contents (C_{con})	Unit contents cost × floor area × mean damage index	500 MU/m^2
Rental (C_{ren})	Rental rate × gross leasable area × loss of function	10 MU/month/m^2
Income (C_{inc})	Rental rate × gross leasable area × down time	2,000 MU/year/m^2
Minor Injury ($C_{inj,m}$)	Minor injury cost per person × floor area × occupancy rate* × expected minor injury rate	2,000 MU/person
Serious Injury ($C_{inj,s}$)	Serious injury cost per person × floor area × occupancy rate* × expected serious injury rate	2 × 10^4 MU/person
Human fatality (C_{fat})	Human fatality cost per person × floor area × occupancy rate* × expected death rate	2.8 × 10^6 MU/person

* Occupancy rate 2 persons/100 m^2

The limit state cost (C_{LS}), for the i-th limit state, can thus be expressed as follows

$$C_{LS}^{i} = C_{dam}^{i} + C_{con}^{i} + C_{ren}^{i} + C_{inc}^{i} + C_{inj}^{i} + C_{fat}^{i} \tag{10.4a}$$

$$C_{con}^{i} = C_{con}^{i,\theta} + C_{con}^{i,acc} \tag{10.4b}$$

where C_{dam}^{i} is the damage repair cost, $C_{con}^{i,\theta}$ is the loss of contents cost due to structural damage that is quantified by the maximum inter-storey drift while $C_{con}^{i,acc}$ is the loss of contents cost due to floor acceleration (Elenas and Meskouris 2001), C_{ren}^{i} is the loss of rental cost, C_{inc}^{i} is the income loss cost, C_{inj}^{i} is the cost of injuries and C_{fat}^{i} is the cost of human fatality. These cost components are related to the damage of the structural system. A more detailed description of the different cost evaluation for each limit state cost can be found in Table 10.2 (Wen and Kang 2001a). The values of the mean damage index, loss of function, down time, expected minor injury rate, expected serious injury rate and expected death rate used in this study are based on (Lagaros 2007; Ellingwood and Wen 2005). Table 10.3 provides the ATC-13 (1985) and FEMA-227 (1992) limit state dependent damage consequence severities.

Based on a Poisson process model of earthquake occurrences and an assumption that damaged buildings are immediately retrofitted to their original intact conditions after each major damage-inducing seismic attack, Wen and Kang (2001b) proposed the following formula for the limit state cost function considering N limit states

$$C_{LS} = C_{LS}^{\theta} + C_{LS}^{acc} \tag{10.5a}$$

$$C_{LS}^{\theta}(t,s) = \frac{\nu}{\lambda}\left(1 - e^{-\lambda t}\right)\sum_{i=1}^{N} C_{LS}^{i,\theta} \cdot P_{i}^{\theta} \tag{10.5b}$$

Table 10.3 Limit state parameters for cost evaluation

	Fema-227 (1992)				ATC-13 (1985)	
Limit state	Mean damage index (%)	Expected minor injury rate	Expected serious injury rate	Expected death rate	Lose of Function (%)	Down time (%)
(I) – None	0	0	0	0	0	0
(II) – Slight	0.5	3.0E−05	4.0E−06	1.0E−06	0.9	0.9
(III) – Light	5	3.0E−04	4.0E−05	1.0E−05	3.33	3.33
(IV) – Moderate	20	3.0E−03	4.0E−04	1.0E−04	12.4	12.4
(V) – Heavy	45	3.0E−02	4.0E−03	1.0E−03	34.8	34.8
(VI) – Major	80	3.0E−01	4.0E−02	1.0E−02	65.4	65.4
(VII) – Collapsed	100	4.0E−01	4.0E−01	2.0E−01	100	100

$$C_{LS}^{acc}(t,s) = \frac{v}{\lambda}\left(1 - e^{-\lambda t}\right)\sum_{i=1}^{N} C_{LS}^{i,acc} \cdot P_i^{acc} \tag{10.5c}$$

where

$$P_i^{DI} = P(DI > DI_i) - P(DI > DI_{i+1}) \tag{10.6}$$

and

$$P(DI > DI_i) = (-1/t) \cdot \ln[1 - \bar{P}_i(DI - DI_i)] \tag{10.7}$$

P_i is the probability of the ith limit state being violated given the earthquake occurrence and C_{LS}^i is the corresponding limit state cost; $P(DI - DI_i)$is the exceedance probability given occurrence; DI_i, DI_{i+1} are the damage indices (maximum inter-storey drift or maximum floor acceleration) defining the lower and upper bounds of the ith limit state; $\bar{P}_i(DI - DI_i)$is the annual exceedance probability of the maximum damage index DI_i; v is the annual occurrence rate of significant earthquakes modelled by a Poisson process and t is the service life of a new structure or the remaining life of a retrofitted structure. Thus, for the calculation of the limit state cost of Eq. 10.5c the maximum inter-storey drift DI is considered, while for the case of Eq. 10.5b the maximum floor acceleration is used. The first component of Eqs. 10.5b or 10.5c, with the exponential term, is used in order to express C_{LS} in present value, where λ is the annual monetary discount rate. In this chapter the annual monetary discount rate λ is taken to be constant and equal to 5%, since considering a continuous discount rate is accurate enough for all practical purposes according to Rackwitz (2006) and Rackwitz et al. (2005). Various studies have proposed values of the discount rate λ in the range of 3–6% (Ellingwood and Wen 2005).

Each limit state is defined by drift ratio limits or floor acceleration, as listed in Table 10.1. When one of the DIs is exceeded the corresponding limit state is assumed to be reached. The annual exceedance probability $\bar{P}_i(DI>DI_i)$ is obtained from a relationship of the form

$$\bar{P}_i(DI > DI_i) = \gamma(DI_i)^{-k} \tag{10.8}$$

where the parameters γ and k are obtained by best fit of known $\bar{P}_i - DI_i$ pairs for each of the two *DIs*. According to Poisson's law the annual probability of exceedance of an earthquake with a probability of exceedance p in t years is given by the formula

$$\bar{P} = (-1/t) \cdot \ln(1 - p) \tag{10.9}$$

This means that the 2/50 earthquake has a probability of exceedance equal to $\bar{P}_{2\%} = -\ln(1\text{–}0.02)/50 = 4.04 \times 10^{-4}$ ($4.04 \times 10^{-2}\%$).

10.5 Heuristic Design of Reinforce Concrete Structures

The ultimate objective of our study is to compare lower-bound designs, or in other words comparing the designs that satisfy the code requirements in the most cost-effective way, i.e. those with minimum cross section and reinforcement dimensions. For this reason, a structural optimization problem is formulated and the designs obtained are then assessed. The formulation of a structural optimization problem, which constitutes the basis for the performance evaluation of different reinforced concrete building designs, is defined as follows

$$\begin{aligned} \min_{\mathbf{s}\in \mathsf{F}} \quad & C_{IN}(t, \mathbf{s}) \\ \text{subject to} \quad & g_j^{SERV}(\mathbf{s}) \leq 0 j = 1,\ldots,m \\ & g_j^{ULT}(\mathbf{s}) \leq 0 j = 1,\ldots,k \end{aligned} \tag{10.10}$$

where $\mathbf{s}$ represents the design vector with the cross-section dimensions of all columns and beams, F is the feasible region where all the serviceability and performance-based constraint functions (g^{SERV} and g^{ULT}) are satisfied. The objective function considered is the initial cost C_{IN} of the design. For the solution of the optimization problem at hand an Evolutionary type of algorithm (EA) is employed. In all formulations the initial construction cost is the objective function to be minimized. The columns and beams are of rectangular cross-sectional shape, and are separated into groups. The two dimensions of the columns/beams along with the longitudinal, transverse reinforcement and its spacing are the five design variables that are assigned to each group of the columns/beams.

10.6 Implementation of LCCA Procedure

The limit-state cost calculation procedure requires the assessment of the structural capacity in at least three hazard levels of increased intensity with the definition of at least three pairs of annual probability of exceedance $(\bar{P}_i)$ and maximum value of the damage index in question (DI_i). In this study the *abscissa* values of the $(\bar{P}_i - DI_i)$ pairs, corresponding to the maximum values of the damage index for the number hazard levels in question, are obtained either by means of CSM or MSDA, while the *ordinate* values correspond to the annual probabilities of exceedance. In both cases eight pairs are considered corresponding to eight hazard levels (100/50, 83.95/50, 59.47/50, 38.28/50, 24.82/50, 16.35/50, 10.19/50, 5.90/50). These probabilities correspond to discrete values of annual probabilities of exceedance obtained from the US Geological Survey (USGS) ground motion parameter calculator (US Geological Survey (USGS)). In the case of CSM the damage index considered is the maximum inter-storey drift, while in the case of MSDA two damage indices are employed, namely the maximum interstorey drift and maximum floor acceleration. In both cases the hazard levels are defined in accordance to the hazard curve of the city of San Diego, California (Latitude (N) 32.7°, Longitude (W) −117.2°, Fig. 10.1).

The selection of the proper external loading for performing life-cycle cost assessment is not an easy task due to the uncertainties involved in the seismic characteristics. For this reason a rigorous treatment of the seismic loading is to assume that the structure is subjected to a set of records that are more likely to occur in the region where the structure is located. In our case for both implementations

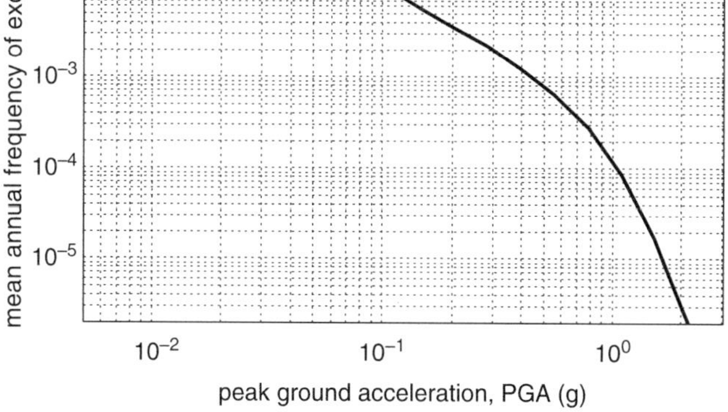

Fig. 10.1 Hazard curve of the city of San Diego, California (Latitude (N) 32.7°, Longitude (W) − 117.2°)

(CSM and MSDA) a series of 10, 20, 40 and 60 seismic records per hazard level are implemented. The records were randomly selected from the three lists of records given in (Mitropoulou et al. 2011). These records have been selected from the PEER strong-motion database (Pacific Earthquake Engineering Research (PEER) 2010) according to the following features: (i) Events occurred in specific area (longitude –124° to –115°, latitude 32–41°). (ii) Moment magnitude (M) is equal to or greater than 5. (iii) Epicentral distance (R) is smaller than 150 km. To make sure that the randomly selected number of records *no_recs* (where *no_recs* = 10, 20 or 40) of both *CSM* and *MSDA* implementations are not dominated by a few events, equal number of records from the each earthquake was kept. This was performed by means of Latin Hypercube Sampling (LHS) selecting one, two or more records from the same earthquake depending on the number of records used. Since the records of Mitropoulou et al. (2011) belong to 12 earthquakes. LHS is a strategy for generating random sample points ensuring that every part of the random space is represented. Latin hypercube samples are generated by dividing each random variable into N non-overlapping segments of equal probability. Thus, if M random variables are considered the random variable space is partitioned into N^M cells. For each random variable, a single value is randomly selected from each segment, producing a set of N values. The values of each random variable are randomly matched with each other to create N samples. The analysis procedure adopted (CSM or MSDA) for the implementation of the LCCA is described below.

10.6.1 Implementation of the Capacity Spectrum Method

Non-linear static analysis (NSA) is based on the assumption that the response of the structure is described by an equivalent single degree of freedom system with properties proportional to the first mode of the structure. For the implementation of CSM a lateral load distribution that follows the fundamental mode is adopted. The analysis is terminated when 150% of the target displacement that correspond to the 2% in 50 years (2/50) hazard level is reached, or earlier if the structure has collapsed (FEMA-356 2000). According to the capacity spectrum method, adopted in this chapter, the capacity of a structural system to resist lateral forces is compared to the demand of an earthquake response spectrum and results with the calculation of the maximum inter-storey drift (θ_i). In order to calculate the maximum inter-storey drift in multiple hazard levels the corresponding earthquake response spectra are used. These spectra are the median ones of 10, 20, 40 and 60 records when scaled to the corresponding hazard level based on the $S_A(T_1,5\%)$ resulting to four implementations: CSM(i), i = 10,20,40,60. In Fig. 10.2 the spectra corresponding to the longitudinal and translational components of the records scaled to the 10/50 hazard level are shown.

CSM is limited with regard to evaluation of the simultaneous response to ground shaking in different directions. In this chapter the recommendation of FEMA-350 (2000) is employed where multidirectional excitation effects are accounted for by

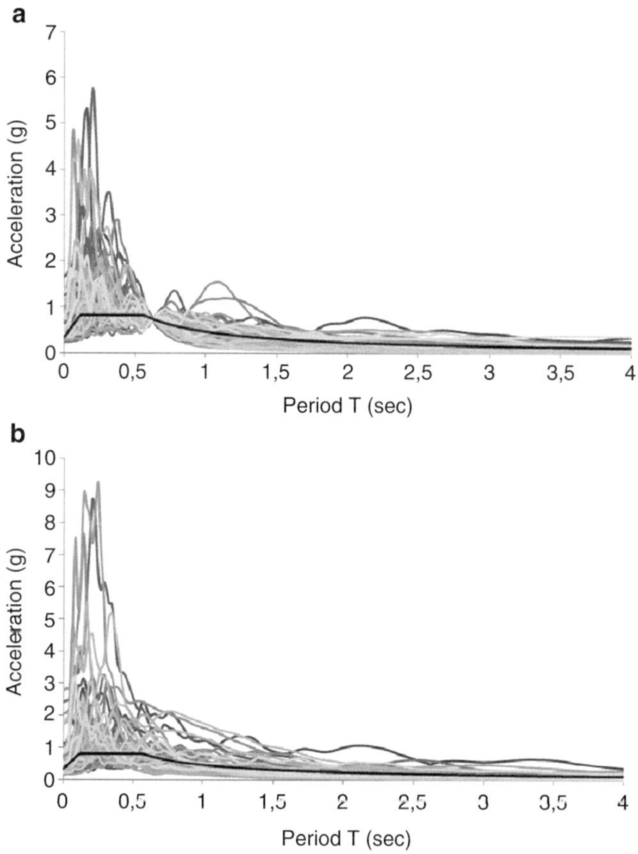

Fig. 10.2 Response spectra of the group of 60 records scaled to the 10/50 hazard level, of the city of San Diego, according to the $S_A(T_1,5\%)$ (**a**) longitudinal and (**b**) transverse direction

combining 100% of the response due to loading in the longitudinal direction with 30% of the response due to loading in the transverse direction, and vice versa. The worst of these two combinations in each hazard level is used in order to assess the structural performance in the corresponding performance levels.

10.6.2 Implementation of the Multistripe Dynamic Analysis Method

For the implementation of the MSDA, multiple non-linear dynamic analyses have to be performed in order to assess its performance in all eight hazard levels. For each hazard level the median response among the records is used.

Therefore, the application of NDA incorporated into LCCA results in a time-consuming and computationally-demanding procedure compared to the corresponding CSM implementation. From this procedure a scale factor is calculated for each one of the 60 ground motions and for each hazard level. In order to preserve the relative scale of the two components of the records in the longitudinal and transverse directions, the component of the record having the highest $S_A(T_1,5\%)$ is scaled first, while a scaling factor that preserves their relative ratio is assigned to the second component. Similar to the CSM implementation four variants of the MSDA are applied according to the number of the records used.

10.7 Numerical Study

In this chapter one 3D RC building has been considered in order to perform LCCA and study the influence of four factors on the LCCA procedure. Non-linear Static Analysis and Multiple-Stripe Dynamic Analysis were performed in the 3D RC building for four groups of seismic events each composed by 10, 20, 40 and 60 records, respectively. In order to study the influence of the *DI* considered, for the case of MSDA, the performance of the 3D RC building was assessed with reference to the maximum inter-storey drift and floor acceleration induced by eight hazard levels for each group of ground motions. The final part of this study deals with the evaluation of the influence of four factors: the type of the analysis (*CSM, MSDA*), the number of seismic records (*10, 20, 40, 60*), the type of the damage index (maximum interstorey drift-θ_{max} and combination of maximum interstorey drift-θ_{max} and maximum floor acceleration-a_{max}), into the life-cycle cost assessment procedure of 3D RC buildings.

10.7.1 Structural Models and Numerical Simulation

The multi-storey 3D RC building is shown in Fig. 10.3. The test example considered is an eight storey RC building having symmetrical plan view. The building has been designed for minimum initial cost following an optimization strategy proposed by Mitropoulou et al. (2010). The cross-sections of the beams and the columns of the test example are provided in Table 10.4, where hl × bl and ht × bt correspond to the cross sectional dimensions of horizontal and vertical beams. Concrete of class C20/25 (nominal cylindrical strength of 20 MPa) and steel of class S500 (nominal yield stress of 500 MPa) are assumed. The slab thickness for the test example is equal to 15 cm. In addition to the self weight of beams and slabs, a distributed permanent load of 2 kN/m^2 due to floor finishing-partitions and an imposed load with nominal value of 1.5 kN/m^2, are considered.

The optimum designs achieved for different values of the q factor are presented in Table 10.4, where the initial costs are given in monetary units (MU, corresponding to Dollars or Euros). It can be seen that, with respect to total initial cost design $D_{q=1}$ is increased by the marginal quantity of 3% compared to $D_{q=2}$ and by 10% and 15% compared to $D_{q=3}$ and $D_{q=4}$, respectively. In the case when the four designs are compared with reference to the cost of the RC skeletal members alone, design $D_{q=1}$ is increased by 12% compared to $D_{q=2}$ and by 65% and 95% compared to $D_{q=3}$ and $D_{q=4}$, respectively. Confirming the results of the first test example, it can be also seen that the initial construction cost of RC structures designed based on elastic response for the design earthquake is by no means prohibitive.

All structural analyses were performed using the OpenSEES platform (McKenna and Fenves 2001). Each member is modelled with a single force-based, fibre beam-column element. This element provides a good balance between accuracy and computational cost. The modified Kent-Park model (Kent and Park 1971) is employed for the simulation of the concrete fibres. This model was chosen because it allows for an accurate prediction of the demand for flexure-dominated RC members despite its relatively simple formulation. The transient behaviour of the reinforcing bars was simulated with the Menegotto-Pinto model (Menegotto and Pinto 1973), while a nonlinear shear force-shear distortion (V-γ) law is adopted to

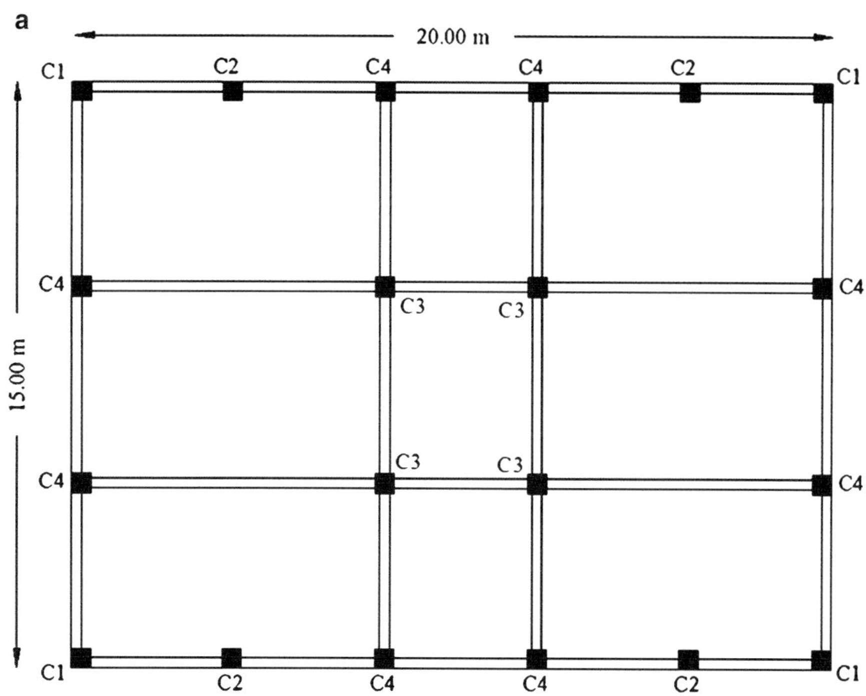

Fig. 10.3 Eight storey test example: (**a**) Plan view

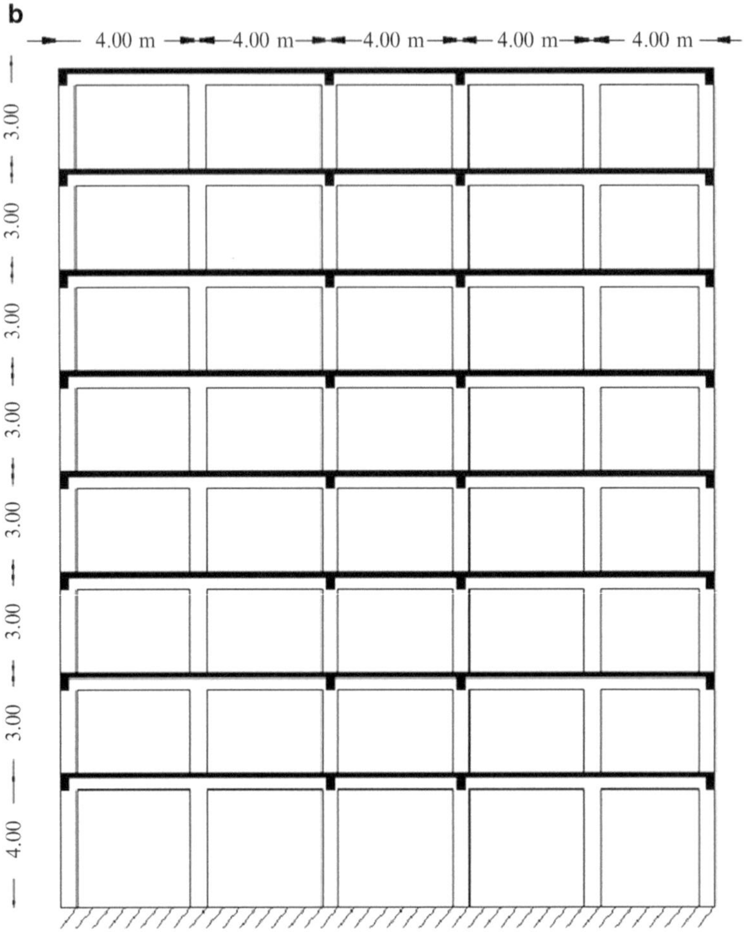

Fig. 10.3 (continued) (**b**) front view

account for shear failure, based on the work of Marini and Spacone (2006). The effect of gravity loads and second-order effects are considered using geometric stiffness matrix.

10.7.2 Multiple Stripe Analysis Results

Figures 10.4 and 10.5 present four groups of MSDA curves that depict the relation of the maximum inter-storey drift and the floor acceleration with reference to eight hazard levels. In particular, the MSDA curves present the first mode spectral acceleration $S_A(T_1,5\%)$ against the maximum inter-storey drift in Fig. 10.4 and

Table 10.4 Eight storey test example – Optimum designs obtained for different values of behaviour factor q

Design variables	Optimum designs q = 1	q = 2	q = 3	q = 4
h1 × b1	1.35 × 1.35, LR: 88Ø30, TR: (5)Ø10/10 cm	0.80 × 0.75, LR: 26Ø32, TR: (4)Ø10/15 cm	0.80 × 0.60, LR: 24Ø28, TR: (4)Ø10/20 cm	0.80 × 0.60, LR: 10Ø24 + 12Ø28, TR: (2)Ø10/20 cm
h2 × b2	0.90 × 0.85, LR: 34Ø30, TR: (4)Ø10/10 cm	0.80 × 0.80, LR: 30Ø32, TR: (4)Ø10/10 cm	0.65 × 0.60, LR: 26Ø28, TR: (4)Ø10/20 cm	0.75 × 0.35, LR: 8Ø22 + 12Ø26, TR: (2)Ø10/20 cm
h3 × b3	1.05 × 1.10, LR: 50Ø30, TR: (4)Ø10/10 cm	0.80 × 0.80, LR: 30Ø32, TR: (4)Ø10/10 cm	0.75 × 0.65, LR: 28Ø28, TR: (4)Ø10/20 cm	0.75 × 0.70, LR: 10Ø22 + 12Ø26, TR: (2)Ø10/20 cm
h4 × b4	1.10 × 1.05, LR: 60Ø30, TR: (4)Ø10/10 cm	0.80 × 0.85, LR: 32Ø32, TR: (4)Ø10/10 cm	0.75 × 0.65, LR: 30Ø28, TR: (4)Ø10/15 cm	0.70 × 0.65, LR: 8Ø26 + 12Ø32, TR: (2)Ø10/20 cm
h5 × b5	0.85 × 0.85, LR: 36Ø30, TR: (4)Ø10/10 cm	0.65 × 0.65, LR: 8Ø22 + 12Ø26, TR: (4)Ø10/10 cm	0.65 × 0.55, LR: 8Ø24 + 4Ø26, TR: (2)Ø10/20 cm	0.60 × 0.55, LR: 8Ø20, TR: (2)Ø10/20 cm
h6 × b6	0.80 × 0.80, LR: 36Ø30, TR: (4)Ø10/10 cm	0.65 × 0.65, LR: 22Ø32, TR: (2)Ø10/10 cm	0.55 × 0.55, LR: 8Ø24 + 12Ø28, TR: (2)Ø10/20 cm	0.55 × 0.40, LR: 6Ø22 + 12Ø26, TR: (2)Ø10/20 cm
h7 × b7	0.85 × 0.85, LR: 36Ø30, TR: (4)Ø10/10 cm	0.65 × 0.65, LR: 24Ø32, TR: (2)Ø10/10 cm	0.75 × 0.50, LR: 24Ø28, TR: (2)Ø10/20 cm	0.75 × 0.55, LR: 10Ø22 + 12Ø26, TR: (2)Ø10/20 cm
h8 × b8	0.85 × 0.85, LR: 40Ø30, TR: (4)Ø10/10 cm	0.70 × 0.70, LR: 24Ø32, TR: (4)Ø10/10 cm	0.55 × 0.60, LR: 22Ø28, TR: (2)Ø10/20 cm	0.55 × 0.55, LR: 8Ø24 + 12Ø28, TR: (2)Ø10/20 cm
h9 × b9	0.60 × 0.60, LR: 8Ø26 + 12Ø30, TR: (4)Ø10/10 cm	0.65 × 0.65, LR: 8Ø22 + 4Ø26, TR: (4)Ø10/10 cm	0.55 × 0.55, LR: 8Ø18 + 4Ø22, TR: (2)Ø10/20 cm	0.60 × 0.55, LR: 8Ø20 + 4Ø24, TR: (2)Ø10/20 cm
h10 × b10	0.75 × 0.75, LR: 28Ø30, TR: (4)Ø10/10 cm	0.65 × 0.65, LR: 8Ø22 + 12Ø26, TR: (4)Ø10/10 cm	0.55 × 0.50, LR: 6Ø20 + 12Ø28, TR: (2)Ø10/20 cm	0.50 × 0.35, LR: 4Ø26 + 4Ø32, TR: (2)Ø10/20 cm
h11 × b11	0.75 × 0.75, LR: 28Ø30, TR: (4)Ø10/10 cm	0.65 × 0.65, LR: 8Ø26 + 12Ø32, TR: (4)Ø10/10 cm	0.50 × 0.50, LR: 4Ø22 + 12Ø26, TR: (2)Ø10/20 cm	0.45 × 0.45, LR: 4Ø22 + 12Ø26, TR: (2)Ø10/20 cm
h12 × b12	0.80 × 0.80, LR: 34Ø30, TR: (4)Ø10/10 cm	0.65 × 0.65, LR: 8Ø26 + 12Ø32, TR: (4)Ø10/10 cm	0.55 × 0.55, LR: 11Ø18 + 10Ø20, TR: (2)Ø10/20 cm	0.55 × 0.55, LR: 8Ø24 + 4Ø28, TR: (2)Ø10/20 cm

(continued)

Table 10.4 (continued)

Design variables	Optimum designs q = 1	q = 2	q = 3	q = 4
h13 × b13	0.60 × 0.60, LR: 26Ø20 + 35Ø18, TR: (2)Ø8/15 cm	0.65 × 0.65, LR: 31Ø20, TR: (2)Ø8/ 15 cm	0.55 × 0.55, LR: 11Ø18 + 10Ø20, TR: (2)Ø8/15 cm	0.60 × 0.55, LR: 6Ø18 + 5Ø20, TR: (2)Ø8/15 cm
h14 × b14	0.75 × 0.75, LR: 26Ø20 + 35Ø18, TR: (2)Ø8/15 cm	0.65 × 0.65, LR: 33Ø20, TR: (2)Ø8/ 15 cm	0.55 × 0.50, LR: 9Ø18 + 10Ø20, TR: (2)Ø8/15 cm	0.50 × 0.35, LR: 6Ø18 + 3Ø16, TR: (2)Ø8/15 cm
$C_{IN, Frame}$	3.92E + 02	3.51E + 02	2.40E + 02	1.99E + 02
C_{IN}	1.59E + 03	1.55E + 03	1.44E + 03	1.40E + 03

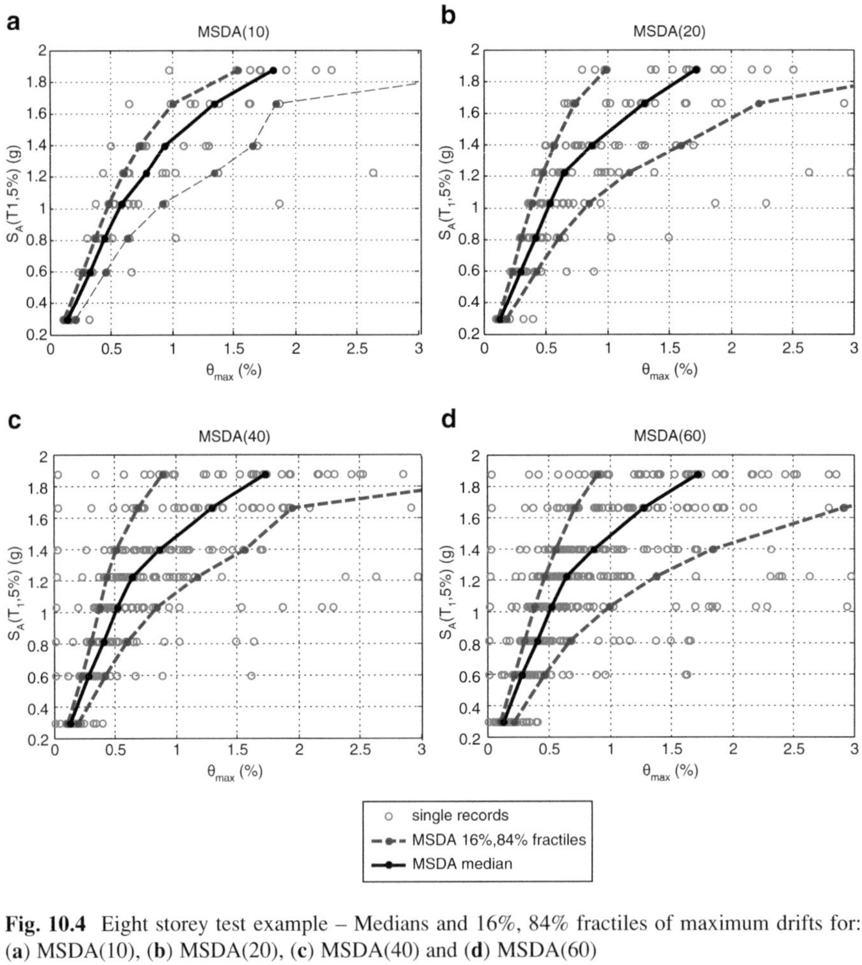

Fig. 10.4 Eight storey test example – Medians and 16%, 84% fractiles of maximum drifts for: (**a**) MSDA(10), (**b**) MSDA(20), (**c**) MSDA(40) and (**d**) MSDA(60)

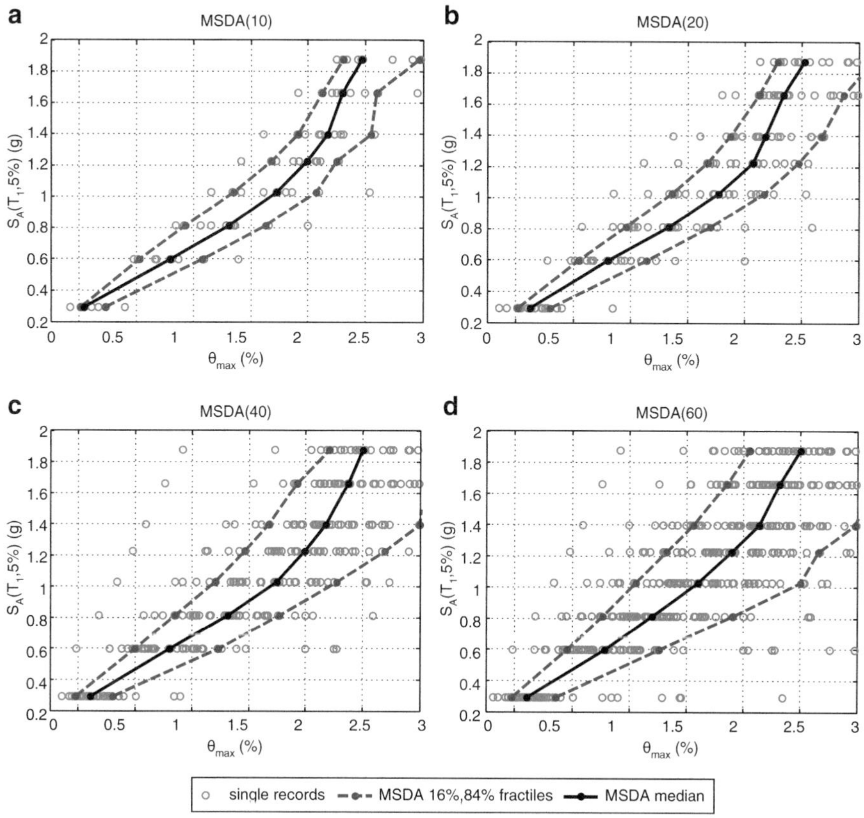

Fig. 10.5 Eight storey test example – Medians and 16%, 84% fractiles of maximum floor acceleration for: (**a**) MSDA(10), (**b**) MSDA(20), (**c**) MSDA(40) and (**d**) MSDA(60)

the maximum floor acceleration in Fig. 10.5. In these graphs the medians along with the 16% and 84% fractile curves are shown.

As it can be seen in Fig. 10.4 the four median curves almost coincide in the lower and the higher hazard levels. The curves corresponding to MSDA(20), MSDA(40) and MSDA(60) are almost identical with respect to the maximum inter-storey drift. On the contrary, deviation in the values of the upper hazard levels is observed for the curves of MSDA(10) with reference to the MSDA(20), MSDA(40) and MSDA(60). More significant variation is noticed with reference to the 16% and 84% fractile curves. In Fig. 10.5 the three out of four MSDA median curves differ significantly in all hazard levels. On the other hand, the curves corresponding to MSDA(20) and MSDA(40) almost coincide while they overestimate the capacity of the structure in comparison with the other two curves. The MSDA curve of the 60 records differs significantly from the other three curves showing greater deviation with higher hazard levels. In Fig. 10.5 the three curves of MSDA(20), MSDA(40) and MSDA(60) are close while the MSDA(10) curve present a more intensive exponential trend and differ

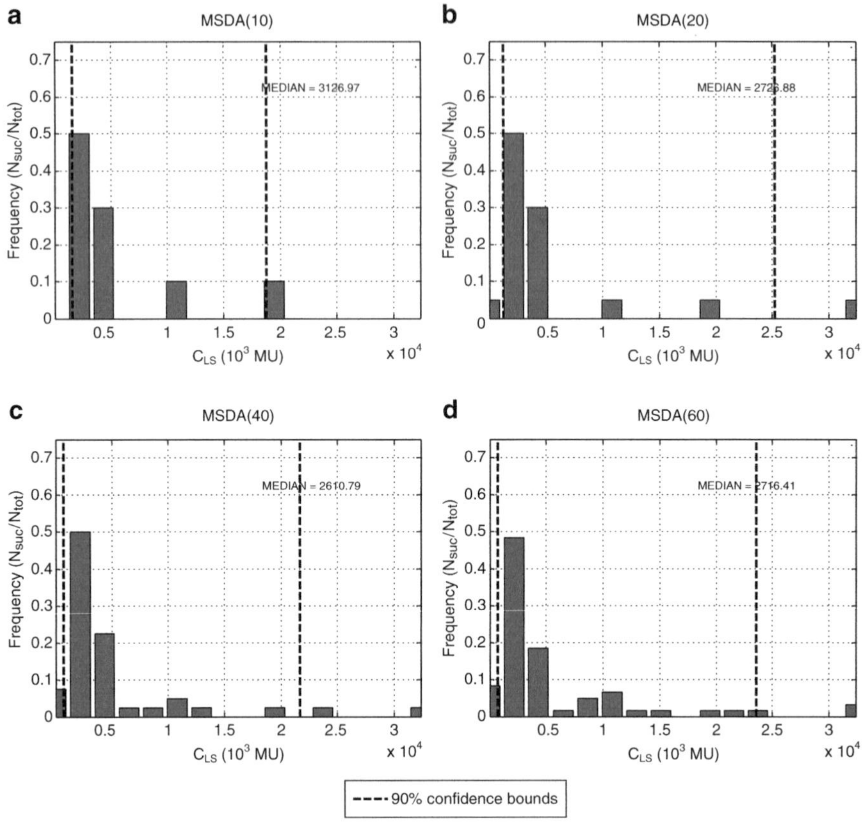

Fig. 10.6 Eight storey test case – Frequency of occurrence for the case of: (**a**) MSDA(10), (**b**) MSDA(20), (**c**) MSDA(40), (**d**) MSDA(60)

in the third to the sixth and in the eighth hazard levels by underestimating the first four levels and overestimating the level related to the capacity of the structure. Comparing the MSDA curves of the test example it can be noticed that in all graphs the curves of MSDA(10) underestimate the capacity of the structure in contrast to those of MSDA (60). Furthermore, the curves with respect to the maximum floor acceleration (Fig. 10.5) increase exponentially and almost coincide in all hazard levels.

10.7.3 Sensitivity Analysis of LCCA

In the second group of figures the variability of the life-cycle cost with respect to the number of the seismic records is demonstrated. The histograms in Fig. 10.6 show the probabilistic distribution of the life-cycle cost values due to the number of seismic records implemented in the MSDA for the test example, respectively, along

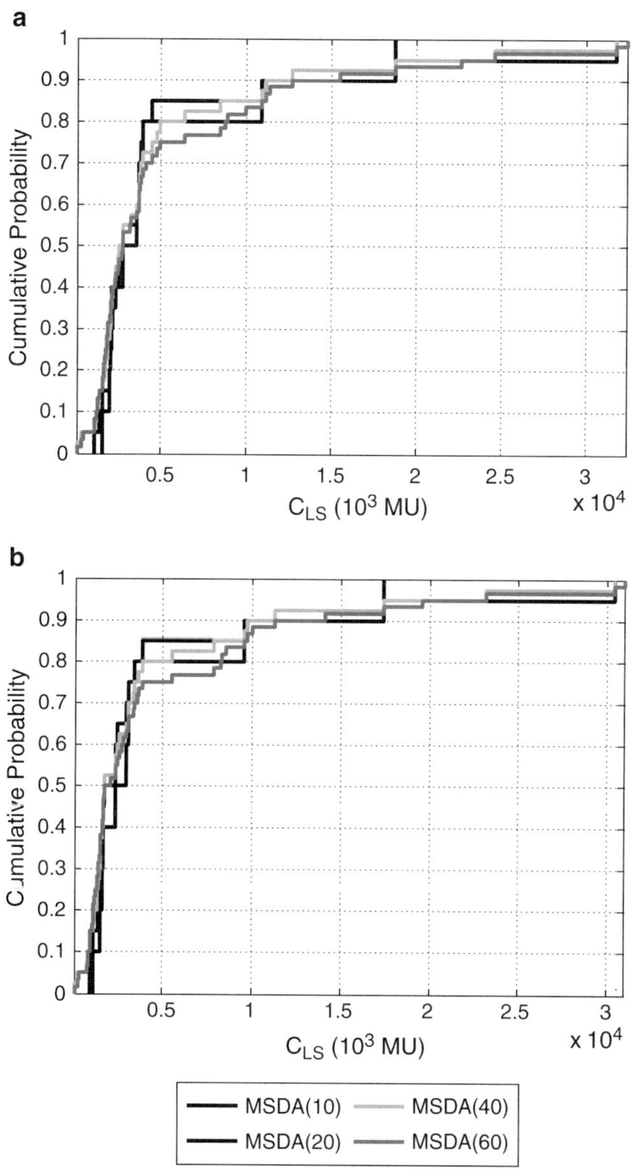

Fig. 10.7 Eight storey test example – Cumulative density function: (**a**) drift plus floor acceleration, (**b**) drift

with the 90% confidence bounds for each group of seismic records. The frequency of C_{LS} value occurrence is defined as the ratio of the number of simulations (N_{suc}), corresponding to limit state cost values in a specific range, over the total number of simulations (N_{tot}), where N_{tot} is equal to 10, 20, 40 and 60 for the respective number of 10, 20, 40 and 60 groups of seismic records respectively.

Comparing the histograms of Fig. 10.6 it can be noticed that in Fig. 10.6b–d histograms the width of the confidence bounds is almost the same. On the other hand, the width of the 90% confidence bounds of the life-cycle cost values when 10 records are implemented for the MSDA is 20% narrower compared to the other three confidence bounds (Fig. 10.6b–d). Furthermore, the mean value of the life-cycle cost is almost the same for the MSDA with 20, 40 and 60 records and 15% higher for the MSDA with 10 records. The conclusions that can be derived from the histograms of the probabilistic distribution of C_{LS} values are similar to those observing the MSDA curves with respect to the maximum inter-storey drift. In fact the performance of the test example defined by the value of the inter-storey drift at each hazard level has a direct effect on the value of the life-cycle cost. Thus the maximum inter-storey drift plays a more important role in the calculation of the life-cycle cost than the maximum floor acceleration.

The cumulative density function (CDF) curves, corresponding to the 10, 20, 40 and 60 groups of records are depicted in Fig. 10.7. In Fig. 10.7a, b the CDF for groups of 10, 20, 40 and 60 records coincide significantly in the range up to 5.0 million monetary units (MU, corresponding to Dollars or Euros) where a sharp slope is noticed up to the 80% cumulative probability. In this range small increase of C_{LS} causes large increase of probability of occurrence. On the other hand in the range 5.0–15.0 millions MU there is a markedly variation, while a smooth slope is encountered. Finally, for greater values of C_{LS} the three CDF curves for the groups of 20, 40, 60 records reach asymptotically the 100% cumulative probability, while the CDF curves of the group of 10 records has already reached the 100% cumulative probability. It can also be seen that Fig. 10.7a are slightly shifted compared to the corresponding Fig. 10.7b. This is because in the first ones the loss of contents due to floor acceleration was taken into account.

10.7.4 Nonlinear Static Analysis Versus Nonlinear Dynamic Analysis

In addition to the MSDA the CSM is also undertaken. In Table 10.5 the life-cycle cost values are presented following MSDA and CSM procedures imposing *10*, *20*, *40* and *60* number of seismic records. In the case of MSDA both DIs of maximum inter-storey drift (θ) and the combination of θ and maximum floor acceleration ($\theta + \alpha$) are considered. In Table 10.5 it can be seen that $MSDA_{(\theta)}$ gives a more favourable life-cycle cost value compared to the $MSDA(\theta + \alpha)$. For both $MSDA_{(\theta)}$ and $MSDA_{(\theta+\alpha)}$ the value of life-cycle cost increases as the number of the records increases except of the case of MSDA(40) of Table 10.5. The *CSM* case gives extremely lower values when records *10* and *20* are used and an almost double life-cycle cost value in the case of *60* records compared to the values of the $MSDA(\theta)$ and $MSDA_{(\theta+\alpha)}$ cases.

Comparing the values obtained for the $MSDA_{(\theta)}$ and $MSDA_{(\theta+\alpha)}$ cases, it can be observed that the life-cycle cost values of MSDA appear to have similar trend. The comparison between the $MSDA_{(\theta)}$, $MSDA_{(\theta+\alpha)}$ and *CSM* leads to the same

Table 10.5 Eight storey test example: Total life-cycle cost (10^3 MU)

Procedure	Number of records 10	20	40	60
$MSDA_{(\theta)}$	4.43E + 03	4.55E + 03	4.31E + 03	4.87E + 03
$MSDA_{(\theta+a)}$	5.10E + 03	5.34E + 03	5.07E + 03	5.65E + 03
CSM	1.17E + 03	1.21E + 03	5.10E + 03	8.57E + 03

conclusions as in Table 10.5 with two important exceptions: the life-cycle cost value of *60* records with *CSM* is 2 times higher than the corresponding values of $MSDA_{(\theta)}$ and $MSDA_{(\theta+\alpha)}$, while the life-cycle cost value of *40* records with *CSM* test case is one order of magnitude less than the corresponding of $MSDA_{(\theta)}$ and $MSDA_{(\theta+\alpha)}$ value obtained for the symmetrical test case.

10.8 Concluding Remarks

In this chapter a procedure is proposed to calculate the life-cycle cost of reinforced concrete buildings subjected to seismic actions. The numerical study was performed on a 3D RC building structure with regular plan view. The life-cycle cost estimation is examined with respect to the adopted analysis procedure, the number of seismic records imposed, the performance criterion used and the type of the building structure. Multiple stripe dynamic analysis, a variant of the incremental dynamic analysis, and the nonlinear static (pushover) analysis were applied to compute the performance criteria such as maximum inter-storey drift and floor acceleration. The initial cost is related to the material and the labour cost for the construction of the building that includes both structural and non-structural component cost. The life-cycle cost refers to the potential damage cost from earthquakes that may occur during the life of the structure. In construction industry decision making for structural systems situated in seismically active regions, requires consideration of the damage and other losses costs resulting from earthquakes occurring during the lifespan of the structure. Thus, the life-cycle cost assessment becomes an essential component of the design process in order to control the initial and the future cost of building ownership. The life-cycle cost was calculated on the basis of the damage repair cost and loss of contents cost, due to structural damage that is correlated to the maximum inter-storey drift, the loss of rental cost, the income loss cost, the cost of injuries as well as the cost of human fatality and the loss of contents cost due to floor acceleration.

The most important findings of this study can be summarized as follows: (i) From the examination of the multiple-stripe dynamic analysis (MSDA) curves, it can be concluded that 10–20 records are not enough to obtain reliable life-cycle cost analysis prediction results. (ii) The results obtained from the capacity curves and the corresponding life-cycle cost values, as well as from their variation between $MSDA_{(\theta)}$ and $MSDA_{(\theta+\alpha)}$ cases, one would suggest that the use of 40 records is a sufficient number of seismic records for a reliable performance assessment.

(iii) The use of the maximum inter-storey drift as performance criterion, instead of using the combination of both maximum inter-storey drift and maximum floor acceleration, leads to an underestimation of the life-cycle cost of a structure. The effect of inter-storey drift appears to be more critical in the calculation of the life-cycle cost than the maximum floor acceleration. (iv) The nonlinear static analysis procedure is not recommended as a performance estimation tool compared to the nonlinear dynamic analysis. (v) The structural type of the building affects its structural performance. It has been verified that a symmetrical structure sustains less damage and therefore less repair cost during its life compared to a non symmetric structure.

Acknowledgment The first author acknowledges the financial support of the John Argyris Foundation.

References

Arditi DA, Messiha HM (1996) Life-cycle costing in municipal construction projects. J Infrastruct Syst 2(1):5–14

Asiedu Y, Gu P (1998) Product life cycle cost analysis: state of the art review. Int J Prod Res 36:883–908

ATC-13 (1985) Earthquake damage evaluation data for California. Applied Technology Council, Redwood City

Beck JL, Porter KA, Shaikhutdinov RV (2003) Simplified estimation of seismic life-cycle costs. In: Frangopol DM, Bruhwiler E, Faber MH, Adey B (eds) Life-cycle performance of deteriorating structures: assessment, design, and management, ASCE, Reston, Virginia, USA, pp 229–236

Chopra AK, Goel RK (2002) A modal pushover analysis procedure for estimating seismic demands for buildings. Earthquake Eng Struct Dyn 31(3):561–582

Elenas A, Meskouris K (2001) Correlation study between seismic acceleration parameters and damage indices of structures. Eng Struct 23:698–704

Ellingwood BR, Wen Y-K (2005) Risk-benefit-based design decisions for low-probability/high consequence earthquake events in mid-America. Prog Struct Mater Eng 7(2):56–70

Fajfar P (2000) A nonlinear analysis method for performance-based seismic design. Earthquake Spectra 16(3):573–592

FEMA 227 (1992) A benefit–cost model for the seismic rehabilitation of buildings. Federal Emergency Management Agency, Building Seismic Safety Council, Washington, DC

FEMA 273 (1997) NEHRP guidelines for seismic rehabilitation of buildings. Federal Emergency Management Agency, Washington, DC

FEMA-350 (2000) Recommended seismic design criteria for new steel moment-frame buildings. Federal Emergency Management Agency, Washington, DC

FEMA-356: Prestandard and commentary for the seismic rehabilitation of buildings. Federal Emergency Management Agency, Washington DC, SAC Joint Venture, 2000

Frangopol DM, Liu M (2007) Maintenance and management of civil infrastructure based on condition, safety, optimization, and life-cycle cost. Struct Infrastruct Eng 3(1):29–41

Freeman SA (1998) Development and use of capacity spectrum method. In: Proc.6 US National Conference on Earthquake Engineering

Ghobarah A (2004) On drift limits associated with different damage levels. In: International workshop on performance- based seismic design, Bled, June 28–July 1 2004

Ghobarah A, Abou-Elfath H, Biddah A (1999) Response-based damage assessment of structures. Earthquake Eng Struct Dyn 28(1):79–104

Giovenale P, Cornell CA, Esteva L (2004) Comparing the adequacy of alternative ground motion intensity measures for the estimation of structural responses. Earthquake Eng Struct Dyn 33 (8):951–979

Jalayer F, Cornell CA (2009) Alternative non-linear demand estimation methods for probability-based seismic assessments. Earthquake Eng Struct Dyn 38(8):951–972

Kappos AJ, Dimitrakopoulos EG (2008) Feasibility of pre-earthquake strengthening of buildings based on cost-benefit and life-cycle cost analysis, with the aid of fragility curves. Nat Hazards 45(1):33–54

Kent DC, Park R (1971) Flexural members with confined concrete. J Struct Div 97(7):1969–1990

Lagaros ND (2007) Life-cycle cost analysis of construction practices. Bull Earthquake Eng 5:425–442

Liu M, Burns SA, Wen YK (2003) Optimal seismic design of steel frame buildings based on life cycle cost considerations. Earthquake Eng Struct Dyn 32:1313–1332

Marini A, Spacone E (2006) Analysis of reinforced concrete elements including shear effects. ACI Struct J 103(5):645–655

McKenna F, Fenves GL (2001) The OpenSees Command Language Manual – Version 1.2, Pacific Earthquake Engineering Research Centre, University of California, Berkeley

Menegotto M, Pinto PE (1973) Method of analysis for cyclically loaded reinforced concrete plane frames including changes in geometry and non-elastic behavior of elements under combined normal force and bending. In: Proceedings, IABSE symposium on resistance and ultimate deformability of structures acted on by well defined repeated loads, Zurich, pp 15–22

Mitropoulou ChCh, Lagaros ND, Papadrakakis M (2010) Building design based on energy dissipation: a critical assessment. Bull Earthquake Eng 8(6):1375–1396

Mitropoulou ChCh, Lagaros ND, Papadrakakis M (2011) Life-cycle cost assessment of reinforced concrete buildings subjected to seismic actions, Reliability Engineering and System Safety, (to appear)

Pacific Earthquake Engineering Research (PEER) (2010) http://peer.berkeley.edu/smcat/search.html/. Last accessed February 2010

Pei S, Van De Lindt JW (2009) Methodology for earthquake-induced loss estimation: an application to woodframe buildings. Struct Saf 31(1):31–42

Rackwitz R (2006) The effect of discounting, different mortality reduction schemes and predictive cohort life tables on risk acceptability criteria. Reliab Eng Syst Saf 91(4):469–484

Rackwitz R, Lentz A, Faber M (2005) Socio-economically sustainable civil engineering infrastructures by optimization. Struct Saf 27(3):187–229

Sanchez-Silva M, Rackwitz R (2004) Socioeconomic implications of life quality index in design of optimum structures to withstand earthquakes. J Struct Eng 130(6):969–977

Takahashi Y, Der Kiureghian A, Ang AH-S (2004) Life-cycle cost analysis based on a renewal model of earthquake occurrences. Earthquake Eng Struct Dyn 33:859–880

US Geological Survey (USGS). http://earthquake.usgs.gov/hazards/designmaps/

Vamvatsikos D, Cornell CA (2002) Incremental dynamic analysis. Earthquake Eng Struct Dyn 31 (3):491–514

Wen YK, Kang YJ (2001a) Minimum building life-cycle cost design criteria. II: Applications. J Struct Eng 127(3):338–346

Wen YK, Kang YJ (2001b) Minimum building life-cycle cost design criteria. I:Methodology. J Struct Eng 127(3):330–337

Chapter 11
Simplified Estimation of Seismic Risk for Buildings with Consideration of Structural Ageing

Daniel Celarec, Dimitrios Vamvatsikos, and Matjaž Dolšek

Abstract A simplified method for estimating the seismic risk of deteriorating buildings is presented utilizing a probabilistic framework and a simplified nonlinear method for seismic performance assessment of structures. Firstly, the probabilistic methodology with the extension to deteriorating structures is briefly explained. Then the methodology is applied to the example of a four-storey RC frame building with corroded reinforcement in order to estimate the influence of corrosion on seismic risk for the near-collapse limit state. The results reveal that after 50 years from the initiation of corrosion, the peak ground acceleration that causes the structure to violate the defined near collapse limit state decreases by 17% and the seismic risk for the near-collapse limit state increases by 7%, compared to the case where corrosion is neglected. It is also shown that degradation due to corrosion may change the collapse mechanism from ductile to brittle shear failure, raising an important question on the seismic safety of the existing buildings.

Keywords Performance-based earthquake engineering • Seismic risk • Capacity degradation • Corrosion • Reinforced concrete frame • Pushover analysis

11.1 Introduction

Structures are subject to deterioration over time due to extreme actions and environmental conditions. These usually refer to the varying negative effects of the outdoor environment, such as severe climate conditions or exposure to

D. Celarec (✉) • M. Dolšek
Faculty of Civil and Geodetic Engineering, University of Ljubljana,
Jamova 2, Ljubljana, SI 1000, Slovenia
e-mail: dcelarec@ikpir.fgg.uni-lj.si; mdolsek@ikpir.fgg.uni-lj.si; mdolsek@gmail.com

D. Vamvatsikos
Metal Structures Laboratory, School of Civil Engineering,
National Technical University of Athens, Athens, Greece
e-mail: divamva@central.ntua.gr

DOI 10.1007/978-94-007-1448-9_11,

aggressive atmospheric agents. The deterioration effects become even more important for structures subjected to seismic loads, since material corrosion is usually accompanied by a reduction of the ductility and strength of the structures (Stewart and Rosowsky 1998; Jin and Zhao 2001). For these reasons, it is important to provide seismic risk estimates with consideration of deterioration effects over time.

Seismic risk assessment with consideration of the structural ageing process is a complicated task, since it is highly uncertain to model corrosion attack mechanisms, especially, if it is assumed that the corrosion is spatially distributed over the structure. The extremely complicated deterioration phenomena, for example the localized (pitting) corrosion (Val et al. 1998), the cracking of concrete cover (Pantazopoulou and Papoulia 2001), and the reduction of bond strength between the steel reinforcement and the concrete (Val et al. 1998; Berto et al. 2009), are still not yet completely understood, at least, as far as their effects on the seismic response of structural elements are concerned. Moreover, because of the complexity of corrosion phenomena, its modelling is usually accompanied with a great level of uncertainty.

For these reasons, significant simplifications are needed in the process of modelling and analysis of deteriorating structures, in order to provide a practical basis for the probabilistic seismic risk assessment of structures with consideration of the structural ageing process. Recently, such a methodology was introduced by Vamvatsikos and Dolšek (2010), which represents an extension of the SAC/FEMA and correspondingly the Pacific Earthquake Engineering Research Center (PEER) probabilistic framework (e.g. Cornell et al. 2002). However, the method requires a comprehensive analysis of the seismic response of the structure, typically conducted by using the computationally demanding Incremental Dynamic Analysis (IDA) (Vamvatsikos and Cornell 2002). Thus, it is more practical to replace IDA with a simplified nonlinear static procedure for seismic performance assessment of structures, e.g. the N2 method (Fajfar 2000), which involves pushover analysis and inelastic response spectra. The application of the PEER probabilistic framework, in which the most demanding part, the IDA analysis, is replaced by the N2 method, was introduced by Dolšek and Fajfar (2008a). However, in that work the deterioration of the structure over time was neglected.

In the following, the methodology for seismic risk assessment of ageing structures is summarized with emphasis on those parameters that are determined with the N2 method or have to be predefined. Then, the methodology is demonstrated by means of an example of a four storey RC frame structure that is affected by corrosion.

11.2 Summary of the Probabilistic Framework

The extension of the PEER probabilistic framework, which enables the seismic risk assessment of deteriorating structures, has been formulated in a previous study by Vamvatsikos and Dolšek (2010). The authors provided a closed-form solution for the probabilistic treatment of structures with consideration of the deterioration effects over time and the applicability of the method has been illustrated using an existing multi-storey building. In this Section, the methodology (Vamvatsikos and Dolšek 2010) for seismic risk assessment of ageing structures is briefly summarized, and the basic assumptions are given that make the methodology applicable in practice. However, a full explicit derivation of the methodology is given elsewhere (Vamvatsikos and Dolšek 2010).

The probabilistic framework introduced by Vamvatsikos and Dolšek (2010) relies on a broader SAC/FEMA (Cornell et al. 2002), or PEER formulation, that represents a probabilistic basis for a performance-based seismic design and assessment guidelines. The formulation is based on the philosophy that the probability problem can generally be represented by the three basic variables. The first variable is related to the ground motion intensity I_M; the second variable is the seismic demand D, which is defined as a measure for the response of the structure, and the third variable describes the seismic capacity C, presumed as a measure for structural seismic performance at the defined limit state. Using these three random variables, the expected Mean Annual Frequency (MAF) of the exceedance of limit state becomes (Cornell et al. 2002):

$$\lambda_{ls} = \int_{0}^{+\infty} P(C < D | I_M = i_m) |dH(i_m)|, \tag{11.1}$$

where $|dH(i_m)|$ is the absolute value of the differential of the seismic hazard curve, i.e. the probability that $I_M = i_m$, and $P(C<D|I_M = i_m)$ is the probability that the seismic demand D exceeds the capacity C given the intensity $I_M = i_m$.

The integral in Eq. 11.1 can be solved numerically for any given probabilistic distribution of the random variables. However, in order to bring the formulation more closely to practical application, it is preferred to provide the expression in Eq. 11.1 in closed-form. This can be done in two different ways, depending on whether the structural seismic capacity C is defined via the intensity measure (IM-based method), or via the Engineering Demand Parameter (EDP-based method), for example using the storey drifts or displacements of the structure.

Theoretically speaking, both of the above mentioned methods are equivalent and so will give the same result. However, the problem associated with the EDP-based method is that structural instability in the region near collapse may mean that even a small increase in the seismic intensity results in practically “infinitely” large values of the displacements. This may consequently lead to problems with estimating the first two moments of a probability distribution of EDP-based parameters in the near-collapse region of the structure. On the other hand, the extreme values of EDP in the near-collapse region are automatically taken into account within the IM-based method. Also, the latter requires no simplifications on the analytical form of the EDP given the seismic intensity. This makes the IM-based method clearer and less error-prone. For these reasons, the IM-based method is adopted in this chapter as more convenient method for the practical representation of the PEER methodology.

11.2.1 The IM-Based Method

The analytical solution of the integral in Eq. 11.1 in terms of the IM-based method may be achieved under the two basic assumptions. Let us consider that seismic intensity is defined by the peak ground acceleration a_g, although other intensity measures can be used. Further it is assumed that seismic hazard curve may be approximated by the form (Vamvatsikos and Dolšek 2010):

$$H(s) = k_0 \cdot a_g^{-k}, \tag{11.2}$$

where k_0 and k are parameters of the approximating power-law function (Cornell et al. 2002). The seismic hazard curve has to be fitted in the region of interest, that is, in the range of the intensities corresponding to the structural seismic performance at the defined limit state. It is also assumed that the peak ground acceleration that causes the structure to violate the defined limit state $a_{g,ls}$ is distributed log-normally with the median $\tilde{a}_{g,ls}$ and the standard deviation of the natural logarithm β_{CRU}. Here, the notation β_{CRU} emphasizes that the total dispersion of the peak ground acceleration that causes the structure to violate the defined structural capacity (C) results from the aleatoric uncertainty usually referred to the "record-to-record" variability (R), and from the epistemic uncertainty (U) stemming from the incomplete knowledge of the actual physical and modelling parameters of the structure. The dispersions in the peak ground acceleration that causes the structure to violate the defined limit state due to aleatoric and epistemic uncertainty, β_{CR} and β_{CU}, respectively, are considered to be independent. Therefore, the total dispersion β_{CRU} can be approximately calculated by using the SRSS combination, i.e. with the square root of the sum of the squares (Vamvatsikos and Dolšek 2010):

$$\beta_{CRU}^2 = \sqrt{\beta_{CR}^2 + \beta_{CU}^2}. \tag{11.3}$$

Considering the above assumptions, the expected MAF (Eq. 11.1) can be formulated with a closed-form solution as in (Vamvatsikos and Dolšek 2010):

$$\lambda_{ls} = H(\tilde{a}_{g,ls}) \cdot exp\left(k^2/2 \cdot \left(\beta_{CR}^2 + \beta_{CU}^2\right)\right). \tag{11.4}$$

In Eq. 11.4, $\tilde{a}_{g,ls}$ is approximately estimated by using the N2 method (Fajfar 2000). Thus, the mean value of seismic capacity is estimated and not the median one, as it is assumed in Eq. 11.4. However, the estimation of the median via the mean value is conservative for lognormal distributions. Alternatively, the median seismic capacity can be calculated by assuming the log-normal distribution of structural seismic capacity and the corresponding dispersion.

Several methods for direct determination of median and the dispersion values for randomness and uncertainty in seismic capacity, β_{CR} and β_{CU}, exist. For example, they can be estimated by using the SPO2IDA tool (Vamvatsikos and Cornell 2006), a simple-to-use tool that provides the median, the 16th and 84th fractile values of the demand or the capacity for an equivalent single-degree-of-freedom (SDOF) system. Alternatively, dispersions can be estimated directly by performing IDA (Vamvatsikos and Cornell 2002) either for the equivalent SDOF or for the entire structure, for example as shown in Dolšek (2009). In that study it was assumed that the dispersion for randomness and uncertainty in seismic capacity, at least for the structure presented in this chapter, are $\beta_{CR} = 0.68$ and $\beta_{CU} = 0.28$ (Dolsek 2009). Note that the value of $\beta_{CU} = 0.28$ also agrees with the dispersion measures for the epistemic uncertainty that have been estimated for two RC buildings (Celarec and Dolsek 2010) by using the N2 method in conjunction with the Latin Hypercube Sampling (LHS) technique (Vořechovský and Novák 2009). However, stochastic analysis with consideration of epistemic uncertainties is not within the scope of this chapter. Thus, for a simplified approach, it is convenient to define the dispersion measures in advance. The dispersions in intensity measure due to randomness and modelling uncertainties for different types of structural systems and for the IM-based approach were recently proposed also in FEMA P695 (2009). The values for record-to-record and modelling-related variability in the intensity are in range of 0.35–0.45 for randomness and 0.2–0.65 for uncertainties.

The expression for λ_{ls} in Eq. 11.4 is used under the assumption that the median peak ground acceleration $\tilde{a}_{g,ls}$ and the dispersion measures, β_{CR} and β_{CU}, are time-independent. As a result, the calculated λ_{ls} is also a constant value over time, which is a basic assumption of a Poisson process. According to the definition of a Poisson process, the probability P that some event would occur n-times in the considered time interval equals to:

$$P = \frac{v^n}{n!} \cdot exp(-v), \tag{11.5}$$

where it is assumed that the occurrences of an event in a given time interval are independent of those in any other non-overlapping interval, and the expected number of occurrences of the event in the selected time interval is given by v.

The disadvantage of the Poisson distribution defined with Eq. 11.5 is that the parameter v, which defines the expected number of the occurrences of the event, depends on the length of the considered time interval. Thus, if the conditions are not changed, the parameter v in Eq. 11.5 can be defined as a linear function over time. In this case, the probability mass function for the random variable defined as the number of the occurrences of the event in time interval Δt can be expressed as:

$$P = \frac{(\lambda \cdot \Delta t)^n}{n!} \cdot exp(-\lambda \cdot \Delta t), \tag{11.6}$$

where λ is defined as the expected number of the occurrences of the event per unit of time, in other words being the time-independent rate of events.

Let us now translate the above terminology to that used in earthquake engineering. In engineering language, the parameter λ in Eq. 11.6 represents the expected MAF λ_{ls}. Then, by drawing a parallel to the Poisson process, we can see that if the median seismic capacity and the parameters regarding the dispersion measures do not change over time, then the expected number of the limit-state violations (or capacity-exceedance events) over the arbitrary selected time interval can be simply evaluated as λ_{ls} multiplied by the length of the considered time interval Δt. This is a typical case of using Eq. 11.4, which considers that median seismic capacity does not change over time. Thus, Eq. 11.4 can only be used under the very limited condition that λ_{ls} stays constant over the entire time interval under consideration.

11.2.2 Extension of the IM-Based Formulation to Deteriorating Structures

In the case if unfavourable environmental or other effects deteriorate the structural seismic capacity over time, then the MAF of exceeding the defined limit state is not constant over time. Consequently, the expected number of seismic events exceeding the defined limit state over the considered time interval, which is herein denoted as η_{ls}, cannot be estimated by simply using the Poisson distribution. A more comprehensive solution involves defining the time-dependent parameters as function of time τ, and integrating Eq. 11.1 over the considered time interval (t_0, t_0 + Δt) to determine the total number of exceedance events. In that case, η_{ls} is given by:

$$\eta_{ls}(t_0, \Delta t) = \int_{t_0}^{t_0+\Delta t} \int_{0}^{+\infty} P(C(\tau) < D | I_M = i_m) |dH(i_m)| d\tau, \tag{11.7}$$

where t_0 and Δt are, respectively, the initial time and time interval, $P(C(\tau) < D|$ $I_M = i_m)$ is the probability that the seismic demand D exceeds the capacity $C(\tau)$ given the intensity $I_M = i_m$and $|dH(i_m)|$ is the absolute value of the differential of the seismic hazard curve.

The analytical solution of the integral in Eq. 11.7 is possible if additional approximations are assumed for the time-dependent capacity expressed in terms of intensity measure (e.g. median peak ground acceleration $\tilde{a}_{g,ls|\tau}$ and corresponding dispersion measures $\beta^2_{CR|\tau}$ and $\beta^2_{CU|\tau}$). However, the detailed discussion on solving analytically the time integral in Eq. 11.7 is out of the scope of this chapter. Thus, on the following, only the basic assumptions and the final solution for an averaged MAF introduced by Vamvatsikos and Dolšek (2010) are briefly presented. For a more detailed discussion the reader is referred to Vamvatsikos and Dolšek (2010) and Torres and Ruiz (2007).

Basically, there are three assumptions that have to be considered. First, it is assumed that the same power-law approximation of the seismic hazard curve can be considered over the entire integration time interval. This assumption can be adopted if the power-law parameters k and k_0 in Eq. 11.2 are fitted within an extended seismic intensity region, in order to include the lower values of the median seismic intensities at the performance limit state of the structure, which decrease due to the unfavourable effects of the degrading process over time. An appropriate interval of seismic intensity over which the seismic hazard curve is to be approximated is a matter of the discussion. Earlier research, e.g. in Cornell et al. (2002), proposed that the fitting process be performed locally by fitting only at the intensities of structural seismic performance. However, a recent study conducted by Dolšek and Fajfar (2008a) revealed that better results can be achieved through an asymmetrically extended interval which in our case can be determined to be within $\left[0.25 \times \tilde{a}^{0}_{g,ls}, 1.25 \times \tilde{a}^{\Delta t}_{g,ls}\right]$, where $\tilde{a}^{0}_{g,ls}$ and $\tilde{a}^{\Delta t}_{g,ls}$ are the median peak ground accelerations at the beginning and at the end of the considered time interval, respectively.

The other two approximations are concerned with the time-dependent seismic response parameters, namely, the median peak ground acceleration $\tilde{a}_{g,ls|\tau}$ and the dispersion measures for randomness and uncertainty $\beta^{2}_{CR|\tau}$ and $\beta^{2}_{CU|\tau}$, which change over time τ. In contrast to the median seismic capacity, the dispersion measures increase over time, since the uncertainty in predicting the corrosion parameters in the future becomes extremely uncertain. Using the power-law and the linear fit, respectively, the median peak ground acceleration $\tilde{a}_{g,ls|\tau}$ and the dispersion measures for randomness and uncertainty may be approximated as functions of time (Vamvatsikos and Dolšek 2010):

$$\tilde{a}_{g,ls|\tau} = \tilde{a}^{0}_{g,ls} - \gamma\tau^{\delta} = \tilde{a}^{0}_{g,ls}\left(1 - \gamma\tau^{\delta} \big/ \tilde{a}^{0}_{g,ls}\right), \tag{11.8}$$

$$\beta_{CRU}|\tau^{2} = \beta^{0}_{CRU}{}^{2} + c_{\beta}\tau, \tag{11.9}$$

where constants γ, δ and c_{β} define the functions of median peak ground acceleration $\tilde{a}_{g,ls|\tau}$ and that of the dispersion measure $\beta_{CRU_{|\tau}}$ over time. The parameters $\tilde{a}^{0}_{g,ls}$ and β^{0}_{CRU} refer, respectively, to the initial median seismic capacity (herein expressed in terms of the median peak ground acceleration) and corresponding total dispersion measure, which are calculated for the initial time t_0.

Considering the above approximations, the final solution of the integral in Eq. 11.7, i.e. the expected number of events exceeding the defined limit state (η_{ls}) over the time interval (t_0, $t_0 + \Delta t$), can be formulated as follows (Vamvatsikos and Dolšek 2010):

$$\eta_{ls}(t_0, \Delta t) = \lambda^{0}_{ls}\frac{e^{\phi'\Delta t} - 1}{\phi'}, \tag{11.10}$$

where λ^{0}_{ls} is the expected MAF for the initial structure (Eq. 11.4), and ϕ' is a factor that includes the contribution of the deterioration. The parameter ϕ' can approximately be estimated as (Vamvatsikos and Dolšek 2010):

$$\phi' = \phi + k^{2}c_{\beta}/2, \tag{11.11}$$

$$\phi = -\frac{k}{\rho \cdot \Delta t} \cdot ln\left[1 - \frac{\gamma(\rho \cdot \Delta t)^{\delta}}{\tilde{a}^{0}_{g,ls}}\right], \tag{11.12}$$

where k is the parameter (local log-log slope) of the seismic hazard function (Eq. 11.2). The parameter ρ, introduced in Eq. 11.12, is greater than 0 but less or equal to 1. It controls the boundary values within the considered time interval, e.g. $\tau = 0$ and $\tau = \rho\Delta t$, at which the approximation in Eq. 11.12 gives the same values as the exact solution. A parametric investigation (see (Vamvatsikos and Dolšek 2010)) showed that the total error of the approximation in Eq. 11.12 remains within reasonable bounds when using a value for ρ within the interval $\rho \in [0.85, 1]$, e.g. $\rho = 0.9$.

11.3 Application to an Example Frame Building

In this section, an application of the proposed probabilistic formulation is illustrated. The methodology is applied to an existing four-storey reinforced concrete frame building, for which it is assumed that the corrosion of reinforcement has just started. The structure is located in a region of high level of carbonation-induced reinforcement corrosion (Somerville et al. 1992) and of moderate seismic hazard, the later being typical for the South-East part of Slovenia (Dolšek and Fajfar 2008a). The objective of this case study is to estimate the seismic risk for the near-collapse (NC) limit state of the example frame structure as the structural seismic capacity deteriorates over time and compare it with that where corrosion is neglected. Herein, the NC limit state is defined by the roof displacement in the softening part of the pushover curve that corresponds to 20% reduction of the maximum base shear.

In order to estimate the median peak ground acceleration $\tilde{a}_{g,ls|\tau}$, the seismic response parameters of the structure were estimated at different time instants between times t_0 and $t_0 + \Delta t$. Note that the t_0 relates to the initial condition, where the corrosion process just started and the structure is still intact, and Δt is the duration of the considered time interval during which the propagation of corrosion is taking place. Since the evaluation interval is taken to be [0, 50 years], t_0 was set to zero (e.g. in Eqs. 11.8–11.12), and the considered time interval with the duration of $\Delta t = 50$ years was divided in five time instants at which the structure is analyzed. These time instants coincide with 10, 20, 30, 40 and 50 years after the initiation of structural deterioration.

11.3.1 Description of the Structure and Mathematical Model

The example structure has been adopted from one of the previous studies conducted by Dolšek and Fajfar (2008b). The structure was designed to reproduce the design practice in European and Mediterranean centuries about 40–50 years ago (Carvalho and Coelho 2002). However, the structure may also be typical of more recent buildings, but without the application of the capacity principles and without

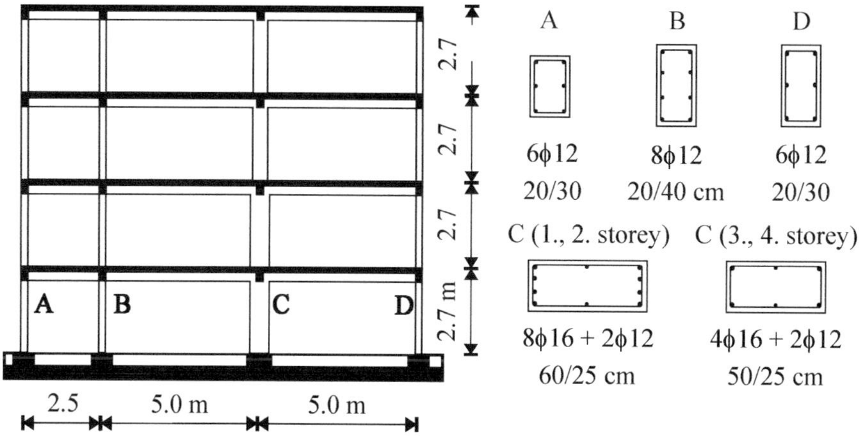

Fig. 11.1 The elevation and typical cross-sections of the columns of the example structure

up-to-date detailing. The elevation and the reinforcement of typical cross sections of the columns are shown in Fig. 11.1.

The structure was modelled by one-component lumped plasticity elements, which consist of an elastic beam-column element and two inelastic rotational hinges at the ends. Envelopes, describing the moment-rotation relationships of the hinges, were modelled with an equivalent tri-linear relationship using the effective initial stiffness of elements. The yield and maximum moments were calculated by section analysis with consideration of axial forces in columns due to gravity loads. The axial forces in beams were considered to be zero. The near-collapse rotations of hinges in columns were determined by using the CAE method (Peruš et al. 2006) and the rotation capacities of the hinges in beams were estimated according to Eurocode 8-3 (CEN 2005). The remainder of the parameters were determined as described in Dolšek (2010). Note that the nonlinear structural model consists only of plastic rotational hinges in columns and beams, and that the effects of shear deformations and interaction between the shear and the moment are not considered in the model. Still, the flexural resistance of columns was limited by the corresponding shear strength, where the shear strength of the columns was approximately estimated according to Eurocode 8-3 (CEN 2005). In this case, whenever the shear force in one of the hinges in a column exceeded the estimated shear strength, the flexural resistance of the column was considered to be decreasing. In doing so, the same negative stiffness of plastic hinges in columns was considered as it was calculated for the original moment-rotation envelope. Such an approach enables the approximate consideration of the nonlinear shear behaviour.

The resulting tri-linear moment-rotation relationship of the plastic hinge at the base of the column C in the third storey is shown in Fig. 11.2. It is presented for the initial conditions at time instant t_0 with the reinforcement bars of the original diameters (the continuous line), and additionally for the degraded column after

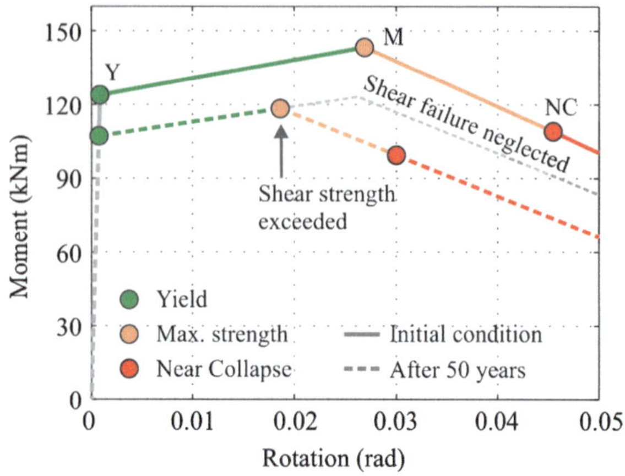

Fig. 11.2 The tri-linear moment-rotation relationship for the plastic hinge at the base of column C in the third storey

50 years of corrosion initiation (the dashed line). In the latter case, the shear force in the column reaches the corresponding shear strength at a rotation of 0.019 rad. From here onwards, the flexure resistance of the column is decreasing as the rotation in plastic hinge increases.

Rigid diaphragms were assumed at the floor levels. The masses, which were lumped at the mass centres, amounted to 46 t for the first three stories, and 40 t for the top storey. The strengths of the concrete and of the reinforcement steel are 16 and 343 MPa, respectively.

Deterioration of the capacity over time is modelled by the simplified model of the corrosion of longitudinal and transverse reinforcement in columns and beams. It is assumed that the corrosion affects only the diameter of reinforcement. The influence of corrosion on the bond stress between the concrete and steel bars as well as the spalling of concrete cover were neglected in this stage of the study. The reduced diameter $D_{rb}(t)$ of a reinforcing steel bar with initial diameter of D_b (mm), which is subjected to corrosion for a time period (years) $\Delta t = t - t_0$ (t_0 relates to the initial condition of the building and is equal to 0) is defined according to the procedure suggested by Pantazopolou and Papoulia (2001):

$$D_{rb} = D_b - 0{,}023 \cdot i_{\mathrm{corr}} \cdot \Delta t, \tag{11.13}$$

where i_{corr} represent the mean annual corrosion current per unit anodic surface area of steel (μA/cm^2). In our analysis the corrosion current $i_{\mathrm{corr}} = 2.0$ μA/cm^2 was considered, which corresponds to a high level of the carbonation-induced reinforcement corrosion. This value is related to the laboratory data of on-site measurements provided by Somerville et al. (1992), and represents more or less the upper value of the corrosion current under the extreme condition of carbonated concrete of low strength, and, at the same time with relatively high humidity content.

The adopted value for the corrosion current $i_{\mathrm{corr}} = 2.0$ μA/cm^2 is, however, typical of the carbonation-induced corrosion and is rather underestimated for the concrete contaminated with chlorides, a typical case of the de-icing salts on bridge

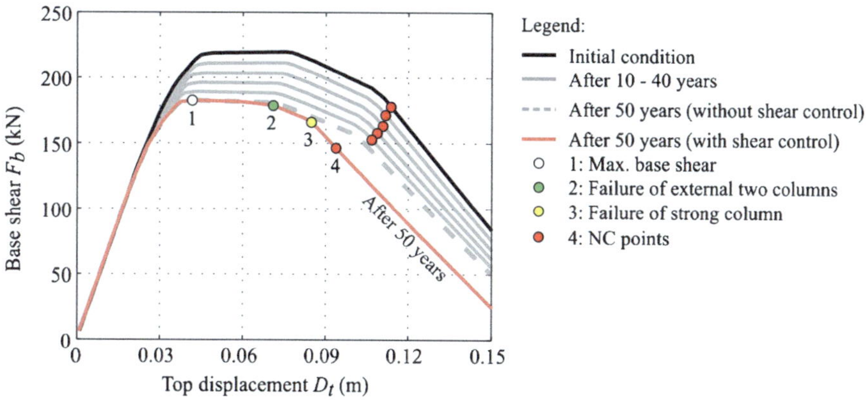

Fig. 11.3 The pushover curves for the initial conditions and for the structure attacked by corrosion. The pushover curves are calculated successively every 10 years after the corrosion initiation

decks or that of sea spray chlorides contaminating the surface of buildings in a coastal environment (Stewart and Vu 2000). In those cases, the corrosion current could be much larger, e.g. $i_{corr} = 10\ \mu A/cm^2$ or more (Somerville et al. 1992).

In order to determine the median seismic capacity as a function of time, six structural models M_0, M_{10}, M_{20}, M_{30}, M_{40} and M_{50} were prepared by using the PBEE toolbox (Dolšek 2010) in conjunction with the OpenSees platform (McKenna and Fenves 2004). These models correspond to the initial condition of the building at time t_0 and to the degraded structure (affected by corrosion) after 10, 20, 30, 40 and 50 years of corrosion propagation, respectively.

11.3.2 Pushover Analyses and Estimation of Seismic Capacity Using the N2 Method

Firstly, pushover analysis was performed for the initial conditions, indicating no signs of the negative effects of the corrosion attack, so that the structure still had its full seismic capacity. After that, the structure was analyzed successively for every decade, i.e. for time instants 10, 20, 30, 40 and 50 years, after initiation of corrosion. In general, any number of analyses could be performed over the considered time interval. It is only required that the results of the performed analyses provide information about deterioration of seismic capacity over time with sufficient accuracy.

The lateral loads for the pushover analyses were determined by multiplying the first mode shape by the storey masses. The first modal period of the initial structure ($t = 0$) is $T_1 = 0.85$ s and only slightly varies with time. Therefore, the increase of the natural periods and mode shapes over time is negligible.

The pushover curves are presented in Fig. 11.3. The black pushover curve represents the relationship between the top displacement and corresponding base shear at the initial conditions ($t = 0$). The maximum base shear and the ductility are gradually reduced for the deteriorated structures at time instants 10, 20, 30 and 40 years. A much larger decrease in ductility is observed for the structure that has been affected by corrosion for 50 years (red line). For previous cases, where the structure has been affected by corrosion for less than 50 years, the shear forces in columns never exceeded its shear strength. Therefore, the failure of columns was ductile. However, for the case when $t = 50$ years, shear failures occurred in the column A and C in the third storey.

As already mentioned, the shear failure in columns was modelled by reducing their flexural strength after the shear strength is exceeded. Since we couldn't know in advance which columns would be subject to shear failure, the pushover analysis for the time instant $t = 50$ years has been iteratively repeated two times. First, the pushover analysis was performed for the structural model with the columns with full available flexural capacities. Then, based on the results of the pushover analysis, the moment-rotation envelopes of the columns with shear strength exceeded were recalculated as described in the sect. 11.3.1. Lastly, the pushover analysis was repeated again, this time for the corrected structural model, representing the final analysis with consideration of the shear failure in columns. The resulting pushover curve with consideration of shear failure of columns is presented in Fig. 11.3 (red line) and compared with the pushover curve calculated in the first try of pushover analysis neglecting potential shear failures in column (grey dashed line).

Based on the results of the pushover analyses, it is revealed that the corrosion has the greatest impact on the maximum base shear and the displacement capacity. For example, the strength of the structure, which initially amounts to 220 kN, decreases over time and after 50 years it is reduced by about 17%. In contrast to the maximum base shear, the displacements only slightly decrease over the first four decades, but suffer a considerable drop at $t = 50$ years. The reason is the deteriorating effect of the corrosion on the transverse reinforcement. It is predicted that columns A and C in the third storey would fail in shear if the structure would be exposed to corrosion for about 50 years (see Fig. 11.3). In this case, the displacement capacity decreases by about 17.5%.

More comprehensive information on the seismic behaviour of the structure, than that given by pushover plots, can be provided through a visual representation of the damage on the structure (Fig. 11.4). The damage is presented for the initial conditions and for the degraded structures after 20 and 50 years, respectively. In the latter case, the damage is presented for the two different cases with and without consideration of shear failure in the columns. It can be observed that the selected structure typically collapses under the storey sway collapse mechanism developed in the third storey. The damage in the hinges of columns and beams, which slightly varies from case to case as the corrosion propagates over time, should be interpreted in accordance with the limit states shown for the moment-rotation relationship of the plastic hinges (Fig. 11.2).

However, the IM-based type of the probabilistic formulation requires the seismic capacity to be expressed with the intensity measure (Eqs. 11.8–11.12), which was in the case of the presented example estimated with the N2 method.

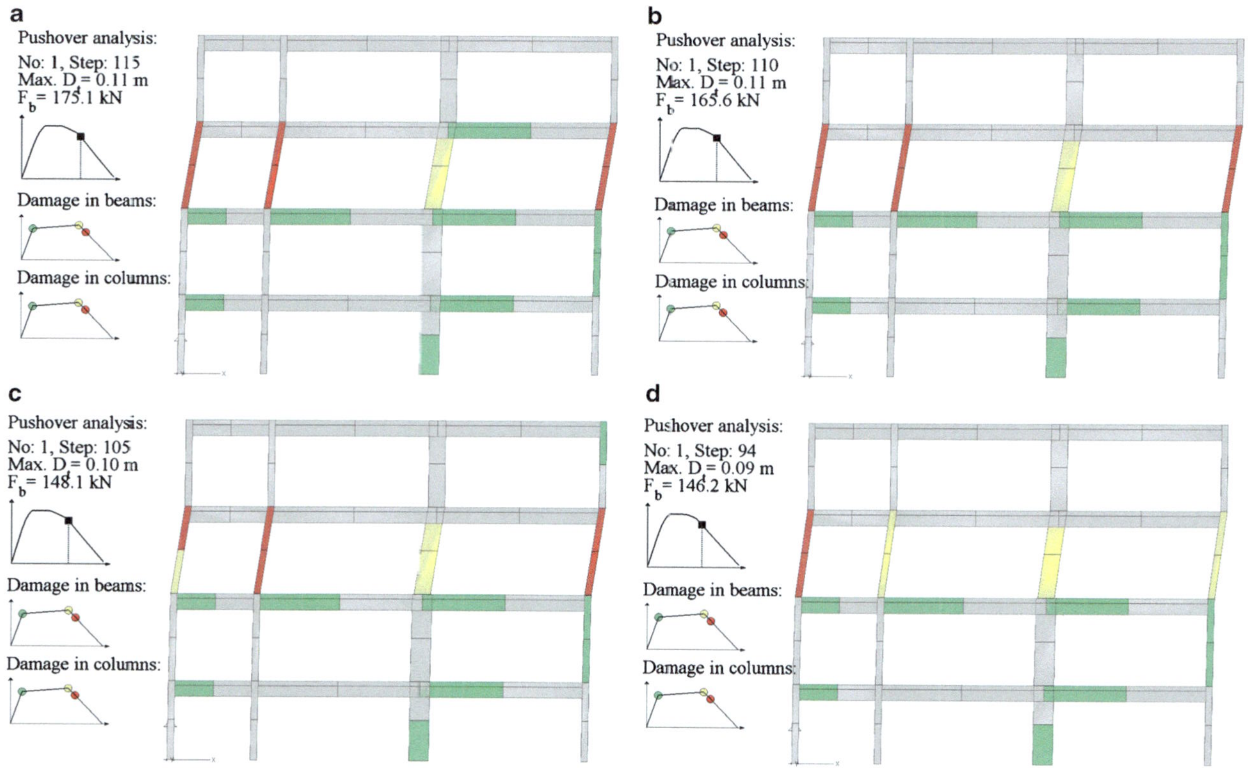

Fig. 11.4 The visual representations of structural damage for the NC limit state. The damage is presented for (**a**) the initial conditions with full structural capacity and for the corroded condition after (**b**) 20 and (**c-d**) 50 years of the degrading process. In the case (**c**) and (**d**) the damage is visualized, respectively, without and with consideration of shear failure

Table 11.1 The top displacement and associated median peak ground accelerations at NC limit state. The results are presented for the initial conditions ($\Delta T = 0$ years) and for the corroded state, successively every 10 years after the corrosion initiation ($\Delta T = 10$, 20, 30, 40 and 50 years)

Condition	ΔT (y)	$D_{\mathrm{nc}}(m)$	$\tilde{a}_{g,nc}(g)$
Initial	0	0.114	0.321
Corroded	10	0.112	0.316
	20	0.111	0.313
	30	0.109	0.308
	40	0.107	0.302
	50	0.094	0.265
$*(x_{50} - x_0)/x_0$		−17.5%	−17.5%

For that reason, the pushover curves were approximated by the elastic-perfectly plastic force-displacement relationship according to the procedure prescribed in Eurocode 8 (CEN 2004). In doing so, the yield force, which represents the strength of the idealized system, was taken equal to the maximum base shear of the structure. Then, the idealized force-displacement relationship was transformed to that of the SDOF system and the median peak ground acceleration $\tilde{a}_{g,nc}$ at the NC limit state was determined through the acceleration response spectra. The equal displacement rule was employed, which assumes that the mean displacement of the structure analyzed by using either the elastic or the nonlinear model is the same. Such an approximation is acceptable since the first modal periods of all structural models were in the medium period range of the acceleration response spectrum, i.e. the period corresponding to the first mode shape exceeds the characteristic period T_C of the elastic acceleration spectrum.

The advantage of the N2 method in comparison to the dynamic analysis is the simple definition of the seismic load on the structure since it can be easily defined via the acceleration spectrum. In the example building, Eurocode's elastic response spectrum for soil class *C* (CEN 2004) was adopted for the seismic load. The corresponding soil factor amounts to $S = 1.15$ and the characteristic periods are $T_{\mathrm{B}} = 0.2$ s, $T_{\mathrm{C}} = 0.6$ s and $T_{\mathrm{D}} = 2.0$ s.

The results of the N2 method and the displacements D_{nc} corresponding to the NC limit state are presented in Table 11.1. The seismic capacity in terms of the median peak ground acceleration for the initial structure amounts to $\tilde{a}_{g,nc} = 0.321g$, but decreases over time as well as the displacements D_{nc}. Fifty years after the corrosion initiation $\tilde{a}_{g,nc}$ drops considerably due to the shear failure of critical columns. Due to using the equal displacement rule, which suggests that the mean ductility and the reduction factor of the idealized system have the same values, the degradation of the median seismic capacity is equal to that of the displacements D_{nc} and amount to around 17.5% if compared to the initial conditions.

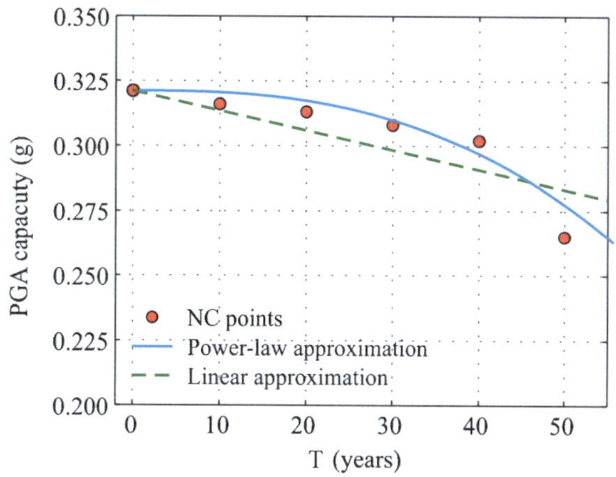

Fig. 11.5 The median peak ground acceleration at the NC limit state over time. The figure illustrates the calculated NC points versus the linear and power-law approximations of the median peak ground acceleration $\tilde{a}_{g,nc}$ over time

11.3.3 Evaluation of the Fitting Parameters γ, δ and k

In the study, only the median peak ground acceleration $\tilde{a}_{g,nc}$ was taken as time-dependent, while the dispersion measure β_{CRU} was considered constant over time. The parameters γ and δ, which define the median peak ground acceleration $\tilde{a}_{g,nc}$ as a function of time $\tilde{a}_{g,nc}(t)$, Eqs. 11.8 and 11.9, were determined by assuming two different types of approximation. The first choice was related to the simplest approach in which, $\tilde{a}_{g,nc}(t)$ is approximated by a linear function. In this case $\delta = 1$, while parameter γ was determined by using the least square approach and amounts to $\gamma = 7.56 \times 10^{-4}$. The linear approximation of $\tilde{a}_{g,nc}(t)$ significantly overestimates the calculated $a_{g,nc}$ for time instants below 50 years and underestimates it for the case of 50 years (Fig. 11.5). At that time, the shear failures of the two critical columns contributed to the considerable drop in the seismic capacity as described before. Thus, a higher-order approximation of $\tilde{a}_{g,nc}(t)$ is needed in order to model a nonlinear trend in the deterioration of the median seismic capacity over time with sufficient accuracy. The improved type of the approximation is based on the power-law function, as defined in Eq. 11.9. By considering the time interval from 0 to 50 years as the domain of the input time variable, over which the fitting process was performed, the values $\gamma = 0.155 \times 10^{-5}$ and $\delta = 2.62$ have been determined as representative for the power-law function parameters. Also in this case the fitting process has been performed by using the least square approach.

In Fig. 11.5 it can be observed that the power-law approximation of $\tilde{a}_{g,nc}(t)$ is much more accurate than the linear approximation. Nevertheless, both types of the approximations were used and the differences between them have been compared in terms of the calculated seismic risk for the NC limit state.

In addition to the approximation of the median seismic capacity over time, the seismic hazard needs to be approximated by the power-low function in accordance

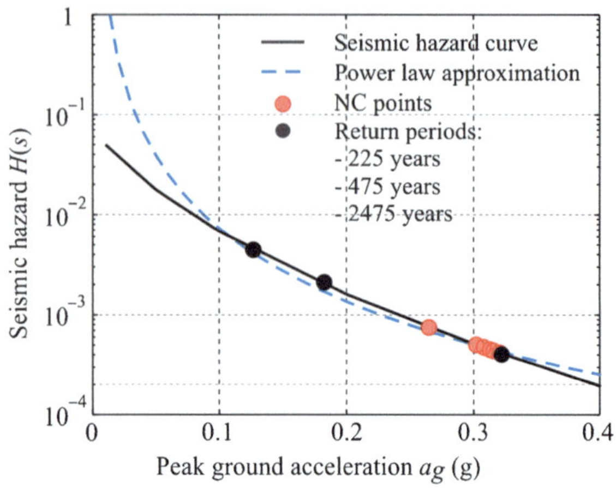

Fig. 11.6 The seismic hazard curve and its power-law approximation. The calculated $\tilde{a}_{g,nc}$ (*red circle marks*) are compared with those corresponding to the return periods of 225, 475 and 2,475 years

to Eq. 11.2. Note that for the adopted probabilistic formulation (Eqs. 11.8–11.12) only the parameter k needs to be explicitly defined, which could be done by a simple tangent line in log-log space (Cornell et al. 2002). Nevertheless, based on the previous study (Dolšek and Fajfar 2008a) it was decided to determine the parameters k and k_0 by fitting the hazard curve over the interval from $0.25 \times \tilde{a}_{g,nc|0}$ to $1.25 \times \tilde{a}_{g,nc|50}$, where $\tilde{a}_{g,nc|0}$ is the median peak ground acceleration at which the NC limit state is violated for the initial structure, and $\tilde{a}_{g,nc|50}$ is the corresponding median peak ground acceleration for the degraded structure after 50 years of degrading. The parameter k, which was determined by using the least square approach, amounted to $k = 2.42$. The value of k was used only when evaluating Eqs. 11.4 and 11.12, while $H(\tilde{a}_{g,nc})$ (see Eq. 11.4) was determined directly from the hazard curve in order to partially eliminate the error which is the consequence of power-law approximation of the seismic hazard curve.

The seismic hazard curve, which corresponds to the South-East part of Slovenia, and its power-law approximation are presented in Fig. 11.6. The peak ground accelerations for return periods 225, 475 and 2,475 years are 0.127, 0.182 and 0.323 g, respectively, while the calculated median peak ground acceleration at the NC limit state for the initial structure is 0.321 g and decreases with time.

11.3.4 The Seismic Risk Estimates

The seismic risk was determined in terms of the expected number of exceedance of the NC limit state (η_{nc}) and the instantaneous MAFs (λ_{nc}). The values of η_{nc} and λ_{nc} were calculated by assuming the dispersion measures β_{CR} and β_{CU} equal to 0.68 and 0.28, respectively, for both the closed-form and the "exact" solution. These values have been preliminary estimated in one of the studies conducted by Dolšek (2009) as discussed in Sect. 11.2.1.

Table 11.2 The expected number of exceedance events of the NC limit state. Estimated η_{nc} according to Eq. 11.4 is compared to the "exact" solution based on numerical integration of Eq.11.7

Condition	Time Δt (y)	Expected number of exceedance over time interval η_{nc} (10^{-2})			MAF λ_{nc} (10^{-2})
		"Exact solution"	Power-law fit	Linear fit	Eq. 11.4
Corroded	1	0.16	0.20	0.20	0.20
	10	1.62	2.02	2.08	0.21
	20	3.27	4.06	4.28	0.22
	30	4.97	6.16	6.62	0.23
	40	6.72	8.37	9.11	0.25
	50	8.56	10.8	11.8	0.36
No deterioration	50	7.96	10.1	10.1	0.20
*$(x_{deg,\,50} - x_{nodeg,\,50})/x_{nodeg,\,50}$		7.5%	6.9%	16.8%	80%

$x_{deg,50}$ and $x_{nodeg,50}$ x_{50} relate to the conditions after 50 years for the cases with and without consideration of the corrosion effects

The results for η_{nc} and λ_{nc} are presented in Table 11.2 for different time instants of the corroded structure, for the case of $t = 50$ years without corrosion effects. Further, the results for η_{nc} are additionally compared between the "exact" (Eq. 11.7) and the two variants of the closed-form solutions that incorporate corrosion, i.e. by means of Eqs. 11.10–11.12 for the linear and power-law approximations of $\tilde{a}_{g,nc}(t)$. Comparing η_{nc} for the "exact" solution for $t = 50$ years with the corresponding η_{nc} without consideration of corrosion, it can be observed that the impact of corrosion increases η_{nc} by about 7.5%. In this case it is predicted that the NC limit state would be reached, on average, 8.56×10^{-2} times in 50 years.

Slightly different results are observed if η_{nc} is estimated by using the closed-form solution based on the linear and power-law approximation of $\tilde{a}_{g,nc}(t)$. In that case, the values for η_{nc} are equal to 11.8 and 10.8×10^{-2}, respectively, for the linear and power-law approximation.

The differences between the results can be more clearly presented by plotting the expected number of exceedances as a function of time. In Fig. 11.7 it can be observed that η_{ls}, if estimated according to Eq. 11.4 and the power-law approximation of $\tilde{a}_{g,nc}(t)$, is overestimated by about 26% with respect to the value calculated by the numerical integration of Eq. 11.7. However, a detailed analysis showed that the differences in the results between the "exact" and the closed-form solution become negligible if the hazard curve in the process of the numerical integration is defined by the approximated power-law form and not with the actual seismic hazard curve. This means that the major part of the error of the closed form solution results from the approximate representation of the seismic hazard curve. Nevertheless, the power-law based approximation of $\tilde{a}_{g,nc}(t)$ is sufficiently accurate for the prediction of the expected number of failures due to the deterioration effects. This, however, cannot be declared for the linear-based approximation, since it does not allow approximating the nonlinearities in the median peak ground acceleration $\tilde{a}_{g,nc}$ over time.

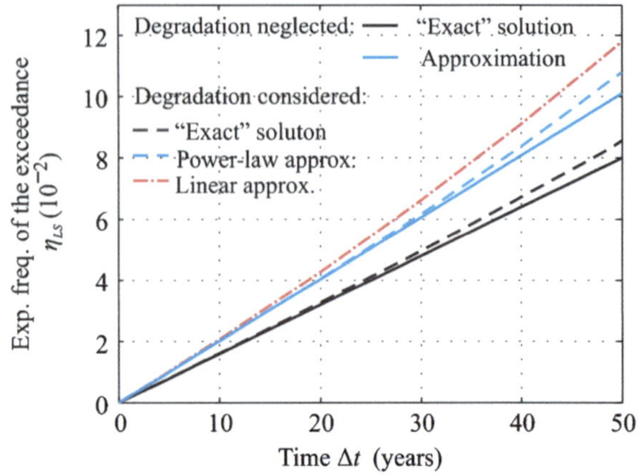

Fig. 11.7 The expected number of exceedances of the NC limit state over time. The results are presented for the approximate and the "exact" solution, with and without consideration of the capacity deterioration

Now, if we take a closer look at the results of the closed form solution based on the power-law approximation of $\tilde{a}_{g,nc}(t)$, it can be observed that the corrosion only slightly increases the calculated expected number of exceedances, i.e. by 6.9%. The explanation for that follows the fact that a great part of the structural loss happened after the 48th year. Thus, for the major part of the integration time interval, except for the last 2 years, only the lower seismic hazard levels are taken into account due to the higher seismic capacity (see Fig. 11.6). The associated increase in the expected number of the NC limit state violations represents a total sum of the structural failures over the entire considered time interval (50 years), and thus cannot be directly compared to the "instantaneous" reduction of seismic capacity or, for example, with the significant increase in the instantaneous MAF λ_{nc}. Note that MAF λ_{nc} at time instant 50 years increases for 80% if compared to the case where corrosion is neglected.

Therefore, the corrosion would have much larger effect on the seismic risk if evaluated for the next period of 50 years, what would be reasonable to do, since the use of the buildings is often extended over their design life. In doing so, the lower (un-conservative) estimate of the η_{nc} for the next period of 50 years can be based on a constant MAF $\lambda_{nc} = 0.36 \times 10^{-2}$ calculated at $t = 50$ years. Then the corrosion has significant effect since η_{nc} for the next 50 years exceeds η_{nc} for the first 50 years for about 67% and amounts:

$$\eta_{nc} = 0.0036 \cdot 50 = 0.18. \tag{11.14}$$

The accuracy of the results of the closed-from probabilistic formulation, which is adopted in this chapter, is likely to be very sensitive to the analytical approximations of the basic input variables. For example, the expected number of the exceedance events of the NC limit state over the time interval of 50 years equals $\lambda_{nc} = 10.8 \times 10^{-2}$ if the seismic hazard curve is approximated for the interval from $0.25 \times \tilde{a}^{0}_{g,nc}$ to $1.25 \times \tilde{a}^{\Delta t}_{g,nc}$. The value for λ_{nc} is estimated with a relative error of 26%, if compared to the "exact" solution obtained by the numerical integration. It is revealed, however, that a better accuracy can be achieved, by shifting the upper bound of the interval to $1.00 \times \tilde{a}_{g,nc}^{\Delta t}$. In this case the expected number of the exceedance events over the time interval 50 years is estimated with relative error of 0.5% and amounts to $\lambda_{nc} = 8.51 \times 10^{-2}$. On the other hand, the type of the approximation of $\tilde{a}_{g,nc}$ over time has only a minor influence on the results. For example, the results for λ_{nc} calculated based on the linear or the power-law-based approximations of $\tilde{a}_{g,nc}$ over time differ only for about 9%.

11.4 Conclusions

A simplified methodology for probabilistic seismic performance assessment of buildings with consideration of the structural ageing process has been summarized in this chapter. The methodology is an extension of the SAC/FEMA and PEER probabilistic formulation for seismic risk estimation to make them applicable to deteriorating structures and can be easily applied within the N2 method. Generally, there are two basic simplifications that are considered. The first relates to the computational analysis of the structure, which is the most demanding part of the probabilistic analysis. Herein, the IDA was replaced by the N2 method, a relative simple nonlinear method, which involves pushover analysis in combination with the nonlinear response spectra. The second simplification in the presented methodology concerns the mathematical model of the corrosion. The model of the corrosion proposed is relatively simplified, since it involves only reducing the steel bar cross-sectional areas. Thus, the method allows quick estimates of the influence of the corrosion effects on the global seismic response parameters rather than using very accurate mathematical models of the corrosion phenomenon.

Within the presented case study, it was shown that the corrosion has only a moderate influence on the seismic risk estimates for the NC limit state. Still, the corrosion simply cannot be ignored if structures contain shear-critical members, something that tends to be the norm in older RC buildings. It is envisaged that further refinement of the adopted corrosion model with inclusion of concrete spalling and bond degradation will additionally increase the estimated seismic risk of ageing RC structures.

Acknowledgements The research presented in this chapter represents the continuation of the bilateral project with cooperation of the University of Ljubljana and University of Cyprus and it is based on the work supported by the Slovenian Research Agency within the framework of the project High-throughput computing environment for seismic risk assessment (http://ice4risk.slo-projekt.info/) (J2-0845-0792-08) and the young researcher program (1000-07-310190). This support is gratefully acknowledged.

References

Berto L, Vitaliani R, Saetta A, Simioni P (2009) Seismic assessment of existing RC structures affected by degradation phenomena. Struct Saf 31:284–297

Carvalho EC, Coelho E (eds) (2002) Seismic assessment, strengthening and repair of structures. ECOEST2-ICONS report no. 2, European Commission-Training and Mobility of Research Programme

Celarec D, Dolsek M (2010) A simplified method for seismic performance assessment of RC frames with consideration of epistemic uncertainties. In: Proceedings of the international symposium on reliability engineering and risk management (ISRERM 2010), Beijing

CEN (2004) Eurocode 8: Design of structures for earthquake resistance. Part 1: General rules, seismic actions and rules for buildings. EN 1998-1. European Committee for Standardisation, Brussels

CEN (2005) Eurocode 8: Design of structures for earthquake resistance. Part 3: Strengthening and repair of buildings. EN 1998-3. European Committee for Standardisation, Brussels

Cornell CA, Jalayer F, Hamburger RO, Foutch DA (2002) Probabilistic basis for 2000 SAC federal emergency management agency steel moment frame guidelines. J Struct Eng 128(4):526–533

Dolsek M (2009) Incremental dynamic analysis with consideration of modeling uncertainties. Earthquake Eng Struct Dyn 38(6):805–825

Dolšek M (2010) Development of computing environment for the seismic performance assessment of reinforced concrete frames by using simplified nonlinear models. Bull Earthquake Eng 8 (6):1309–1329

Dolšek M, Fajfar P (2008a) The effect of masonry infills on the seismic response of a four storey reinforced concrete frame-a probabilistic assessment. Eng Struct 30:3186–3192

Dolšek M, Fajfar P (2008b) The effect of masonry infills on the seismic response of a four-storey reinforced concrete frame-a deterministic assessment. Eng Struct 30:1991–2001

Fajfar P (2000) A nonlinear analysis method for performance-based seismic design. Earthquake Spectra 16(3):573–592

FEMA, (2009), Quantification of building seismic performance factors, FEMA P-695 report, prepared by the Applied Technology Council for the Federal Emergency Management Agency, Washington, DC

Jin W, Zhao Y (2001) Effect of corrosion on bond behaviour and bending strength of reinforced concrete beams. J Zhejiang Univ Sci 2(3):298–308

McKenna F, Fenves GL (2004) Open system for earthquake engineering simulation. Pacific Earthquake Engineering Research Center, Berkeley, http://opensees.berkeley.edu

Pantazopoulou SJ, Papoulia KD (2001) Modeling cover-cracking due to reinforcement corrosion in RC structures. J Eng Mech 127(4):342–351

Peruš I, Poljanšek K, Fajfar P (2006) Flexural deformation capacity of rectangular RC columns determined by the CAE method. Earthquake Eng Struct Dyn 35:1453–1470

Somerville G, Andrade C, Fagerlund G, Lagerblad B, Rodriguez J, Tuutti K (1992) The residual service life of reinforced concrete structures. http://www.epicuro.co.uk/uploads/cr1-3.pdf

Stewart MG, Rosowsky DV (1998) Time-dependent reliability of deteriorating reinforced concrete bridge decks. Struct Saf 20:91–109

Stewart MG, Vu KAT (2000) Structural reliability of concrete bridges including improved chloride-induced corrosion models. Struct Saf 22:313–333

Torres MA, Ruiz SE (2007) Structural reliability evaluation considering capacity degradation over time. Eng Struct 29:2183–2192

Val DV, Stewart MG, Melchers RE (1998) Effect of reinforcement corrosion on reliability of highway bridges. Eng Struct 20(11):1010–1019

Vamvatsikos D, Cornell CA (2002) Incremental dynamic analysis. Earthquake Eng Struct Dyn 31:491–514

Vamvatsikos D, Cornell CA (2006) Direct estimation of the seismic demand and capacity of oscillators with multi-linear static pushovers through Incremental dynamic analysis. Earthquake Eng Struct Dyn 35(9):1097–1117

Vamvatsikos D, Dolšek M (2010) Equivalent constant rates for performance-based seismic assessment of ageing structures. Struct Saf 33(1):8–18. doi:10.1016/j.strusafe.2010.04.005

Vořechovský M, Novák D (2009) Correlation control in small-sample Monte Carlo type simulations I: A simulated annealing approach. Probab Eng Mech 24(2):452–462

Chapter 12
A Toolbox and Web Application for the Seismic Performance Assessment of Buildings

Matjaž Dolšek, Robert Klinc, Matevž Dolenc, Marko Brozovič, and Iztok Peruš

Abstract A software tool, known as a PBEE toolbox, for the seismic performance assessment of buildings, which was developed in Matlab in conjunction with the software framework OpenSees, and a web application for the prediction of approximate IDA curves are presented in this chapter. Although, in its present version, the PBEE toolbox supports only simple nonlinear models, its capabilities exceed usual software tools for computational simulation, since it enables seismic performance assessment of buildings with various procedures and adopts an open-source philosophy so that it can be easily extended or modified to suit specific user requirements. The capabilities of the PBEE toolbox and the web application, which involves the response database of a single-degree-of-freedom system with a quadrilateral force-displacement relationship, are demonstrated by means of an assessment of the seismic response parameters of an eight-storey reinforced concrete frame, using incremental dynamic analysis, progressive incremental dynamic analysis, approximate incremental dynamic analysis, and the N2 method. It is shown that, for the case of the presented example, all the methods produce similar results, although each method has some advantages and some disadvantages.

Keywords Performance-based earthquake engineering • PBEE toolbox • Web application • Progressive incremental dynamic analysis • Approximate incremental dynamic analysis • Reinforced concrete frames

M. Dolšek (✉) • R. Klinc • M. Dolenc • M. Brozovič • I. Peruš
Faculty of Civil and Geodetic Engineering, University of Ljubljana,
Jamova 2, Ljubljana SI 1000, Slovenia
e-mail: mdolsek@ikpir.fgg.uni-lj.si; mdolsek@gmail.com; robert.klinc@itc.fgg.uni-lj.si; matevz.dolenc@itc.fgg.uni-lj.si; Mbrozovic@ikpir.fgg.uni-lj.si; iperus@ikpir.fgg.uni-lj.si

M. Dolšek (ed.), *Protection of Built Environment Against Earthquakes*,
DOI 10.1007/978-94-007-1448-9_12,

12.1 Introduction

The response of structures to severe earthquakes is nonlinear, since they are usually designed to withstand large seismic demands without local or global collapse, but selected structural elements may be significantly damaged. From this point of view, the seismic response parameters of a structure, which are estimated in the design or assessment process, should be determined by using nonlinear methods of analysis. Unfortunately, in the case of practical applications such an approach is rare, for many reasons. The main reason probably lies in the fact that nonlinear analysis is far more complex than linear analysis. This means that the performance assessment of structure based on nonlinear analysis requires much more knowledge, input data, and work. The complexity of nonlinear analysis is also probably the reason why the use of nonlinear methods of analysis is not required in structural codes, but only defined as an alternative to linear methods of analysis. Consequently, structural engineers do not check the response of structures by using nonlinear methods of analysis, since it is not necessary to do so according to the structural codes. However, they are aware of the fact that structural response will be nonlinear in the case of the design earthquake.

In order to facilitate the use of nonlinear methods of analysis in practice, without forcing such use into structural codes, it is necessary to extend the knowledge of engineers and to develop user-friendly and reliable software tools that are able to support various nonlinear models and methods for the seismic performance assessment of structures.

Some commercially available software already supports performance-based methodologies. For example, SAP 2000 (CSI 2008), PERFORM-3D (CSI 2006) and MIDAS (2008) provide nonlinear structural analysis methods, as well as different types of simplified procedures for the performance assessment of structures. In addition to commercial software, free (SEISMOSOFT 2007; Elnashai et al. 2008) and open-source software (McKenna and Fenves 2007) is also available. This kind of software has certain advantages in comparison with commercially available software, since it provides a comprehensive library of nonlinear elements, material models, analysis types and solvers. From this point of view OpenSees (McKenna and Fenves 2007) is probably the most comprehensive software framework in the field of earthquake engineering. However, it does not usually support the performance-based assessment methods prescribed by various structural codes, since it is mostly focused on the research community, in order to support researchers to develop applications for the simulation of the performance of structural systems subjected to earthquakes.

In order to further extend the applicability of softwarc for computational simulation, a PBEE toolbox for the seismic performance assessment of reinforced concrete frames and a web application for the prediction of approximate IDA (incremental dynamic analysis) curves have been developed. They are briefly presented and their use is demonstrated by means of the seismic performance assessment of an eight-storey frame, whose performance has been analysed with various methods.

12.2 Overview of the Procedures for the Seismic Performance Assessment of Buildings

Many different procedures for the seismic performance assessment of buildings have been developed in past decades, not only because of different needs but also because the fact that different researchers and structural engineers have had different philosophies for the solving of such problems. Consequently, seismic performance assessment procedures can be classified according to many different criteria.

The most general classification of seismic performance assessment procedures can be based on the type of analysis. In linear elastic analysis it is assumed that the material remains in the elastic range. Since earthquakes often cause damage to structures, this assumption is not sufficiently accurate from the theoretical point of view, but it is often applied in the case of seismic analysis. For example, the European standard Eurocode 8-3 (CEN 2005) for the assessment and retrofitting of buildings prescribes that seismic response parameters can also be determined by lateral force or modal response spectrum analysis, which are both types of linear elastic analysis. However, the advantage of the linear elastic analysis in comparison to the nonlinear analysis is its numerical stability, relatively short computational time, the simple definition of earthquake loading, and the broad availability of user-friendly software. Nonlinear analysis is, of course, computationally much more demanding and therefore still rarely suitable for practical use.

Linear elastic and nonlinear analysis can be further classified into static and dynamic analysis. In the case of static analysis the equilibrium equations are independent of time, whereas in the case of dynamic analysis the seismic response parameters are computed for each time step. Since nonlinear dynamic analysis considers the nonlinear behaviour of structures, and takes into account the dynamic effects, which are important in the case of earthquake actions on a structure, this analysis method is the most advanced for the simulation of structural response under seismic action. However, it is the subject of many uncertainties, and its use is not trivial. Firstly, nonlinear dynamic analysis is computationally extremely demanding, and, in the case of complex structural models, can also become the subject of numerical instability. Furthermore, earthquake action has to be defined by means of ground motion records. Although simple guidelines for the selection of ground motion records exist, different engineers/researchers may select different ground motion records for the analysis. Consequently, estimated seismic response parameters may vary from case to case. Additionally, predicted seismic response parameters can differ due to different nonlinear models used in the analysis. This is because many different nonlinear models exist, and there is no consensus as to which model is the most appropriate for the seismic analysis. Last but not least, nonlinear dynamic analysis involves the numerical integration of the equations of motion. It is well-known that results of nonlinear dynamic analysis more or less depend on the parameters of integration schemes. Even the level of critical damping and the corresponding model of viscous damping can significantly

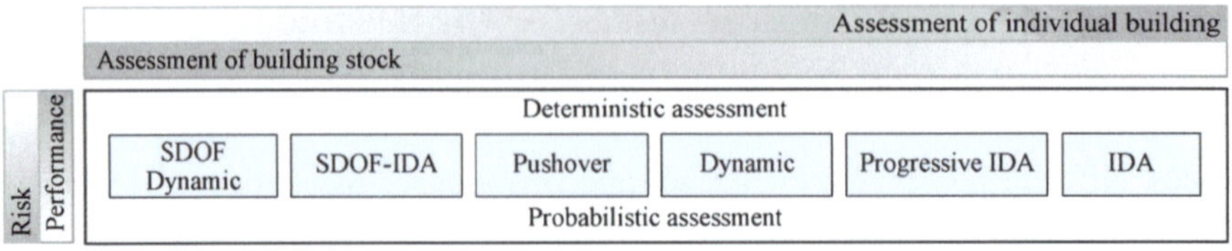

Fig. 12.1 List of the different types of analyses used for the seismic performance assessment of buildings

affect the results of nonlinear dynamic analysis. Because of all the issues encountered in nonlinear dynamic analysis, practice-oriented nonlinear analysis procedures, which typically involve nonlinear static (pushover) analysis and prediction of the target displacement, have emerged over the last two decades, and become popular among both researchers and engineers. Many different procedures have been developed (Fajfar 2000; Priestley and Kowalsky 2000; Chopra and Goel 2002; Aydinoğlu 2003; Antoniou and Pinho 2004), and some of them have become a part of standards and guidelines, e.g. CEN (2004a), FEMA (2000), and FEMA (2005).

Seismic performance assessment procedures can be also classified as deterministic or probabilistic procedures, which can involve different type of analysis (Fig. 12.1). In deterministic procedure all the input parameters are assumed according to the best knowledge and/or often conservatively. For example, the mean/median or other characteristic value (fractile) is adopted for the strength of a material, although it can vary in different parts of the structure. The opposite is in the case, if the seismic performance assessment of building is based on probabilistic procedure where some or all of the input parameters are considered to be uncertain, and are most frequently treated by appropriate random variables. In this case, the uncertainties are usually classified into two principal categories (Ellingwood and Kinali 2009): the aleatoric uncertainties, which are related to the random nature of earthquakes and are, by definition, irreducible, and the epistemic uncertainties, which are knowledge-based and are most often related to the physical properties of the structure and its modelling parameters.

Most seismic performance assessment procedures were developed to be used for the assessment of a single building structure, but many procedures have been simplified, and can be used for the seismic performance assessment of building stock. Such classification is therefore also viable.

Some of the analyses used in the process of the seismic performance or risk assessment of buildings are illustrated in Fig. 12.1. In this case all the presented analyses are nonlinear, starting with the dynamic (SDOF Dynamic, see Fig. 12.1) and incremental dynamic analysis (IDA) (Vamvatsikos and Cornell 2002) of an equivalent single-degree-of-freedom (SDOF) system (SDOF-IDA, see Fig. 12.1). Since this type of analysis involves a simple nonlinear model, which can be arbitrarily defined based on some structural parameters, it can be used for the seismic performance assessment of building stock. The IDA curves of the equivalent SDOF system can also be used for determination of the target displacement, if combined with pushover analysis (Pushover, see Fig. 12.1), which is the first analysis from the list of analyses presented in Fig. 12.1, which require structural model of the entire building.

The next analysis from the list Fig. 12.1 is the nonlinear dynamic analysis (Dynamic, see Fig. 12.1), which has already been described, and the last is incremental dynamic analysis, which is a well-known parametric analysis method, and involves nonlinear dynamic analysis and subjecting a structural model to a number of ground motion records, each scaled to multiple levels of intensity. IDA is powerful analysis method, but is computationally extremely demanding. An alternative to IDA is progressive IDA (Azarbakht and Dolšek 2011) (Progressive IDA, see Fig. 12.1). A new element of progressive IDA, when compared to the elements of IDA, is the precedence list of ground motion records. In general, IDA curves are calculated for all the ground motion records in a set of such records, whereas in progressive IDA, the IDA curve is first calculated for the first ground motion record from the precedence list, and then progressively for the other ground motion records from the precedence list of records. After several IDA curves have been calculated, the analysis can be terminated, since the acceptable tolerance is achieved. Such an approach facilitates practical application, with the aim of selecting the most representative ground motion records for incremental dynamic analysis.

The methods of analysis presented on the left hand side of Fig. 12.1, as well as the corresponding simplified versions of these methods, are not computationally demanding, and are therefore often used for the seismic performance assessment of building stock, whereas the methods of analysis on the right hand side of Fig. 12.1 require more computational time and they can be afforded only for the seismic performance assessment of individual building structures. Additionally, all the listed analyses can be used for deterministic or probabilistic performance assessment of structures. Deterministic assessment is most often adopted in the case if the seismic performance assessment is expressed in terms of seismic response parameters, such as, inter-storey drift, storey acceleration, or similar, whereas probabilistic assessment is required if seismic performance of a structure is judged based on the mean annual frequency of exceedance of a given limit state, expected monetary or other kind of losses.

12.3 Seismic Performance Assessment of Buildings by Using the PBEE Toolbox and Web Application

An effective approach for the seismic performance assessment of buildings is introduced. It involves the use of the PBEE toolbox (Dolsek 2010) in conjunction with a web application for the prediction of approximate IDA curves. The PBEE toolbox is a software tool, which was developed in order to fill the gap between the user, who needs to evaluate the seismic performance of a structure, and OpenSees (McKenna and Fenves 2007), which consists of comprehensive software for computational simulation in the field of earthquake engineering.

12.3.1 Overview of the PBEE Toolbox

The PBEE toolbox is a set of Matlab (MathWorks 2007) functions, which can be used for the seismic performance assessment of reinforced concrete (RC) frames in conjunction with OpenSees. The aim of the PBEE toolbox is to enable rapid definition of simple nonlinear structural models of RC frames. In this case the most time-consuming part of the work involves the determination of the properties of the plastic hinges. Since the PBEE toolbox automatically generates the properties of plastic hinges, based on data regarding material strength, reinforcement and section properties, the amount of work which has to be performed in order to prepare a structural model is reduced significantly. The current version of the PBEE toolbox combines the Eurocode 8 (CEN 2005) requirements for non-linear modelling and the non-linear seismic analysis of buildings with some other approaches which were used elsewhere (Fajfar et al. 2006; Peruš et al. 2006). However, since the PBEE toolbox is based on open-source philosophy, the user can simply change the function which, for example, is used to determine the moment-rotation relationship of the plastic hinges, and still use other functions of the PBEE toolbox.

Different assumptions which, in general, follow the Eurocode 8 requirements, are used in order to establish as simple as possible but yet adequate structural model. These assumptions are as follows:

- the floor diaphragms are assumed to be rigid in their own planes, and the masses and moments of inertia of each floor are lumped at the corresponding centres of gravity,
- the beam and column flexural behaviour is modelled by one-component lumped plasticity elements, composed of an elastic beam and two inelastic rotational hinges, which are located at the ends of elastic element, and defined by a moment-rotation relationship. The element formulation is based on the assumption of an inflexion point at the midpoint of the element. For beams, a plastic hinge is used for major axis bending only. For columns, two independent plastic hinges for bending about the two principal axes are used.
- the moment-rotation relationship before strength deterioration is modelled by a bi-linear or tri-linear relationship. Zero axial force and the axial load due to gravity loads are taken into account when determining the moment-rotation relationship for beams and columns, respectively. A linear negative post-capping stiffness is assumed after the maximum moment is achieved.
- the gravity load is represented by a uniformly distributed load acting on the beams, and/or by concentrated loads acting at the top of the columns.

Work with the PBEE toolbox involves the Matlab script language. First the user has to define the structural data. In the present version of the PBEE toolbox the data regarding the structure are organized at the structural and element type level. At the structural level the user defines the structural gridlines and the storey masses, whereas at the element level the user separately defines the beams and the columns of the structure. Once the structural data are defined, the user calls functions in order to compute the moment-rotation envelopes in the plastic hinges, and after that the input files for OpenSees can be simply generated by calling functions for the generation of the tcl code. Depending on the type of analysis, the user has to prepare some data regarding the analysis, and runs the analysis in OpenSees through the Matlab environment.

In general there is no need to work with OpenSees since all the results are stored in Matlab data structures after the analysis has been performed in OpenSees. The results are organized based on the so-called global results, which are, for example, the displacements at the mass centre, and the local results, which include the forces and deformations at the plastic hinges. For some type of analyses, e.g. for incremental dynamic analysis (Vamvatsikos and Cornell 2002), the PBEE toolbox provides an iterative process between the results of the non-linear dynamic analysis obtained by OpenSees, the requirements of IDA as defined by the analysis data, and the automatic tcl code generation for OpenSees.

After the analysis has been performed the user can define different limit states at the level of the plastic hinge in order to link the damage in the plastic hinge to the global seismic response parameters. For example, if pushover analysis is performed, the user can simply link the damage at hinge level to the top displacement or base shear, or, based on the results of incremental dynamic analysis, link the damage in the plastic hinges with the intensity measure. Such an approach enables graphical representation of the damage in the plastic hinges on the pushover or IDA curve. Additionally, the user can determine the target displacement and the damage to the structure according to the N2 method (Fajfar 2000), and present the results visually in AD format. The damage in the plastic hinges can also be visually presented by plotting it on a structural drawing.

The PBEE toolbox also includes functions which can be used to determine the IDA curves of equivalent SDOF systems. Such an approach represents an alternative to seismic performance assessment with the N2 method, which involves closed-form expressions for determination of the target displacement that can be developed based on simple parametric studies. However, more sophisticated parametric studies can lead to more accurate predictions of the target displacement, and can also provide new information, such as the dispersion in the displacement demand, or the prediction of collapse capacity, which are needed for the probabilistic risk assessment of structures. The advances in information technology (IT), as well as new software and growing computer power, provide new possibilities for solving the above-described shortcomings, which are embedded in the simplified nonlinear methods for seismic performance assessment. Rather than developing new simplified expressions for inelastic displacement ratios, a web-based application for the prediction of approximate fractile IDA curves has been recently developed, and can be used in conjunction with the PBEE toolbox.

12.3.2 Brief Description of the Methodology for Prediction of Approximate IDA Curves

The methodology which was used to develop the web-based application for determination of approximate IDA curves consists of two independent processes. The first process involves determination of the response database, whereas the second process involves the prediction of approximate IDA curves from the response database. The first process is the parametric study, which is performed for the SDOF model, and involves definition of the input parameters which affect the seismic response, definition of the discrete values of the input parameters, and computation of the IDA curves of the defined SDOF models. The usual input parameters of the parametric study are the period of the system, the parameters of the force-displacement relationship and the hysteretic behaviour, the damping, and the ground motions. It is worth emphasizing that definition of the SDOF model depends on expert judgment, since the selection of a force-displacement relationship or hysteretic behaviour of the SDOF model, which might be appropriate for simulating the global response of a specific structural type, is not trivial. Different experts may select different input parameters or discrete sets of their values. It is therefore important that the process of the parametric study is independent of the second process, since in the future many different databases may be created. The results of the parametric study can be used to create a response database, which can be established for a discrete number of input parameters of the SDOF model, for a set of ground motion records.

Since the IDA curves, which are stored in the response database, are computed for the discrete parameters of the SDOF model, the second process involves prediction of the approximate IDA curves for any input parameter of the SDOF model. Clearly, this process is trivial if the request for the prediction of approximate IDA curves is based on the same input parameters as those for which the IDA curves are available in the response database. However, this is, in general, a rare case. The second process therefore involves two steps: the query for appropriate IDA curves from the database, and the computation of the approximate IDA curves by the selected interpolation method. If the response database is computed for a sufficient number of discrete input parameters, then linear interpolation is a suitable method for determining the approximate IDA curves. Since the IDA curves depends on n input parameters, n-dimensional linear (also known as multi-linear) interpolation is applied by using one-dimensional linear interpolation in each separate coordinate dimension. Such an approach, which is described elsewhere (Peruš et al. 2011), requires the neighbouring data which are mapped into the unit hypercube $[0\ 1]^n$. Consequently, the result of the query from the response database is a set of 2^n IDA curves.

It is known that equivalent SDOF models vary depending on the structural system and the material of the structure. For example, in the case of the prediction of the approximate IDA curves of a reinforced-concrete building, the force-displacement relationship can be described by the four dimensionless parameters $r_v = F_1/F_2$, $r_h = u_1/u_2$, $\mu_u = u_3/u_2$, $\alpha = -k_{pc}/k_i$, where points (u_1, F_1) and (u_2, F_2) represent the first and second characteristic points of the idealized force-displacement relationship and roughly represent, respectively, the cracking of the concrete and, in the case of regular structures, yielding of reinforcement at the base of the columns. The displacement u_3 is related to the displacement where the strength of the structure starts degrading, while k_{pc} and k_i are, respectively, the post-capping and initial stiffness of the idealized force-displacement relationship. With a suitable variation of the four parameters the idealized force-displacement relationship can be fitted to almost any pushover curve. The parameters of the force-displacement relationship are visually explained when the user starts the web application (http://ice4risk.slo-projekt.info/analysis/).

The other two parameters, which describe the SDOF model, are the period T_1 and the critical damping ratio ξ. Note that the Takeda's hysteretic rules (Takeda et al. 1970) can be used for the most basic simulation of the hysteretic behaviour of reinforced concrete buildings. These hysteretic rules were used for determination of the response database of the web application, presented in the next subsection.

The only practical way is that the response database is filled with automated computational procedures, for which OpenSees (McKenna and Fenves 2007) can be used. However, for these computations the SDOF model has to be defined with dimensional quantities. For this purpose the mass and the force F_2 of the SDOF model can be assumed to be constant, e.g. $m = 100$ t and $F_2 = 0.1 \cdot m \cdot g$, where g is the acceleration of gravity. Based on these definitions and the known dimensionless parameters r_v, r_h, μ_u and α, the period T_1 and the critical damping ratio, it is trivial to define the SDOF model. This is because the displacement u_2 can be determined from the relationship between the mass, stiffness and the period of the SDOF model, taking into account the defined parameters r_v and r_h:

$$u_2 = \frac{T_1^2}{4\pi^2} \frac{F_2 \cdot r_v}{m \cdot r_h} = \frac{T_1^2 g}{40\pi^2} \frac{r_v}{r_h}. \tag{12.1}$$

Other dimensional parameters of the SDOF model can be then obtained from the definition of the non-dimensional parameters, while the damping constant c can be calculated from the expression:

$$c = \frac{4\pi\xi}{T_1}. \tag{12.2}$$

It is suitable that the IDA curves in the response database are stored for the relationship between the displacement and the peak ground acceleration, and for the displacement and spectral acceleration at period T_1, which corresponds to the initial stiffness of the SDOF system. The IDA curves can be also expressed with the reduction factor R and the ductility μ, which are used for the definition of inelastic spectra (the R-μ-T relationship). However, the reduction factor should in this case be understood as the normalized spectral acceleration. Note that both dimensionless parameters (R, μ) can be defined based on different assumptions. One possibility is to define the ductility according to the displacement u_2, and the yield acceleration according to the maximum force ($F_2 = F_3$) of the SDOF system. However, elastic spectral acceleration is defined at the period T_1, which corresponds to initial stiffness of the SDOF system. This means that the ductility and reduction factor can be written as $\mu = u/u_2$ and $R = S_{ae}(T_1)/S_{ay}$, respectively, where $S_{ay} = F_2/m$.

The IDA curves of the response database are computed only for the SDOF model, which has the dimensional parameters as described above. Other IDA curves of user-defined SDOF models, if expressed with dimensional quantities, can be determined by means of simple transformations of the computed IDA curves. It can be shown that, in addition to the dimensionless parameters and the period, only one dimensional quantity of the force-displacement relationship of the user-defined SDOF model is needed. Knowing that the IDA curves of the two models are equal if expressed in terms of the reduction factor and ductility, the following relations can be derived:

$$S_{ae}^{U}(T_1) = \frac{m}{m^{U}} \frac{F_2^{U}}{F_2} S_{ae}(T_1), \quad u^{U} = \frac{u_2^{U}}{u_2} u, \tag{12.3}$$

where index U denotes the quantity of the user-defined system. Similarly, as in the case of the derivation of u_2 (Eq. 12.1), it can be shown that:

$$m^{U} = \frac{T_1^2}{4\pi^2} \frac{r_v F_2^{U}}{r_h u_2^{U}}. \tag{12.4}$$

Incorporating Eqs. 12.1 and 12.4 into Eq. 12.3 it can be shown, that the relationship between the IDA curves of the two SDOF models (Eq. 12.3) can be expressed by knowing only one additional parameter, which is, according to our definition of the SDOF model displacement u_2^{U}:

$$S_{ae}^{U}(T_1) = \frac{40\pi^2 r_h}{T_1^2 g\, r_v} \cdot u_2^{U} \cdot S_{ae}(T_1), \quad u^{U} = \frac{40\pi^2 r_h}{T_1^2 g\, r_v} \cdot u_2^{U} \cdot u. \tag{12.5}$$

Since $S_{ae}^{U}(T_1) = \beta \cdot a_g^{U}$ and $S_{ae}(T_1) = \beta \cdot a_g$ the spectral acceleration in Eq. 12.5 can be replaced by the peak ground acceleration

$$a_g^{U} = \frac{40\pi^2 r_h}{T_1^2 g\, r_v} \cdot u_2^{U} \cdot a_g. \tag{12.6}$$

Similar transformations can be used for the case when the approximate IDA curves need to be predicted directly for the top displacement of the building, and not for displacement of the equivalent SDOF model. Since, in this case, the characteristic displacement $u_2^{U,mdof}$ corresponds to the structural model (multi-degree-of-freedom model), the additional parameter, the so-called transformation or modification factor, which relates spectral displacement of an equivalent SDOF system to the top displacement of the MDOF system, is required. The top displacement of the building $u_2^{U,mdof}$ is then computed as the product of u_2^{U} and the transformation factor, which is usually denoted by Γ or C_0, respectively, in the case of the use of the N2 method (Fajfar 2000) or FEMA 356 (2000).

Note that in this case, if the approximate IDA curves are predicted directly for the MDOF model, the dimensionless input parameters are the same as in the case of the SDOF model. The period T_1 corresponds to the SDOF model, but if the quadrilateral force-displacement is used for idealization of the pushover curve, the first mode period of the MDOF model can also be used for T_1, since the initial stiffness of the equivalent SDOF model and that of the MDOF model are practically the same.

12.3.3 *Implementation of the Methodology Through the Web Application*

There are several reasons in favour of the implementation of the presented methodology through a web application. Firstly, the database of the IDA curves can contain huge amount of data, and as such is not appropriate for distribution. Although anyone can create his own database, it is unlikely that many researchers or users will do so, since the determination of the database is computationally and time-demanding and also requires specific knowledge, which is not usually within the domain of potential users. It is therefore most appropriate that the database is stored and maintained in one place, and can be easily accessible through the internet.

The infrastructure of the web application was established by following the characteristics and benefits of cloud computing for the business environment, in order to increase the IT capacity in real time, and "on the fly" with no significant investment in new infrastructure, extensive training of personnel or licensing of the software. Additionally, the concepts introduced through the Web 2.0 revolution were adopted (simplicity, adaptability, remixability, openness, etc.). For the interested reader more on the principles of the newly emerging technologies, which will shape engineering in the future, can be found in Klinc et al. (2009).

The web application was constructed following the traditional three-tier client-server architecture, which enforces a general separation into three parts: the client tier (also named the presentation layer or, more specifically, the user interface), the middle tier (business logic), and the data storage tier. Use of three-tier architecture has some advantages. Firstly, separation of the tiers makes it easier to modify or replace any tier (or part of a tier) without affecting the other tiers. Next, separating the application and the database functionality means better load balancing. For example, the database can reside on a completely different server. Last but not least, adequate security policies can be enforced within the server tiers without significant effects on clients.

The top-most level of the application is the user interface, or the presentation tier of the system. Its main function is to translate tasks and results to something which the user can understand. In the web application the presentation is disseminated through a web browser which handles web pages encoded in (X)HTML language and is generated by a web server on the layer of business logic. Calls between the user interface and the web server are both synchronous and asynchronous. For asynchronous calls, the Ajax web programming approach has been applied.

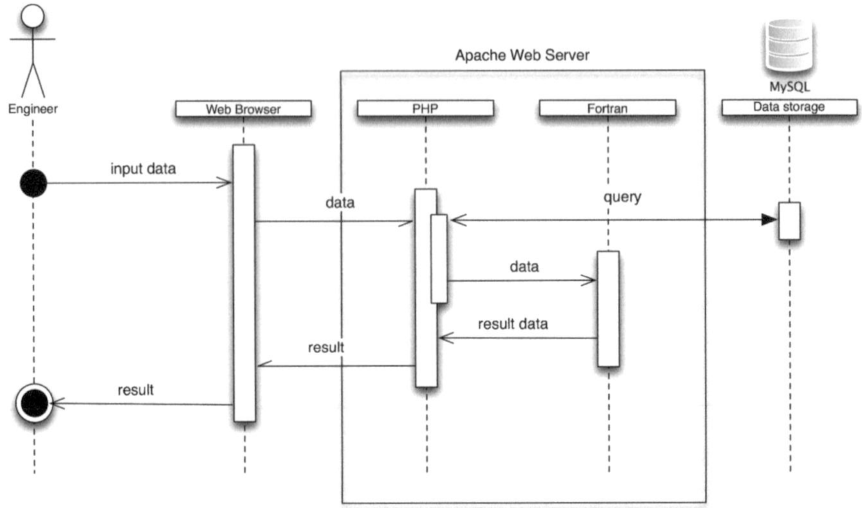

Fig. 12.2 The sequence diagram of the web application

The business logic layer is based on an Apache web server running on a Linux platform. Requests are handled by scripts written by using the PHP programming language, which process the input parameters, interact with the relational database, parse the results and prepare the output (X)HTML pages. For mathematically advanced processing, Fortran script is used.

The data storage layer consists of a relational database, where the IDA curves are stored. On the basis of the excellent connectivity with the PHP programming language and the Apache web server, it was decided to use the MySQL relational database. Another reason for this decision is the fact that this is open-source software, which is freely available. In the test environment, the whole system resides on one server. However, the architecture is scaleable so that the business logic and the data layer can be distributed to different physical servers if such requirements emerge. In the first iteration, the database of the analysis results had approximately five million records spread over two relational tables, and occupied almost 400 MB of space. However, after normalization and optimization the database now has approximately 450,000 records in two relational tables and occupies roughly 170 MB of space, which is enough to store about a quarter of a million IDA curves. As a consequence, the calculation and response time dropped drastically from the initial 30 s to less than 3 s (for combined input and output processing).

Details of the activities and the information passing between and within all the tiers are presented in Fig. 12.2. The events are triggered in the following (simplified) sequence:

1. The engineer navigates to the web site of the developed application and inserts the input parameters (e.g. period, available ductility, damping, etc.) of his structure into the HTML form.

2. The parameters are passed to the PHP script, which builds an appropriate MySQL query and queries the MySQL database. The results of these queries are IDA curves from the database for all the ground motion records and for a combination of the input parameters, which are in the vicinity of the input parameters defined in step 1.
3. After receiving the resulting data, the data is processed and the requested parameters are saved to the file.
4. The Fortran program is called and started, using the saved file as an input. The result is an output file. In this process approximate IDA curves are computed by using n-dimensional linear interpolation.
5. PHP script uses the result data passed through the output file, and processes it.
6. The result is sent to the browser and stored to the result file, which is also available. This file contains approximate IDA curves for all ground motion records, and the 16th, 50th and 84th fractile IDA curves.
7. In parallel, PHP script is used to prepare graphs of the results and passes them to the browser, so that the user can immediately see graphic presentations of the approximate IDA curves.

12.4 Example: Seismic Performance Assessment of an Eight-Storey Reinforced Concrete Frame by Using Different Methods

The use of the PBEE toolbox and the web-application are demonstrated by means of a seismic performance assessment of an eight-storey reinforced concrete frame, which was performed by employing four methods, i.e. the N2 method, web-based approximate incremental dynamic analysis, progressive incremental dynamic analysis, and incremental dynamic analysis. Structural performance was assessed for two limit states, which are defined on the basis of the damage observed in the beams and columns. The top displacement and the peak ground acceleration are adopted as the engineering demand parameter and the intensity measure, respectively.

12.4.1 Description of the Structure, the Structural Model and the Seismic Loading

The structure under consideration is an eight-story reinforced concrete frame building, which was initially designed according to the European standard: Eurocode 8 (Causevic and Mitrovic 2011). After additional design checks, the strength class of the concrete and the amount of reinforcement was slightly changed (Kosič 2010) in order to be in full compliance with Eurocode 8, and because the structure was designed for a peak ground acceleration 0.2 g, soil class B and a behaviour factor of 3.9.

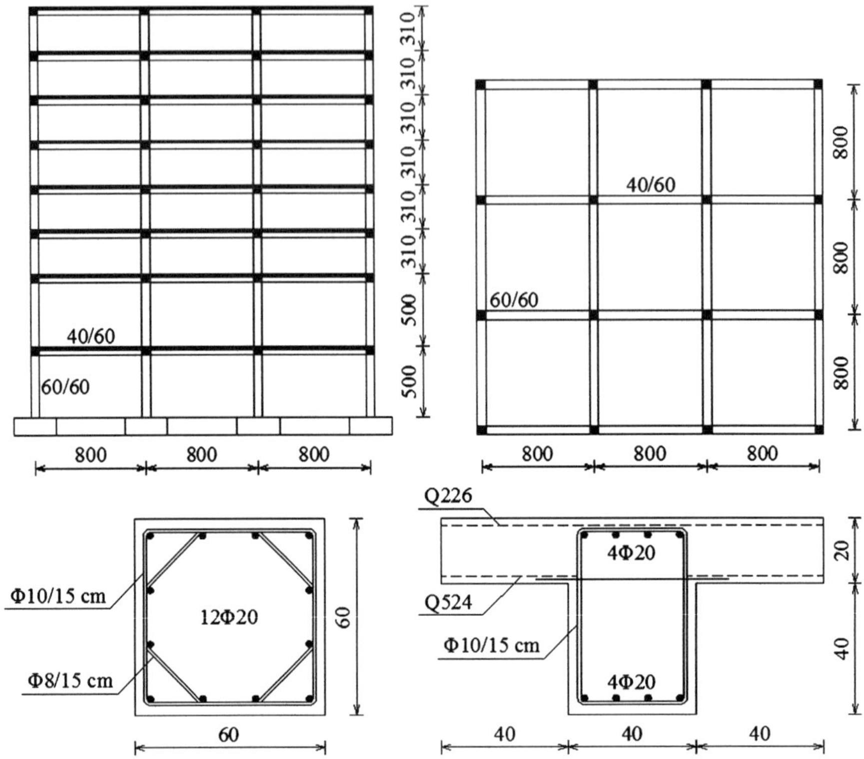

Fig. 12.3 Elevation and plan views of the eight-storey RC frame and the reinforcement in typical cross-sections of the columns and beams

Elevation and plan views of the structure are shown, together with the reinforcement in the columns and beams, in Fig. 12.3, where it can be observed that the height of the first (lowest) and second storeys is 5 m, whereas the other storeys are 3.1 m high. The building has three bays in each horizontal direction, with a distance of 8 m between the centrelines of the columns. All the sections of the columns and beams in the structure have dimensions 60/60 cm and 40/60 cm, respectively. For the columns, the steel reinforcement is the same for all sections, except for the sections at the base, where the density of the stirrups is greater (Φ8/5 cm and Φ10/5 cm). The steel reinforcement for the beams is the same for all cross sections, except for the beams in first two storeys, where at the top of the beams there are 6 instead of 4 Φ20 bars. The concrete cover of the longitudinal reinforcement is 5 cm. The slabs are 20 cm thick, with steel mesh reinforcement Q226 on top and Q524 underneath. The concrete strength of the building is C30/37, and the steel strength is B500.

A structural model of the building was prepared in the PBEE toolbox. Since the height of the stories was determined by the distance between the centrelines of the

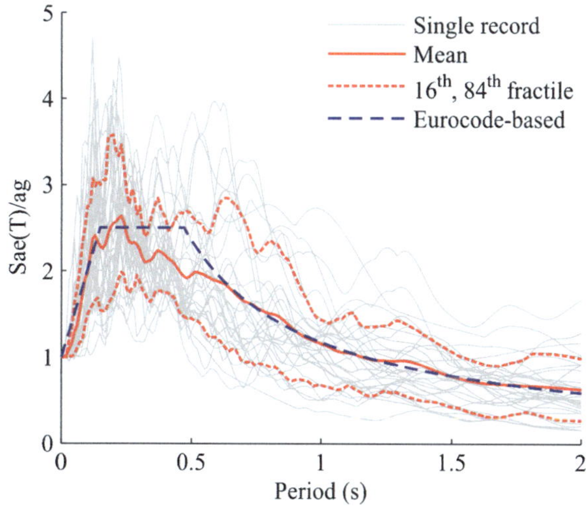

Fig. 12.4 Acceleration spectra for all 30 ground motion records, mean spectrum, 16th, 84th fractile spectrum and the Eurocode-based spectrum

beams, the height of the first storey amounted to 4.7 m. The masses were concentrated at the storey levels, at the centre of gravity. For the vertical loading a uniformly distributed self-weight ($\gamma_c = 25$ kN/m^3) of the structure and a live load ($q = 0.6$ 2kN/m^2) were assumed.

The effective width of the beams was modelled as described in Eurocode 2 (CEN 2004b), assuming zero moment points at the midpoint of the beams. The effective width of beams in the exterior and interior frames was therefore 1.2 m and 2 m, respectively. Beam and column flexural behaviour was modelled by one-component lumped plasticity elements, consisting of an elastic beam and two inelastic rotational hinges. The moment-rotation relationship before strength deterioration was modelled by a bi-linear relationship, whereas the post-capping stiffness was assumed to be linear, with a descending branch. The parameter β_u, which controls the unloading stiffness in the plastic hinges, was assumed to have a value of 0.8.

The seismic loading was defined in two ways, depending on whether the building's performance was assessed by using the N2 method or by nonlinear dynamic analysis. In the later case, a set of 30 ground motion records (Vamvatsikos and Cornell 2006) was used. The peak ground acceleration of the 30 ground motion records varied between 0.05 and 0.52 g. The records were selected within events having a magnitude of between 6.5 and 6.9. All the ground motion records were recorded on firm soil, with a distance range from the epicentre of 12–55 km. Note that the same ground motion records were used to establish the response database for the web-based application for determination of the approximate IDA curves. The acceleration spectra for each of the 30 ground motion records, and the corresponding mean, 16th and 84th fractile spectra are presented in Fig. 12.4.

A Eurocode-based acceleration spectrum was defined for estimating the seismic response parameters when using the N2 method. The estimated parameters of the acceleration spectrum are $S = 1$, $T_B = 0.15$ s, $T_C = 0.47$ s and $T_D = 2$ s.

These parameters were selected in such a way that the mean spectrum of 30 ground motion records and of the Eurocode-based spectrum are compatible in the range near the first period of the structure ($T_1 = 1.76$ s), as can be seen in Fig. 12.4.

12.4.2 *Seismic Performance Assessment Using Different Methods*

The seismic response parameters, the top displacement and the corresponding peak ground acceleration, were estimated for the significant damage (SD) and near collapse (NC) limit states (LS) at the structural level. It is considered that the SD limit state is violated when the rotation in the plastic hinge of the first column or in all the beams in one of the storeys exceeds the rotation which corresponds to the maximum moment in the columns or beams, respectively. Similarly, it is considered that the NC limit state is violated when the rotation in the plastic hinge of first column or all the beams in one of the storeys exceeds the ultimate rotation in the columns or beams, respectively. In dynamic analyses there are two additional criteria, which control the near collapse limit state at the structural level. The near collapse limit state is also violated if the average residual top displacement in last 5 seconds of the analysis exceeds the height of a building by more than 1%, or if global dynamic instability is reached before the near collapse limit state, which is defined based on the ultimate rotations in the columns and beams.

Note that the ultimate rotation Θ_u in the columns corresponds to 80% of the maximum moment measured in the post-capping range of the moment-rotation relationship. It was estimated by means of the Conditional Average Estimate (CAE) method (Peruš et al. 2006). For the beams, the EC8-3 (CEN 2005) formulas were used to compute the ultimate rotations in the plastic hinges, where the parameter γ_{el} is assumed to be equal to 1.5. Note, that computation of the described limit states at the level of the plastic hinge is embedded in the PBEE toolbox, which considers also the P-Δ effects.

12.4.2.1 Pushover Analysis

Nonlinear static (pushover) analysis was performed by assuming a modal distribution of the horizontal forces. Since the structure is symmetric, pushover analysis was performed in one direction only. The pushover curve, together with the points which indicate the defined limit states, is presented in Fig. 12.5. As described in the previous section, in the case of the pushover analysis each limit state is defined by two conditions, which are indicated on the pushover curve. The square indicates the case if the limit state corresponds to the damage in the columns, while the cross indicates the case if the limit state is violated on the basis of damage in the beams. Consequently, the SD limit state at the structural level is controlled by

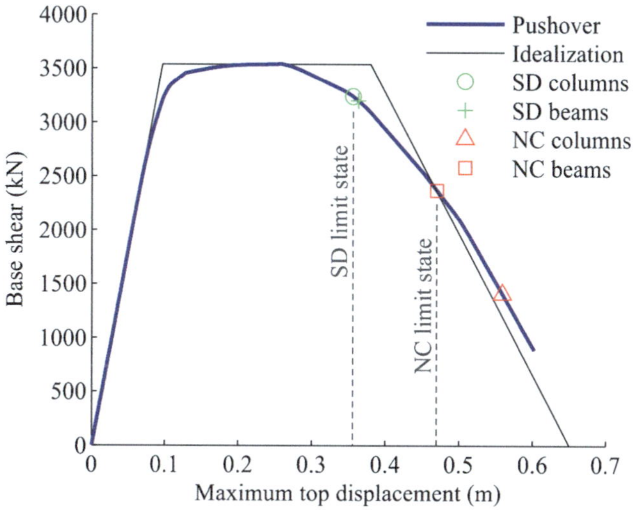

Fig. 12.5 A pushover curve, showing the displacements which correspond to the defined limit states, and the idealized force-displacement relationship

the damage in the columns, whereas the NC limit state is controlled by the damage in the beams (Fig. 12.5).

The damage at the plastic hinges of the beams and columns for the points indicated on the pushover curve are presented in Fig. 12.6. Note that, green, yellow and red colours represent, respectively, yielding of reinforcement, the state of exceeding the maximum moment, and the state of exceeding the ultimate rotation, whereas the grey part of the structure does not suffer any damage. Figures 12.6a, b correspond to the SD limit state, respectively, to the case when the maximum moment is exceeded in the first column, and when the maximum moment is exceeded in all the beams in the first storey, which are oriented in the direction of loading. In the other two figures (Figs. 12.6c, d), the damage in the elements is presented for the NC limit state. More specifically, Fig. 12.6c presents the case when the ultimate rotation is exceeded in all the beams in the first storey, whereas in Fig. 12.6d the ultimate rotation is exceeded in the columns for the first time.

The strength of the structure is, in the case of the NC limit state, reduced by 33% with respect to the maximum strength (Fig. 12.6c). The majority of the damage is concentrated in the first and second storeys. The upper stories remain almost undamaged.

12.4.2.2 The Incremental N2 Method

The first method used to determine seismic demand (maximum top displacement) for different levels of seismic intensity (peak ground acceleration) was the N2 method (Fajfar 2000), which results in an IN2 curve (Dolšek and Fajfar 2007).

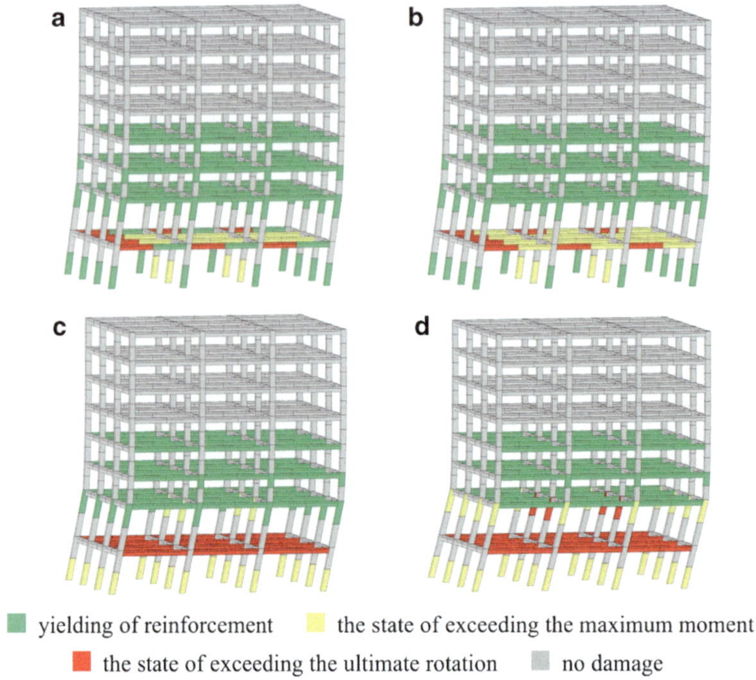

Fig. 12.6 The distribution of damage at different limit states as indicated on the pushover curve in Fig. 12.5 for the case (**a**) "SD columns", (**b**) "SD beams", (**c**) "NC beams" and (**d**) "NC columns"

This method is well-known, and involves the results of pushover analysis and the inelastic response spectrum, which can be easily computed on the basis of the defined Eurocode-based acceleration spectrum shown in Fig. 12.4.

In order to determine the IN2 curve, the pushover curve was idealized as presented in Fig. 12.5. The maximum strength of the idealized force-displacement relationship was set equal to the maximum strength determined by the pushover analysis. Note, that in the case that the structure's performance is assessed by the N2 method, only the ideal elasto-plastic force-displacement relationship of the idealized force-displacement relationship is needed, whereas in the case of the web-based approximate IDA (WIDA) curves and progressive incremental dynamic analysis estimation of the degrading part of the idealized force-displacement relationship is also required. In order to be consistent, the idealized force-displacement relationship is equal for all the analyses, whereas the mass of the equivalent SDOF model m^* and the transformation factor Γ, amounted to 2,860 t and 1.21, respectively.

Since the period of the equivalent SDOF model $T^* = 1.76$ s is longer than the period $T_C = 0.47$ s corresponding to the acceleration spectrum (Fig. 12.4), it can be assumed, according to the N2 method, that the elastic and inelastic displacements are equal, and consequently the IN2 curve is a straight line up to the top displacement, which corresponds to the NC limit state. Definition of the ultimate point on

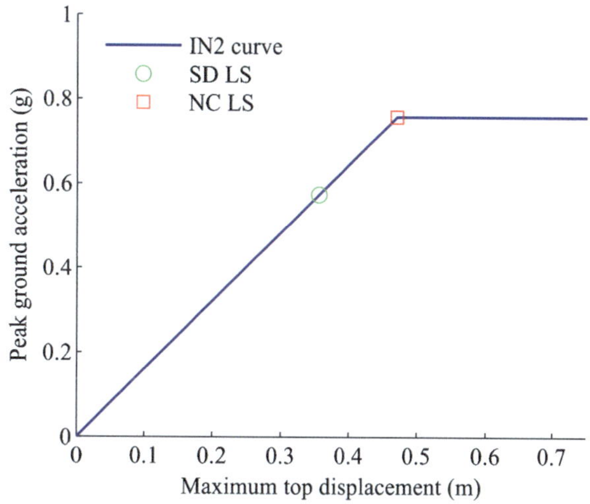

Fig. 12.7 IN2 curve with marked significant damage and near collapse limit states

the IN2 curve is required, since in the case of the N2 method the inelastic spectra are defined on the assumption of unlimited ductility of the equivalent SDOF model. The IN2 curve is presented in Fig. 12.7. It can be seen that the SD and NC limit states are violated if the peak ground acceleration exceeds 0.57 g and 0.76 g, respectively.

12.4.2.3 Web-Based Approximate Incremental Dynamic Analysis (WIDA)

The approximate IDA curves can be determined based on the results of pushover analysis in conjunction with the developed web application (http://ice4risk.slo-projekt.info/analysis). The engineering demand parameter of the approximate IDA curves can be the top displacement of the structure or the top displacement of the equivalent SDOF model, whereas the intensity measure can be the peak ground acceleration and the spectral acceleration corresponding to the first mode period.

The input parameters describing the tri-linear (bi-linear with softening) force-displacement relationship of the pushover curve are the available ductility $\mu_a = 3.92$, and the post-capping stiffness parameter $\alpha = 0.36$, which are dimensionless parameters, and the displacement at yielding u_y for the structural model (0.097 m). The other parameters which are needed to derive the approximate fractile IDA curves are the transformation factor $\Gamma = 1.21$, the period $T_1 = T^*$ $= 1.76$ s and the critical damping ratio, which, in the case of presented example, is assumed to be 0.05.

The approximate IDA curves and the corresponding 16th, 50th and 84th fractile IDA curves are presented in Fig. 12.8 together with highlighted points which indicate the SD and NC limit states. Note, that the limit-state top displacement of the approximate IDA curves (i.e. the WIDA curves), corresponds to the limit-state top displacement presented on the pushover curve (Fig. 12.5).

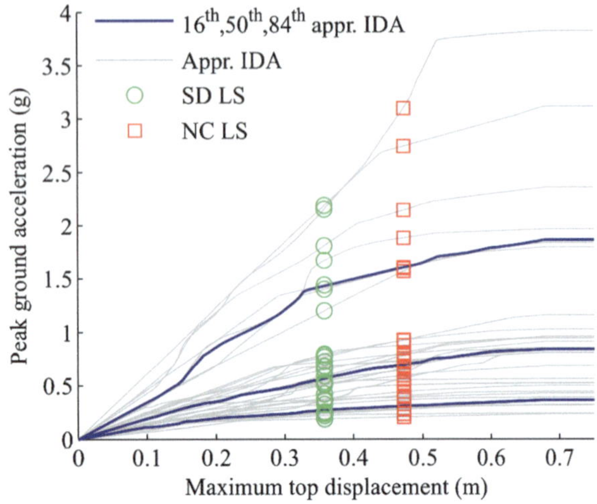

Fig. 12.8 Approximate IDA (i.e. WIDA) curves and corresponding 16th, 50th and 84th fractile IDA curves determined by using the web application, with indication of the significant damage (*SD*) and near collapse (*NC*) limit states

12.4.2.4 Progressive Incremental Dynamic Analysis

Progressive IDA involves a precedence list of ground motion records, which was, for the presented example, determined from the IDA curves for the equivalent SDOF model and the simple optimization procedure (Azarbakht and Dolšek 2011). The same equivalent SDOF model as in the case of the N2 method was used. By using the PBEE toolbox the precedence list was determined and then IDA was performed, firstly, only for the first six records (two subsets ($s = 2$) of three records) from the precedence list. Since the tolerance in this case exceeded the acceptable tolerance, which is 8% in this case, IDA had to be performed for additional subsets of records from the precedence list of records. Finally, the progressive IDA was terminated, since the tolerance became acceptable if IDA was performed for the first four subsets ($s = 4$) of records (12 records). In this case ($s = 4$) the difference between the 16th, 50th and 84th fractile IDA curves with respect to the fractile IDA curves from the previous case ($s = 3$) was minor, and amounted to less than 8%. The IDA curves for the first 12 ground motion records from the precedence list of records and the corresponding 16th, 50th and 84th fractile IDA curves, with indication of the significant damage (SD) and near collapse (NC) limit states, are presented in Fig. 12.9. Note that the top displacements corresponding to the SD or NC limit state, and for a given set of ground motion records, are not equal to those obtained in the case of the WIDA curves, since they are determined directly from the results of dynamic analysis, and not from pushover analysis as in the case of WIDA.

12.4.2.5 Incremental Dynamic Analysis

IDA is the most accurate analysis method used in the case of the presented example. The global dynamic instability of each of the 30 IDA curves was estimated

Fig. 12.9 The IDA curves for first 12 ground motion records (s = 4) from the precedence list of records and the corresponding 16th, 50th and 84th fractile IDA curves, with indication of the significant damage (*SD*) and near collapse (*NC*) limit states

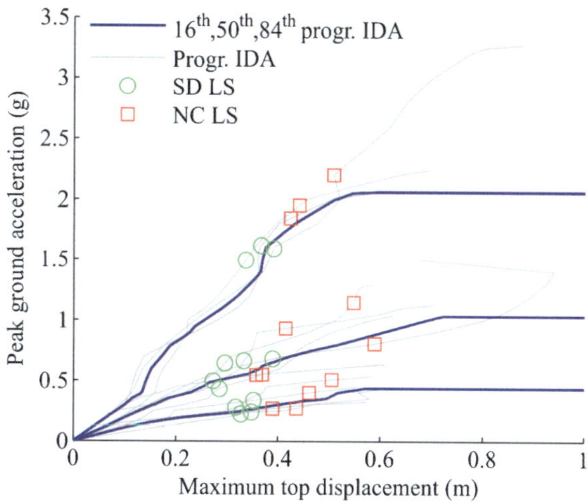

with a precision of 0.02 g. The largest interval between the peak ground acceleration, for which the seismic response parameters were computed, was defined as being equal to 0.05 g. However, if the peak ground acceleration, which corresponds to global dynamic instability, was large, the IDA curves were computed for only 20 points, since the accuracy of predicting these IDA curves is not very important. The same procedure was used in the case of progressive IDA, as well as for determination of the precedence list of ground motion records.

Each nonlinear dynamic analysis within the IDA was calculated by employing the Newmark integration scheme, assuming $\gamma_n - 0.5$, $\beta_n - 0.25$ and an integration time interval of 0.005 s.

The IDA curves for a set of 30 records and the corresponding 16th, 50th and 84th fractile IDA curves, with indication of the significant damage (SD) and near collapse (NC) limit states, are presented in Fig. 12.10.

12.4.2.6 Discussion

The fractile IDA curves, determined by three different approaches and the IN2 curve, are compared in Fig. 12.11. The difference between the IDA curves is minor, even in the case of the NC limit state, where the structure is already severely damaged. This is because the structural damage is concentrated only in the bottom two storeys, which are significantly higher if compared to the height of other stories. However, such mechanism should not occur, since the structure has been designed according to capacity design rules defined in Eurocode 8. The largest difference between the IDA curves is observed for the 16th fractile IDA curves, where the difference between the results of IDA and approximate IDA expressed in terms of peak ground acceleration corresponding to the NC limit state is less than 15%. The points in Fig. 12.11 presents the 16th, 50th and 84th fractiles of the peak ground acceleration, which cause the median top displacement corresponding to the

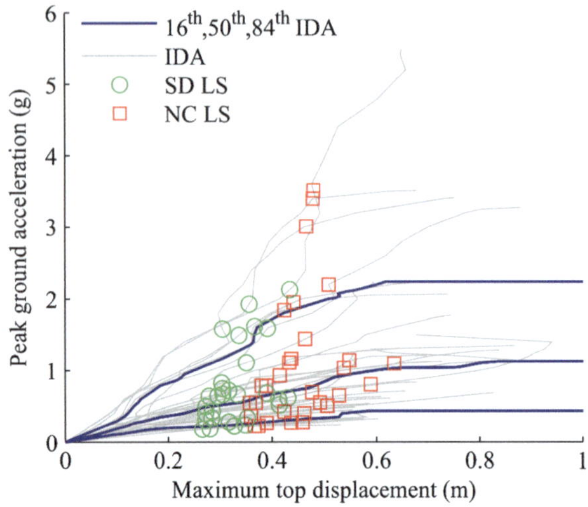

Fig. 12.10 The IDA curves for a set of 30 records and the corresponding 16th, 50th and 84th fractile IDA curves, with indication of the significant damage (*SD*) and near collapse (*NC*) limit states

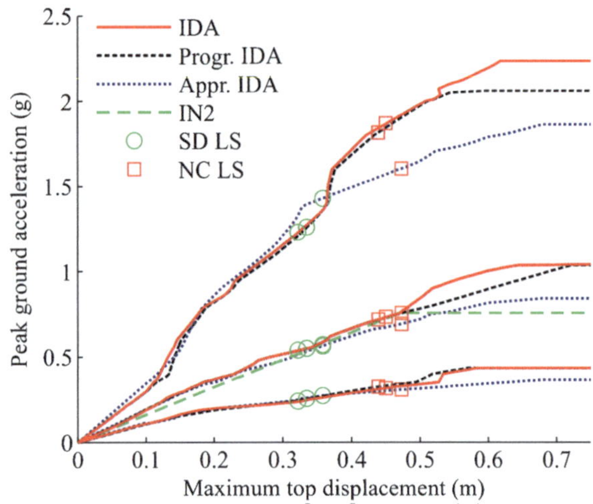

Fig. 12.11 The 16th, 50th and 84th fractile IDA curves determined by IDA, progressive IDA, web-based approximate IDA, and the IN2 curve. The highlighted points on the curves indicate the SD and NC limit states

SD or NC limit state, and are also presented in Table 12.1. It can be observed that the difference between the median estimates based on the simplified methods and that of IDA rarely exceeds 10%. A somewhat larger error is observed in the case of the 16th and 84th fractiles if these are estimated by the web-application, whereas progressive IDA practically produces same results as IDA.

However, the largest error is observed in the case of the prediction of global dynamic instability (Fig. 12.11). On the other hand, the seismic performance assessment by using simplified methods requires significantly less computational time. It is estimated that the time required for assessing the structural performance in the case of the IN2 or WIDA is only about 1% if compared to the computational time needed for IDA.

Table 12.1 The 16th, 50th and 84th fractiles of peak ground acceleration (*PGA*) causing the SD and NC limit states, and the corresponding median top displacement (*MTD*)

Method	Comp. time (%)	Fractile	SD limit state MTD (m)	SD limit state PGA (g)	NC limit state MTD (m)	NC limit state PGA (g)
IN2	1	50	0.36 (11%)	0.57 (6%)	0.47 (5%)	0.76 (3%)
Approximate IDA	1	16	0.36 (11%)	1.43 (16%)	0.47 (5%)	1.61 (14%)
		50		0.57 (5%)		0.69 (6%)
		84		0.27 (13%)		0.31 (3%)
Progressive IDA	50	16	0.33 (4%)	1.26 (2%)	0.44 (2%)	1.82 (3%)
		50		0.55 (2%)		0.72 (2%)
		84		0.26 (6%)		0.33 (3%)
IDA	100	16	0.32	1.23	0.45	1.87
		50		0.54		0.74
		84		0.24		0.32

The values in brackets represent an error with respect to results of the IDA.

The most accurate method among the employed methods presented in this chapter is incremental dynamic analysis, which is widely used for the performance assessment of structures under seismic loading. It is shown that, by using progressive incremental dynamic analysis, computational time can be reduced for about 50% compared to that needed to perform IDA. It was proved, at least for the presented example, that the seismic response parameters are sufficiently accurate when determined by progressive IDA, which requires only 12 IDA curves instead of 30, which was the number of records used in the IDA. Nevertheless there is some additional work with the preparation of the precedence list.

The web-based approximate IDA curves are easily determined based on the parameters of the idealized pushover curve, and can be obtained in a few seconds. However, the use of the web-application is currently limited, since the response database is established for only 30 records.

The IN2 method is the simplest method, which is presented in this chapter. It cannot estimate the global dynamic instability, nor consider record-to-record variability. However, all the simplified methods provided sufficiently accurate results, although they are based on response of an equivalent SDOF model.

12.5 Conclusions

A PBEE toolbox for the seismic performance assessment of RC frames, which consists of Matlab functions and works in conjunction with OpenSees, and a web-application for the prediction of approximate IDA curves, have been briefly presented. The advantage of the web-application in comparison with the traditional approaches which are used for the determination of the approximate seismic response of structures can be found in the systematic separation of the process of the parametric study and the process of the prediction of the approximate IDA curves.

Such an approach enables the creation of many response databases, which would eventually result in a global component for the prediction of approximate IDA curves at any location in the world and for any kind of structural system. On the other hand, the PBEE toolbox can serve for code calibration, as a basis for new applications, e.g. for the design of frames based on nonlinear analysis, or for parametric studies, since it enables rapid definition of simple nonlinear models of RC frames.

In the presented examples it was shown that it is possible to predict the 16th, 50th and 84th fractiles of the seismic response parameters for the SD and NC limit states with sufficient accuracy by using WIDA, which is based on pushover analysis and requires only about 1% of computational time with respect to the computational time needed for IDA. In this case the median peak ground acceleration which causes the defined limit states was estimated with an error of less than 10%, whereas the error in predicting the corresponding 16th and 84th fractiles was slightly greater. However, if such an error is too large, it is possible to use progressive IDA, which predicts the seismic response parameters with minor errors and with only about 50% of the computational time required for IDA.

Acknowledgements The results presented in this paper are based on work supported by the Slovenian Research Agency within the framework of the project High-throughput computing environment for seismic risk assessment (http://ice4risk.slo-projekt.info/) (J2-0845-0792-08). This support is gratefully acknowledged.

References

Antoniou S, Pinho R (2004) Development and verification of a displacement-based adaptive pushover procedure. J Earthquake Eng 8:643–661

Aydinoğlu MN (2003) An incremental response spectrum analysis based on inelastic spectral displacement for multi-mode seismic performance evaluation. Bull Earthquake Eng 1:3–36

Azarbakht A, Dolšek M (2011) Progressive incremental dynamic analysis for first-mode dominated structures. J Struct Eng 137:445–455

Causevic M, Mitrovic S (2011) Comparison between non-linear dynamic and static seismic analysis of structures according to European and US provisions. Bull Earthquake Eng 9:467–489

CEN (2004a) Eurocode 8: Design of structures for earthquake resistance. Part 1: General rules, seismic action and rules for buildings. EN 1998-1, European Committee for Standards, Brussels, Dec 2004

CEN (2004b) Eurocode 2: Design of concrete structures. Part 1-1: General rules and rules for buildings. EN 1992-1-1, European Committee for Standards, Brussels, Dec 2004

CEN (2005) Eurocode 8: Design of structures for earthquake resistance. Part 3: Strengthening and repair of buildings. EN 1998-3, European Committee for Standards, Brussels, Mar 2005

Chopra AK, Goel RK (2002) A modal pushover analysis for estimating seismic demands for buildings. Earthquake Eng Struct Dyn 31:561–582

CSI (2006) PERFORM-3D User guide, Version 4. Computers & Structures Inc., Berkeley

CSI (2008) SAP 2000 User manual, Version 12. Computers & Structures Inc., Berkeley

Dolsek M (2010) Development of computing environment for the seismic performance assessment of reinforced concrete frames by using simplified nonlinear models. Bull Earthquake Eng 8:1309–1329. doi:10.1007/s10518-010-9184-8

Dolšek M, Fajfar P (2007) Simplified probabilistic seismic performance assessment of plan-asymmetric buildings. Earthquake Eng Struct Dyn 36:2021–2041

Ellingwood BR, Kinali K (2009) Quantifying and communicating uncertainty in seismic risk assessment. Struct Saf 31:179–187

Elnashai AS, Papanikolaou VK, Lee DH (2008) Zeus NL – A System for Inelastic Analysis of Structures, User Manual, Version 1.8.7

Fajfar P (2000) A nonlinear analysis method for performance-based seismic design. Earthquake Spectra 16:573–592

Fajfar P, Dolšek M, Marušić D, Stratan A (2006) Pre- and post-test mathematical modelling of a plan-asymmetric reinforced concrete frame building. Earthquake Eng Struct Dyn 35:1359–1379

FEMA (2000) Prestandard and commentary for the seismic rehabilitation of buildings. FEMA publication 356, Washington DC, Nov 2000

FEMA (2005) Improvement of nonlinear static seismic analysis procedures. FEMA publication 440, Washington, DC, June 2005

Klinc R, Dolenc M, Turk Ž (2009) Engineering collaboration 2.0: requirements and expectations. ITcon 14:473–488

Kosič M (2010) Design of eight-storey reinforced concrete frame building according to European standard Eurocode 8. Personal communication, 15 Dec 2010

MathWorks (2007) MATLAB the Language of Technical Computing. http://www.mathworks.com/

McKenna F, Fenves GL (2007) Open system for earthquake engineering simulation, Pacific Earthquake Engineering Research Center, Berkeley. http://opensees.berkeley.edu

MIDAS (2008) Pushover Analysis of RC structures as per EC8:2004, MidasGen Version 7.4.1, MIDAS Information Technologies Co, Ltd, Korea

Peruš I, Poljanšek K, Fajfar P (2006) Flexural deformation capacity of rectangular RC columns determined by the CAE method. Earthquake Eng Struct Dyn 35:1453–1470

Peruš I, Klinc R, Dolenc M, Dolšek M (2011) A web-based methodology for the prediction of approximate IDA curves. Earthquake Eng Struct Dyn (in review).

Priestley MJN, Kowalsky MJ (2000) Direct displacement-based seismic design of concrete buildings. Bull New Zeal Soc Earthquake Eng 33:421–444

SEISMOSOFT (2007) Seismostruct version 4. SeismoSoft Ltd., Messina

Takeda T, Sozen MA, Nielsen NN (1970) Reinforced concrete response to simulated earthquakes. J Struct Division 96:2557–2573

Vamvatsikos D, Cornell CA (2002) Incremental dynamic analysis. Earthquake Eng Struct Dyn 31:491–514

Vamvatsikos D, Cornell CA (2006) Direct estimation of the seismic demand and capacity of oscillators with multi-linear static pushovers through IDA. Earthquake Eng Struct Dyn 35:1097–1117

Chapter 13
Recent Advances in the Seismic Analysis and Design of RC Bridges in Slovenia

Tatjana Isaković and Matej Fischinger

Abstract An overview of the research related to the seismic analysis and design of RC bridges, recently performed at UL FGG is made. Four main topics are addressed: (1) Pushover based analysis of bridges; several recommendations related to the use of different pushover methods are overviewed and criteria which define the applicability of single-mode methods are proposed; (2) Modelling of RC bridge columns; three types of frequently used macro-models are discussed on the example of typical bridge columns; (3) Estimation of the shear strength and shear strengthening of typical RC hollow box bridge columns with substandard construction details; three methods for estimation of the shear strength are compared; the method, proposed at UCSD was found the most appropriate in the investigated case; the concrete jacket and CFRP strips successfully prevented the shear failure of the strengthened column; (4) Seismic isolation of RC bridges using new semi-active device – magnetically controlled elastomer (MCE); the efficiency of the smart MCE bearings in partially isolated bridges, subjected to earthquakes weaker than the design earthquake, is discussed and demonstrated.

Keywords RC bridges • Numerical models • Nonlinear analysis • Seismic strengthening • Shear strength • Seismic isolation • Semi-active isolation

13.1 Introduction

Several topics related to the research of the seismic analysis and design of reinforced concrete (RC) bridges, performed at the University of Ljubljana, Faculty of Civil Engineering (UL FGG) are overviewed. They are: (1) Pushover based

T. Isaković (✉) • M. Fischinger
Faculty of Civil and Geodetic Engineering, University of Ljubljana,
Jamova 2, Ljubljana SI 1000, Slovenia
e-mail: tisak@ikpir.fgg.uni-lj.si; mfischin@ikpir.fgg.uni-lj.si

M. Dolšek (ed.), *Protection of Built Environment Against Earthquakes*,
DOI 10.1007/978-94-007-1448-9_13,

analysis of bridges, (2) Modelling of RC bridge columns, (3) Estimation of the shear strength and shear strengthening of typical RC hollow box bridge columns using different types of jacketing, (4) Seismic isolation of RC bridges using new semi-active device – magnetically controlled elastomer.

The inelastic response history analysis has been used for the research purposes for several decades. However, it is still too complex to be used in the design practice. To simplify the nonlinear seismic analysis, several nonlinear static, or so-called, push-over methods have been developed. They recently became quite popular analysis tool. However, due to the limited understanding of their limitations, these methods are frequently used indiscriminately. Their indiscriminate use is particularly typical for bridges. The principles, rules and procedures which were originally developed for buildings have often been simply extrapolated to bridges, neglecting the major differences between these structural systems and their seismic response. In Sect. 13.2 basic specifics in the application of the pushover methods for the analysis of bridges are briefly summarized, and criteria defining the applicability of the N2 method, which is included into Eurocode standards, are introduced.

To perform the nonlinear analysis, adequate numerical models are needed. Some experiences obtained at UL FGG in modelling RC bridge columns are summarized in Sect. 13.3.

To establish an appropriate numerical model the adequate data about the capacity of columns are needed. While the quantities defining the flexural response are usually well defined, the shear response of columns is much more difficult to predict. The knowledge related to this problem is still incomplete. This is e.g. indicated by quite large differences in the results of different methods proposed for the estimation of the shear strength and stiffness of RC columns. This problem is analyzed on the example of columns of a typical existing viaduct, which includes several construction deficiencies. Several methods for estimation of the shear strength are compared and evaluated by means of the experimental results (see Sect. 13.4).

There are numerous existing bridges, which were designed before the modern principles of the seismic engineering were established. From the nowadays point of view, they include many substandard construction details which demand adequate strengthening and retrofit. One of such examples is analyzed in Sect. 13.4. Different retrofitting techniques, including concrete and FRP jacketing are analyzed on the example of a typical viaduct, built in 1970s.

One of the possibilities for the seismic protection of new bridges as well as for the strengthening of existing structures is the seismic isolation. In the last part of chapter (Sect. 13.5) a new semi-active seismic isolation device, magnetically controlled elastomer (MCE) is briefly introduced and the possibilities for its use in bridges are overviewed.

13.2 Simplified Nonlinear Analysis of Bridges

To simplify the inelastic analysis, and to make it more convenient for everyday design, several simplified nonlinear analysis methods have been developed. They are so called pushover methods. There are several methods of this type available.

The simplest methods are so called single-mode pushover methods. One of the main assumptions of these methods is that the response of a structure is governed mostly by one predominant mode. The typical representative of this group is the N2 method (Fajfar 2000), which is included in the Eurocode standards (CEN 2004a, 2005a). The specifics in the application of this method for the analysis of bridges are described in Sect. 13.2.1.

It is typical for long bridges (with the total length of 500 m and longer) that the response can be considerably influenced by higher modes of vibration. In those cases, the single-mode methods are less accurate, and multi-mode pushover methods can be used instead. Some of the conclusions related to their applicability for the analysis of bridges are shortly overviewed in Sect. 13.2.2.

13.2.1 The N2 Method – Single-Mode Pushover Methods

The N2 method was originally developed for the analysis of buildings. Therefore it should be modified when it is used for the analysis of bridges, since their structural system is considerably different than that of buildings (particularly in their transverse direction). In this Section the appropriate modifications are proposed and summarized. Further details can be found in (Isaković et al. 2003, 2008a).

The proposed modifications of the N2 method for the analysis of bridges are:

1. The distribution of lateral forces along the superstructure;
2. The choice of the point where the displacements are monitored to obtain the force-displacement relationship;
3. Idealization of the force-displacement curve, and calculation of yielding force and yielding displacement.

1. The distribution of the "inertial" forces (lateral load) should be assumed before the nonlinear static analysis is performed. Some of the distributions appropriate for bridges are summarized in Fig. 13.1. Note that two extreme cases of the constraints above the abutments are addressed. In the Annex H of standard Eurocode 8/2 (EC8/2) two possible distributions are proposed: (a) distribution proportional to the 1st mode of the bridge in the elastic range, (b) uniform distribution (see Fig. 13.1(a)1,2). The first distribution can be defined based on a modal analysis with some of the standard programs for elastic modal analysis.

   In the previous research (Isaković and Fischinger 2006; Isaković et al. 2008a) it was found that the parabolic distribution (Fig. 13.1(a)3) is appropriate for bridges that are pinned at the abutments. This distribution is simpler to define than that proportional to the first mode. In many cases, the results of the N2 method and the inelastic response history analysis correspond better when the uniform distribution is replaced by the parabolic one.

   For bridges with roller supports at the abutments, the uniform distribution as well as that proportional to the most important mode, corresponding to certain seismic intensity (see Fig. 13.1(b)1) can be used. For bridges with short stiff central columns, the second solution demands iterations, since the most

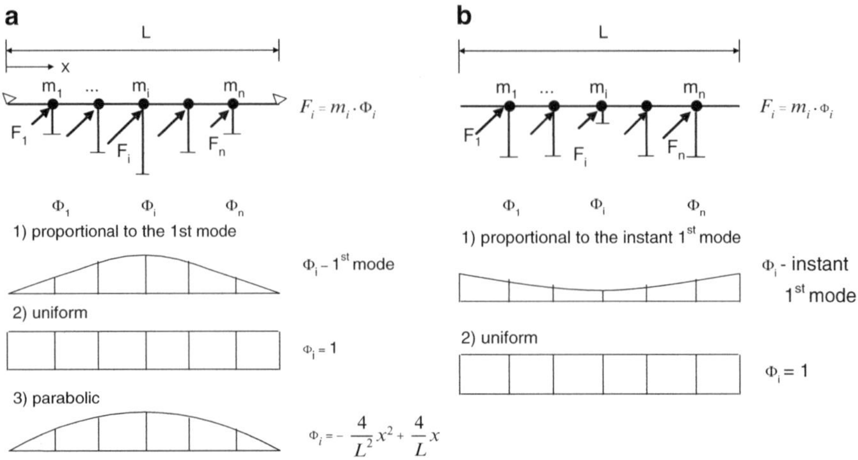

Fig. 13.1 Distributions of the lateral load, appropriate for bridges that are: (**a**) pinned at the abutments, (**b**) with roller supports at the abutments

important mode can change with the intensity of the load. In general it is recommended to use two different distributions of inertial forces and to take into account the envelope of the related response.

The distribution of the lateral load does not influence only the shape of the displacements of the superstructure, but also the value of the maximum displacement. This is illustrated on the example of the bridge, shown in Fig. 13.2. The displacements were determined using three different distributions of the "inertial" forces, shown in Fig. 13.1(a). Results of using the N2 method (dashed line) are compared with the results of the nonlinear time history analysis (solid line). Two seismic intensities were taken into account.

In the central part of the bridge, the largest displacements were obtained when the distribution proportional to the 1st mode is considered. The displacements in the regions close to the abutments were the largest in the case of the uniform load distribution. The parabolic distribution resulted in the deflection line in between.

2. One of the crucial steps in the application of the N2 method is the static nonlinear analysis of the multi-degree-of-freedom (MDOF) system. Based on this analysis the force-displacement relationship is determined, which is further used to define the properties of the equivalent single-degree-of-freedom (SDOF) system.

   In buildings the force-displacement relationship is usually defined monitoring the displacements at the top of the building. In bridges the choice of many researchers is to monitor the displacements at the top of a chosen column. However, in the irregular viaducts it is not clear, which is the appropriate column. The authors propose that the structure should be viewed as a flexibly supported beam and that the maximum displacement of that beam should be monitored.

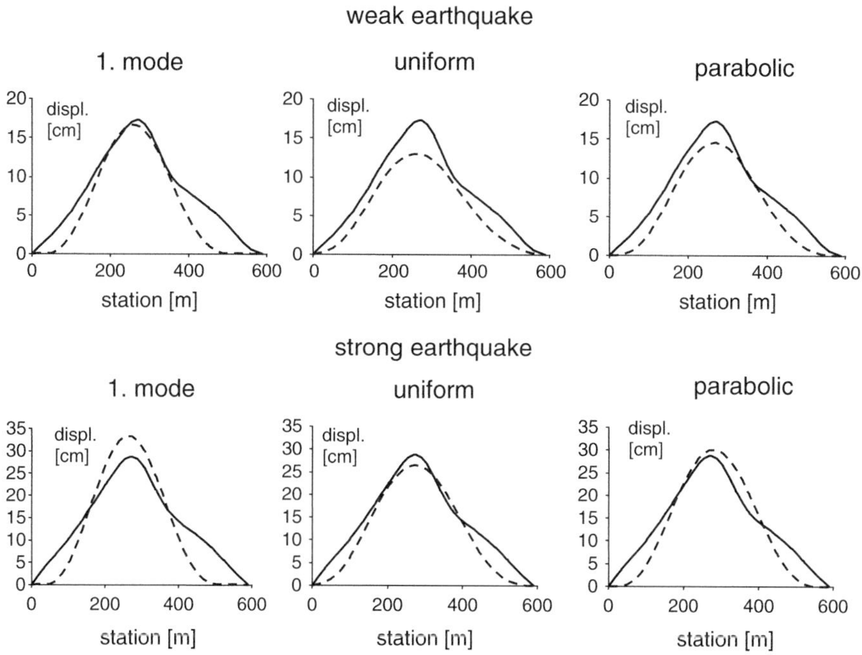

Fig. 13.2 Results of the N2 method (*dashed line*) compared with the results of the NRHA (*solid line*); three distributions of the lateral forces and two seismic intensities (peak ground acceleration of 0.25 and 0.5 g) were taken into account

In bridges supported by very short and stiff columns close to the centre of the superstructure, the monitoring point defined in this way can differ considerably from what is proposed in the e.g. EC8/2. More details on this can be found in Isaković and Fischinger (2006) and in Isaković et al. (2008a).

3. Idealization of the base shear-displacement relationship is one of the basic steps of the N2 method, since it influences the stiffness of the equivalent SDOF model and the value of the maximum displacement. When this stiffness is not adequately estimated, the actual and estimated maximum displacement can be very different (Isaković and Fischinger 2006; Isaković et al. 2008a).

   Elasto-plastic idealisation is typically used. However, this is not appropriate for all types of bridges. Bridges pinned at the abutments, act as a linear beam after all the columns yielded (supposing that the superstructure was designed according to capacity design procedure). Consequently the pushover curve exhibits considerable hardening slope. In such cases bi-linear idealization of the pushover curve is more appropriate.

   The force-displacement relationship is usually simplified using the equal energy principle for idealized and actual curves. Since the energy depends on the reached maximum displacement, which is not known at the moment of the idealization, the authors' opinion is that iterations are necessary. In the majority of cases only one iteration is needed.

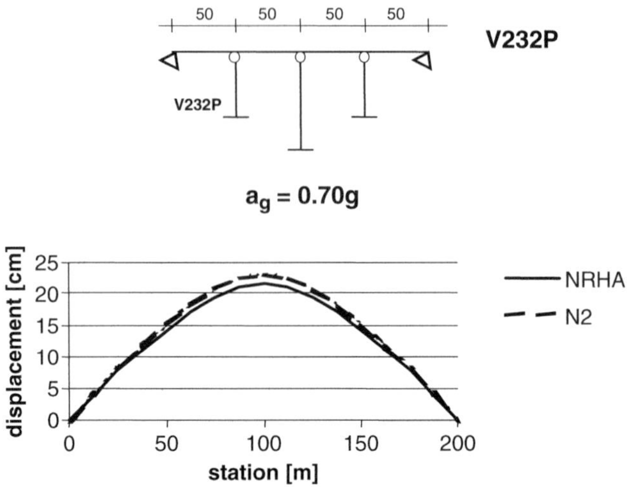

Fig. 13.3 In bridges, where the response is influenced by one predominant mode, the response is estimated well by the N2 method

In the annex H of the EC8/2 it is proposed that the maximum displacement be estimated using the results of the elastic analysis. This solution is very convenient at the first glance. However, to estimate the displacement in the nonlinear range properly, the reduced column stiffness, corresponding to the certain level of the seismic load should be assumed. Often, this procedure also demands iterations, since it is quite difficult to estimate the effective stiffness of columns adequately, particularly in bridges that are supported by columns of very different heights (stiffness). Consequently, the calculation can be more time consuming than that proposed by the authors of this chapter.

The N2 method can be used successfully for analysis of the majority of bridges. An example of a good estimation of the bridge seismic response is illustrated in Fig. 13.3, where the displacements calculated by the N2 method and nonlinear response history analysis (NRHA) are compared. In the presented case these two methods agreed well because the response was influenced by one predominant mode, which did not change considerably with seismic intensity.

The N2 method is, in general, more accurate in the case of short bridges. In the previous research (Isaković and Fischinger 2006) it was found that for long bridges (typically longer than 500 m), because of the flexibility of the superstructure (due to its considerable length) the response is often influenced by higher modes, even if a bridge is supported by relatively flexible columns. The N2 method is less accurate in these cases. Multi-mode pushover methods can be used, or such bridges can be analyzed by the nonlinear response-history methods. The applicability of the N2 method in some cases depends on the seismic intensity. In some cases (e.g. that presented in Fig. 13.4) it increases with the seismic intensity, however the opposite cases (Isaković and Fischinger 2011) were also identified.

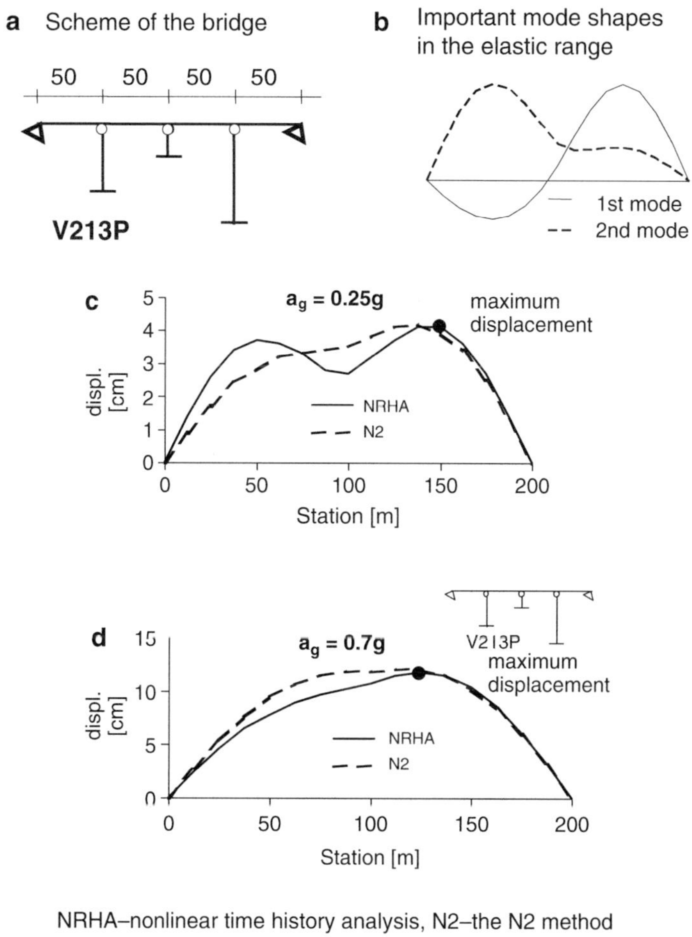

Fig. 13.4 The accuracy of the N2 method in some bridges depends on the seismic intensity

In the previous research (Isaković and Fischinger 2006, 2011) it was found that the N2 method estimates the response well if the response is predominantly influenced by one mode, which has the effective mass at least 80% of the total mass of the structure. To measure the changes of the mode shape a regularity index is proposed (Isaković et al. 2003). The N2 method is accurate enough when this index has value less than 5%. Further details about the applicability of the N2 method can be found in (Isaković et al. 2003; Isaković and Fischinger 2011).

13.2.2 Multi-Mode Pushover Methods

For the analysis of irregular bridges where the response importantly depends on the higher modes of vibration, multi-mode pushover methods are needed. Two typical

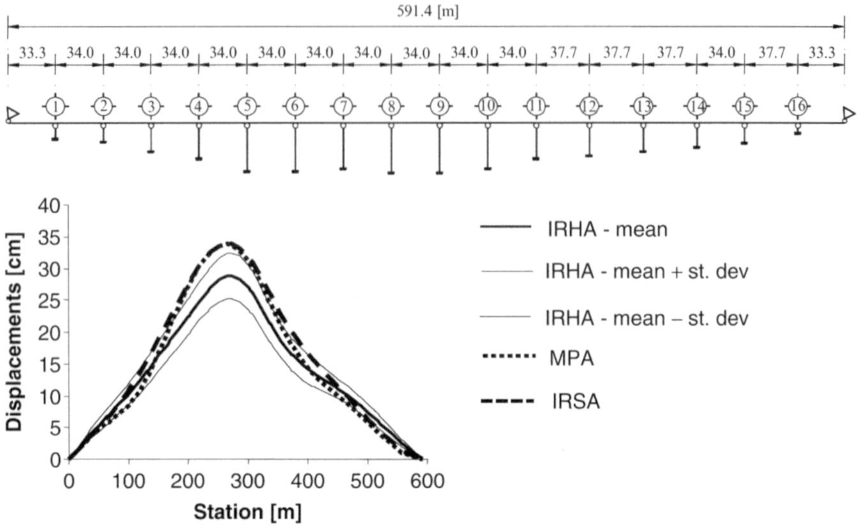

Fig. 13.5 The response of typical long viaducts obtained by MPA, IRSA and IRHA (inelastic response history analysis)

examples of multi-mode pushover methods are the MPA – Modal Pushover Analysis (original version proposed by Chopra and Goel (2002) and modified version proposed by Kappos et al. (Paraskeva et al. 2006)) and the IRSA – Incremental Response Spectrum Analysis (Aydinoğlu 2003) method. Both methods are adequate for the analysis of typical long bridges (see Fig. 13.5).

Both methods, however, have certain limitations and are not universal. Since it is non-adaptive, the MPA method is less accurate for seismic performance of bridges, where the important vibration modes considerably change depending on the intensity of the load. More details about its applicability can be found in (Isaković and Fischinger 2006, 2011). Nevertheless the IRSA method is adaptive (can takes into account changes of the vibration mode shapes) it should be also used with care in highly irregular bridges, particularly those which are torsionally sensitive. Such bridges are typically supported by short, very stiff central columns. For more details see (Isaković and Fischinger 2006, 2011)).

13.2.3 Concluding Remarks

Different types of pushover methods can be used to simplify the nonlinear analysis of bridges in the everyday design practice. Single-mode methods are typically included into the modern standards. One of the representatives of such methods is N2 method.

When the N2 method is used for the analysis of bridges it should be modified. Required modifications are related to the choice of the distribution of the lateral load, the choice of the monitoring point and idealization of the pushover curve.

The N2 method (single-mode methods) can be used for the analysis of bridges where the response is influenced by one mode, which does not considerably change with the seismic intensity. Such bridges are those which have the fundamental mode with the effective mass of at least 80% of the total mass and the value of index of regularity (measure of the changes of the mode shape) of 5% or less. In general the N2 method is accurate enough for the analysis of regular short and medium span bridges.

For longer bridges (with total length in excess of 500 m) multi-mode pushover methods or NRHA should be employed. Typical multi-mode methods are the MPA and the IRSA method. They also have limitations and they should be used with care, particularly in bridges which are torsionally sensitive.

13.3 Numerical Models of RC Bridge Columns

The modern philosophy of seismic design of bridges (which is the basis for most of the modern design codes) includes a consideration that damage of the bridge should be limited to the flexural damage of columns, only. Therefore, the following discussion is focused on the inelastic flexural models of bridge columns. There are several elements, which are suitable for modelling the non-linear flexural behaviour of bridge columns. In general, these elements could be classified as macro- or micro- elements.

Generally, macro-elements are different types of beam-column elements, where the non-linear behaviour is modelled using different hysteretic rules (force-displacement or moment-rotation relationships), and which attempt to capture overall member behaviour. The basis of development of "hysteretic" macro-models is the experiment. The parameters of hysteresis have clearly defined physical meaning, and this makes macro-elements relatively easy to control. Since the hysteretic rules tend to represent the overall member behaviour, macro-models usually include fewer elements than micro-models. This makes macro-models simple and more appropriate for complex dynamic non-linear analyses.

The second group of elements, micro-elements are, in general, plain (2D) or solid (3D) finite elements. Typically, the non-linear behaviour is modelled on the level of stress-strain relationships (using constitutive laws). Compared to the macro-models, more calculations (integrations) are needed, and this makes the complex dynamic response-history analysis more complicated. Compared with the macro-models, the micro-models make the control of the results and their analysis more complex and time-consuming. Some macro-elements, like fibre elements combine the properties of previously described types of elements.

It is the authors' opinion that it is more convenient to use macro-models when the global response of a bridge is analysed. The micro-models are more appropriate when the local responses of some components (e.g. links) are studied. According to this view, the following discussion deals only with macro-elements.

Three types of these elements: (a) beam-column element with lumped plasticity, (b) fibre element and (c) Multiple-vertical-line (MVL) element, are compared

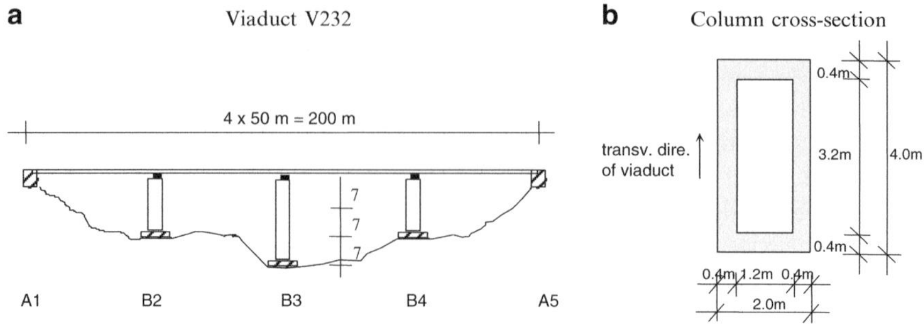

Fig. 13.6 Investigated bridge

using an example of four-span viaduct (see Fig. 13.6). This viaduct was originally investigated experimentally and analytically by Italian researchers, Pinto and Negro (1995). The inelastic dynamic response has been compared with experimental results.

13.3.1 Beam-Column Element with Lumped Plasticity

The beam-column with lumped plasticity, where the response is defined by tri-linear Takeda hysteretic rules, was used to model the columns of the investigated bridge. This element was incorporated into DRAIN-2D program (Kanaan and Powell 1973) and OpenSees (McKenna and Fenves 2007) at UL.

In the initial modelling of the bridge no "tuning" of the element parameters was done. All properties (including hardening parameter) were calculated from the first principles. The common average value of the unloading parameter in the Takeda rules ($\alpha = 0.5$), which determines the rate of the unloading stiffness deterioration, was used.

The correlation between the analytically and experimentally determined displacement response histories, obtained by the initial model, was good and so was the modelling of the predominantly flexural hysteretic behaviour of the tall central column (Fig. 13.7). However, in the case of the design earthquake, the initial model underestimated the actual stiffness degradation on the unloading branch for the short side column (Fig. 13.7a). To account for higher stiffness degradation, unloading parameter $\alpha = 1.0$ should have been used in the modified model. This change improved the calculated response in the case of the design earthquake (Fig. 13.7b), but not in the case of the high-level earthquake (Fig. 13.8), indicating that different α values should be used for different levels of response.

13.3.2 Fibre Element

When modelling columns with fibre elements, the cross-section of a column is divided into certain number of fibres. The nonlinear hysteretic behaviour of the

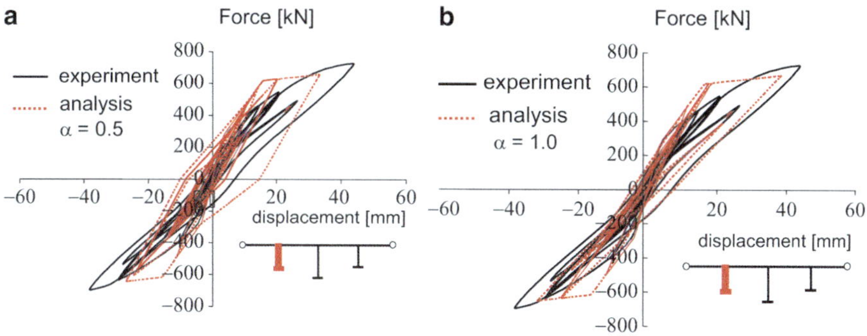

Fig. 13.7 Shear force-displacement diagram for the initial model ($\alpha = 0.5$) and model using $\alpha = 1.0$ (design earthquake)

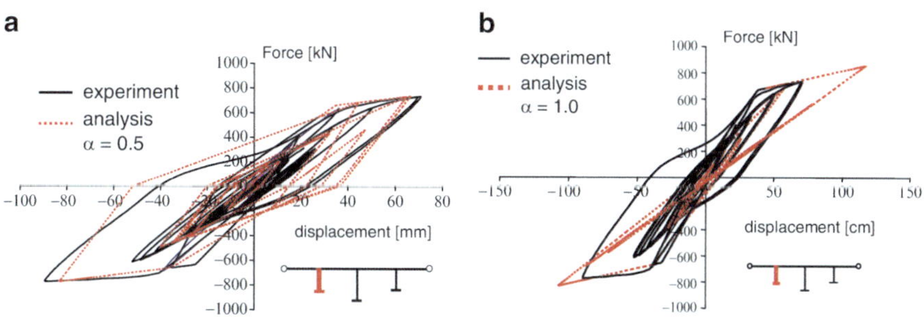

Fig. 13.8 Shear force-displacement diagram for the initial model ($\alpha = 0.5$) and model using $\alpha = 1.0$ (high-level earthquake)

element derives from the constitutive relationship of concrete and reinforcing steel that are associated with each fibre, depending on its material properties. This straightforward approach appears to be natural and simple. However, in practice, this element is complex and sometimes difficult to control. In the case of the element in DRAIN-3DX (Prakash et al. 1993), results are very sensitive to the number and the length of the elements, in particular in the plastic hinge zone.

In the program OpenSees, there are several types of the fibre elements included: (a) Nonlinear beam column element (Spacone et al. 1996a, b), (b) Beam with hinges (Scott and Fenves 2006), and (c) Displacement based beam column element.

In the study, described below, the "Nonlinear beam-column element" was employed. The correlation between the computed displacement response, obtained by the initial model (with typical values of characteristic parameters), and experimentally obtained displacements was good (Fig. 13.9). As the initial beam column element with lumped plasticity, this model failed to predict actual stiffness degradation on the unloading branch (Fig. 13.9). Several improvements were necessary to obtain better results (Fig. 13.10). Model of concrete and model of steel were changed. For example, the strength of the concrete in tension had to be taken into account, and instead of the

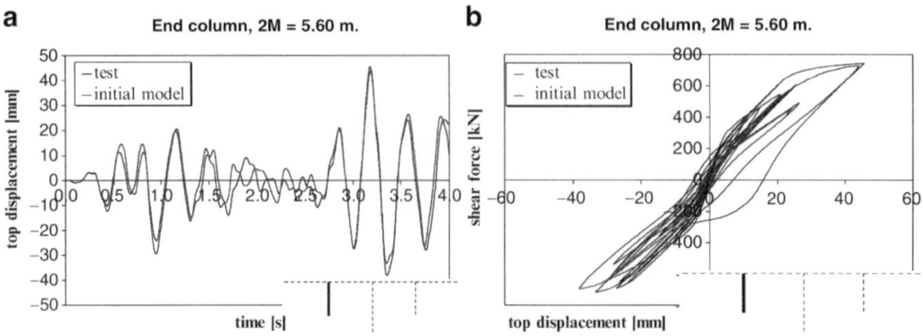

Fig. 13.9 Displacement history and shear force-displacement diagram (initial model)

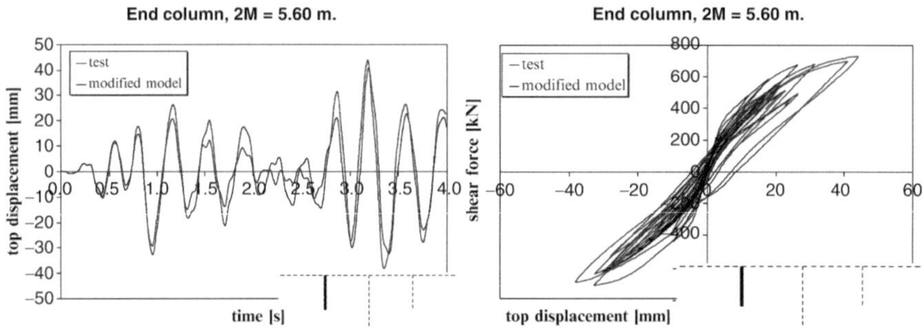

Fig. 13.10 Displacement history and shear force-displacement diagram (modified model)

bilinear stress-strain relationship the Giuffre-Menegotto-Pinto (RC elements under cycling loading – State of the art report 1996) model for steel was used.

The second type of the fibre element the "Beam with hinges element" (Scott and Fenves 2006) is somewhat simpler than the "Nonlinear beam column element". It considers plasticity to be concentrated over specified hinge length at the element ends. Considering an appropriate hinge length it is simpler to take into account different features of the seismic response of RC columns, like shear cracking, and pull out of the reinforcement, which are relatively complicated to model when the nonlinear beam column element is used. Certainly, an important advantage of this element is a stable behaviour in the cases of strain softening (Scott and Fenves 2006), which can be a quite challenging problem for displacement based fibre elements.

13.3.3 Multiple-Vertical-Line (MVL) Element

In terms of the relative level of sophistication, the MVL element falls between the elements presented in Sects. 13.3.1 and 13.3.2. In the MVL element model (Fig. 13.11), the cross section is divided into several springs that are connected by

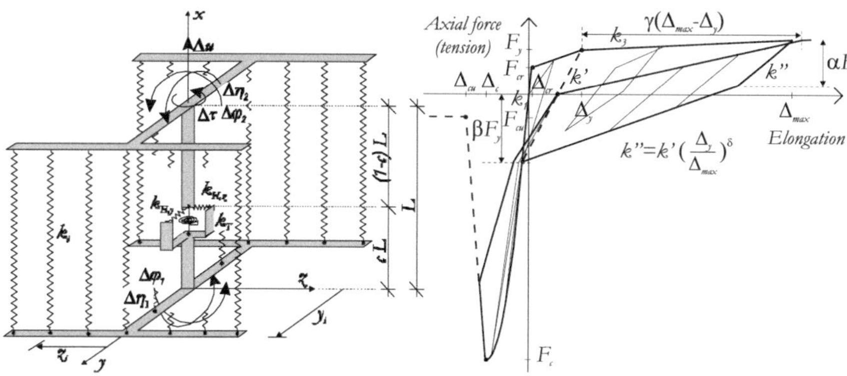

Fig. 13.11 Multiple-Vertical-Line-Element and hysteretic rules of vertical springs

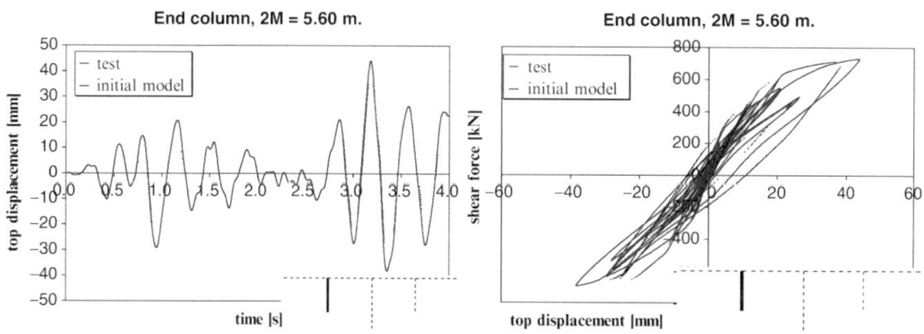

Fig. 13.12 Displacement time-history and shear force-displacement diagram (initial model)

rigid beams at the top and bottom levels of the element. They simulate axial and flexural behaviour of the element using simple hysteretic rules (Fischinger et al. 2004). MVL element includes also a horizontal spring, which models shear behaviour.

This element was originally proposed by Japanese researchers (Kabeyasawa et al. 1983) and later modified in (Vulcano et al. 1989), as well as by the second author of this chapter (Fischinger et al. 1992). All these versions of the element could be used for the analysis of the unidirectional response, only. A UL version (third version) of the element has been extended in (Fischinger et al. 2004), so the bi-directional analysis is also possible. The hysteresis loops have been also improved. The extended version of the element has been incorporated into the program OpenSees. The MVL element has been originally proposed for the analyses of structural walls. Since the larger cross-sections of columns behave similar to the structural walls, the capabilities of this element in modelling the viaducts, has been tested as well.

The displacement response history obtained with the initial model (using standard parameters: $\alpha = 1.0$, $\beta = 1.5$, $\gamma = 1.05$, $\delta = 0.50$), was quite good (Fig. 13.12).

Table 13.1 Advantages and limitations of the presented elements

Type of element	Advantages	Limitations
Beam-column element with lumped plasticity	Simple model with small number of elements (often one per column) Non-linearity defined based on the hysteretic rule with clear physical meaning Easy to control	Cannot be used for the analysis of coupled bi-directional response Unable to estimate stresses and strains
Fibre element	Able to estimate strains and stresses Can be used for the analysis of bi-directional response	Relatively complex analysis Several iterations are necessary to establish the appropriate model Control of results is more complex
MVL element	Relatively simple Non-linearity defined based on the hysteretic rule with clear physical meanings Able to estimate strains and stresses Can be used for analysis of bi-directional response	In general, several elements per column are necessary to obtain acceptable estimation of the response Appropriate number of elements should be defined iteratively

The prediction of the stiffness degradation on the unloading branch, was better than that obtained with the previous two types of elements. Since the prediction was quite good, standard parameters were not changed.

13.3.4 Comparison of the Presented Models and Conclusions

It can be concluded, that all presented models are suitable for modelling the global behaviour of viaduct columns. All the initial models (using standard values of parameters) estimated the maximum displacements as well as the maximum forces quite well. Some discrepancy with the experiment was detected mostly during the unloading phase.

The presented models differ regarding the model sophistication. It can be concluded that although the beam-column element with lumped plasticity is the simplest, it is quite successful in the prediction of the global response. This makes it very suitable for the non-linear response-history analysis, where the simple model is needed to make analysis simple, less time-consuming and easy to control. However, when the strains or stress in some parts of the structure are of the interest, or coupled bi-directional response is investigated, this element cannot be used. In such cases the other two types of elements are more efficient. The advantages and limitations of the presented elements are summarised in Table 13.1.

13.4 Strengthening of Typical Hollow Box RC Columns

13.4.1 Description of the Column

A relatively large number of viaducts in central Europe, which were constructed before the modern principles of seismic design had been established, have non-standard structural details, which are nowadays considered inappropriate for seismic regions. An example of such bridge is shown in Fig. 13.13. It was constructed on one of the main highways in Slovenia. It is a multi-span simply-supported bridge, whose superstructure beams are connected together by means of a continuous deck slab. The superstructure is supported by elastomeric and teflon bearings, located at the top of the single-column piers. The columns have a hollow box cross-section, and are supported by spread footings.

There have been several concerns regarding the seismic safety of this bridge. This Section addresses only those issues which are related to the non-standard reinforcement details in the columns (see Fig. 13.14):

1. The lap splices are constructed near the column foundation, in the region of potential plastic hinges. It was believed that this could considerably reduce the flexural strength of the columns.

Fig. 13.13 Typical existing viaduct with substandard construction details

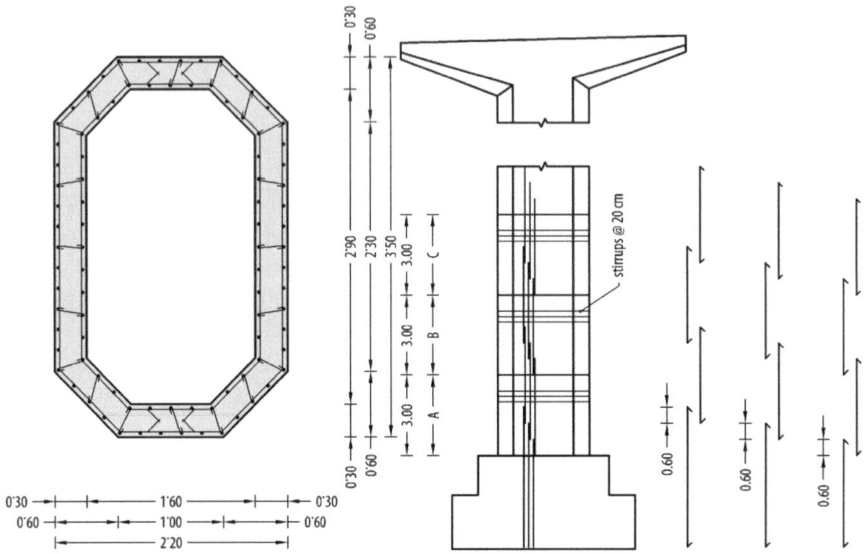

Fig. 13.14 Construction details of the typical column

2. Additional doubts regarding the strength of the columns were caused by the transverse reinforcement which was placed on the inside of the longitudinal bars. Thus there were concerns that column strength could be considerably reduced due to buckling of the longitudinal bars, which could occur prior to their yielding.
3. The amount of the transverse reinforcement gradually reduces from the base to the top of the column. Since the shear force due to seismic loads is constant along the column, the possibility of shear failure at the top of the columns was also considered.
4. Plain bars were used for the longitudinal as well as for the transverse reinforcement.

13.4.2 Short Overview of the Experiment

Two typical columns, with aspect ratios of 4.71 and 1.86, were chosen to be examined experimentally. In Section only the investigation of the short column is presented. The main properties of the 1:4 scale model is presented in Fig. 13.15.

The longitudinal reinforcement ratio at the base was 1.5% of the gross cross-sectional area. At the top of the columns the amount of longitudinal reinforcement was reduced to 0.5%. At the base the transverse reinforcing bars had a diameter of 4 mm, and they were spaced at a distance of 5 cm. At the top the diameter of the transverse reinforcement was reduced to 2.5 mm, but the distance between the bars was kept the same as in the column base (5 cm). The shear reinforcement was placed inside the longitudinal reinforcing bars. The lap splices were constructed

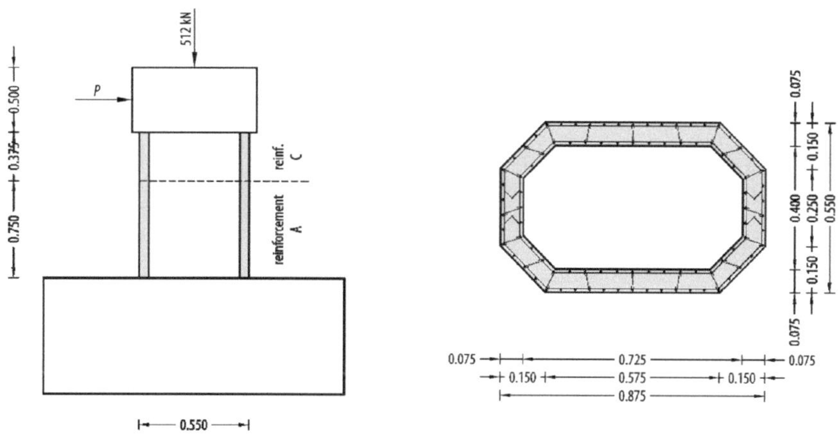

Reinforcement type A:
longitudinal bars 90ϕ6mm (f_y = 324 MPa), transverse bars ϕ4mm/5cm (f_y = 240MPa)
Reinforcement type C:
longitudinal bars 90ϕ3.4mm (f_y = 240 MPa), transverse bars ϕ2.5mm/5cm (f_y = 265MPa)

Fig. 13.15 The 1:4 scale models of the short column

close to the column foundations (see the construction details in the prototype column, shown in Fig. 13.14). The compressive strength of the concrete was 41.6 MPa. The yield stress of the steel was 324 and 240 MPa for the longitudinal and transverse reinforcement, respectively.

The model was subjected to a horizontal cyclic load. During the test, horizontal displacements were imposed cyclically at the middle of the column cap. Their absolute values were increased each time three full cycles had been completed. The axial load was applied at the top of the column, and was kept constant during the whole experiment. The level of the normalized axial forces was equal to about 7% of the compressive strength of the concrete. The column was loaded up to failure. Other basic data about the specimen is given in Fig. 13.15.

Although the column included several sub-standard construction details, its displacement ductility capacity was about 4. This ductility was provided by the favourable hollow box cross-section with its large compression zone, by the low axial forces, and by the high strength of the concrete. A mixed shear-flexural failure mode was observed (see Fig. 13.16). No buckling of the longitudinal bars was observed prior to their yielding.

13.4.3 Shear Strength of the as-Built Column

An estimation of the shear strength of the column was quite complicated and uncertain. Shear strength was calculated using three different procedures:

Fig. 13.16 The investigated column after the failure

Table 13.2 The shear capacity of the as-built column, predicted using different methods

Method	Predicted displ. ductility	V_C	V_N	V_w	V_{tot}	V_{tot}/V_{exp}
EC2 (EC8/2)	–	93	54	171	318 (171)	82% (44%)
EC8/3	3.9	117	110	146	373	96%
UCSD	3.9	83	110	171	364	93%

1. The latest procedure from EC8/2, which is based on the procedure defined in Eurocode 2 (CEN 2004b),
2. The latest procedure proposed in the standard Eurocode 8/3 (EC8/3) (CEN 2005b) and
3. The procedure proposed at University of California, San Diego – USCD (Priestley et al. 1996).

In general, all the procedures listed above determine the shear strength of the columns in the same way taking into account the shear strength of an element without shear reinforcement (i.e. the contribution of the concrete to the shear capacity), the contribution of the compressive stresses to the increase in shear strength, and the contribution of the shear reinforcement.

Although all the considered methods take into account all the important mechanisms contributing to shear strength, the ways in which these mechanisms are considered are quite different. In the investigated case the estimated values of shear strength, particularly the contributions of the concrete, were therefore significantly different. The contribution of all important mechanisms (V_c – the contribution of the concrete to the shear capacity, V_N – the contribution of the compressive stresses to the increase in shear strength, V_w – the contribution of the shear reinforcement) and the total predicted value of the shear strength V_{tot} at the base of the investigated column at the moment of its failure are summarized in Table 13.2. The total predicted shear strength V_{tot} is compared with the experimentally observed value V_{exp} in the last column of Table 13.2 as well as in Fig. 13.17.

The values of the shear strength, determined according to UCSD and EC8/3, matched the experimental data quite well. An estimation of the column's shear strength according to EC2 was less accurate. According to this standard the column

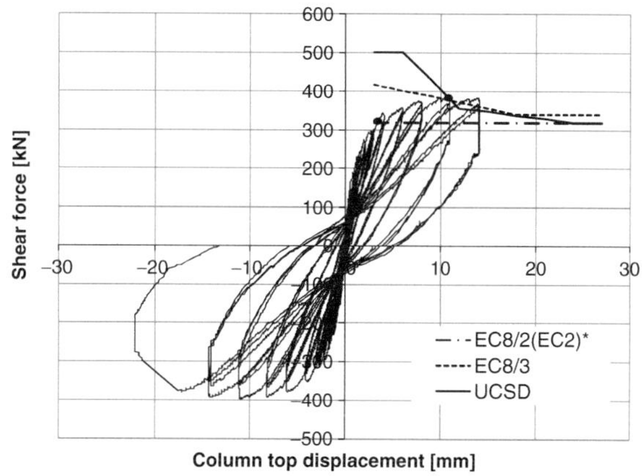

Fig. 13.17 Comparison of the analytically estimated shear strengths and the experimentally determined shear demand

would fail in shear, prior to yielding of the longitudinal bars. When the V_c and V_N are taken into account, the shear strength was estimated to be 318 kN. Actually the value of V_{tot} is significantly smaller (171 kN), since the EC2 neglects the contribution of concrete and compressive stresses when the demand exceeds the sum of this two contributions. In the investigated case this is evidently too conservative, since these two mechanisms contribute almost half of the total shear strength. Based on this and some other observations in similar columns, it can be concluded that the EC2 requirements are not adequate for estimation of the shear strength in hollow box bridge columns and similar structural elements (e.g. RC walls).

The shear strength was estimated also at the upper part of the column. Both of the standards EC8/3 and EC2 were less accurate. The estimated shear strength was quite low (Isaković et al. 2008b). Based on these low values it was concluded that the column would fail in its upper part, and that failure would occur prior to yielding of the longitudinal bars. This was not demonstrated by the experiment.

During the experiment the combined shear-flexural failure mode was observed in the plastic hinge region of the column. This was successfully estimated by UCSD method. According to this method the strength of the upper part of the column was greater than the demand (Isaković et al. 2008b) in spite of the considerably reduced top reinforcement. It was demonstrated by this method that failure could not occur at the top of the column. In the investigated case the method was quite appropriate for the estimation of shear strength and deformability. However, it should be noted that a different value of the angle between the concrete compression strut and the longitudinal column axis (Isaković et al. 2008b) was taken into account than had been proposed by the authors of the method.

13.4.4 Concrete Jacketing of the Investigated Short Column

The main purpose of strengthening was to increase the shear strength of the investigated column. The strengthened column was investigated analytically and experimentally on 1:4 scale model. The layout of the strengthen specimen is

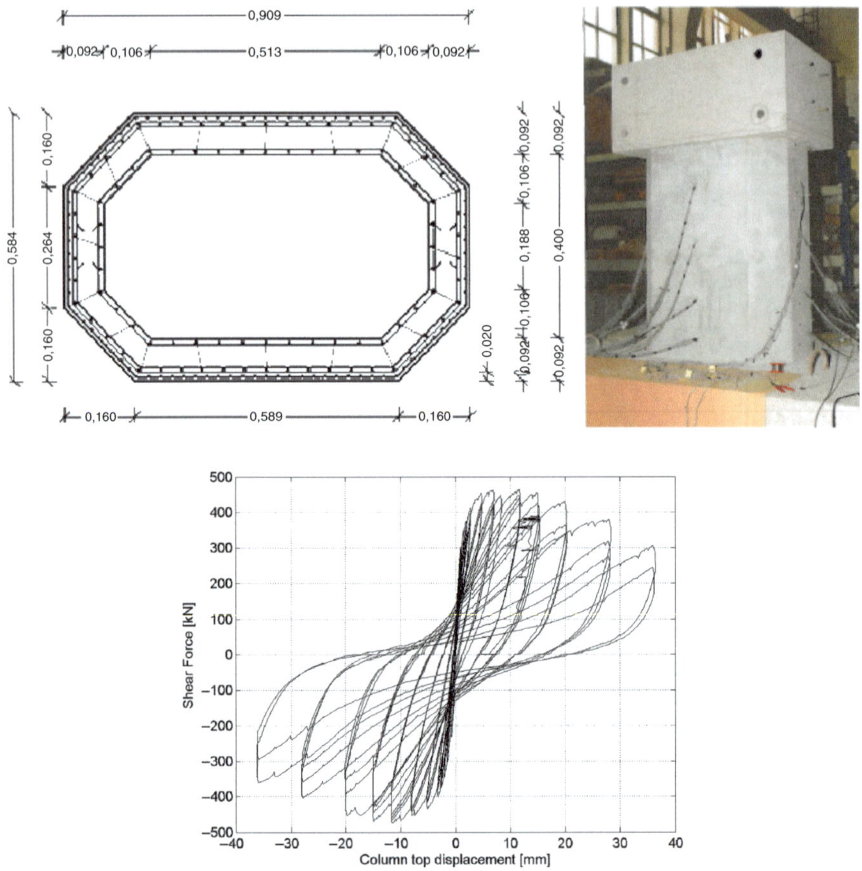

Fig. 13.18 The layout of the strengthened column and its cyclic response

presented in Fig. 13.18. The column was jacketed using an outer layer of concrete. The thickness of this layer was 2 cm. It was reinforced by longitudinal plain bars ϕ3.4 mm, which were placed at distance of 2.5 cm. The transverse reinforcement of the concrete jacket consisted of plain bars ϕ3.4/2.5 cm. The quality of the steel of all reinforcing bars was S240.

The strengthened column was tested cyclically in a similar way as the as-built column. The absolute values of displacements were increased each time three full cycles had been completed. The axial force was somewhat increased (to 780 kN) (compared to the as-built column) to obtain approximately the same compression stresses (normalized stresses of 0.06 f_{ck}, where $f_{ck} = 60$ MPa is the characteristic cylindrical strength of concrete of jacketed column) as in the as-built column.

The shear capacity of the column with and without concrete jacket is compared with the shear demand in Fig. 13.19. The shear capacity was estimated using the analytical methods, described in Sect. 13.4.3.

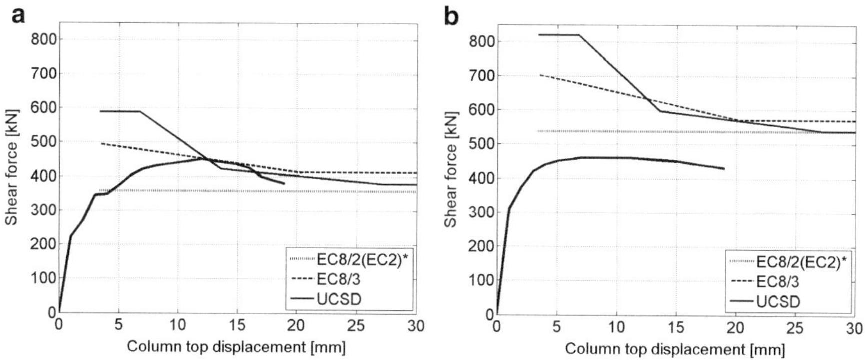

Fig. 13.19 The shear capacity of the column (**a**) without and (**b**) with concrete jacket compared to the shear demand

It can be observed (see Fig. 13.19) that shear failure of the non-jacketed column corresponds to the displacement of approximately 16 mm is reached. The jacketing increased the shear capacity of column and the shear failure was prevented. The failure mechanism was changed. The strength of column was reduced gradually. The 20% strength reduction was observed when the displacement at the top of the column reached the value of 28 mm. The jacketed column failed due to the buckling and fracture of the longitudinal reinforcement.

13.4.5 CFRP Jacketing of the Investigated Short Column

The alternative way of strengthening of the investigated column using carbon fiber reinforced polymer (CFRP) strips was also analyzed analytically and experimentally. The scale of the specimen used in the experiment was kept the same as in the previous cases, as well as the cross-section dimensions and reinforcement details. The axial force was the same as in the case of concrete jacketing (780 kN).

A minimum possible amount of CFRP strips was used to strengthen the column. The one layer 7.5 cm wide strips, which were placed at the distance of 10 cm, and which were overlapped for 20 cm were used for strengthening (see Fig. 13.20). Carbon fibers were oriented only in the horizontal direction, perpendicularly to the vertical axis of the column. The column was wrapped along its total length. The layout of the strengthened column and its cyclic response are presented in Fig. 13.20.

The analytically estimated shear capacity (using the methods described in Sect. 13.4.3) of the as-built and strengthened column is compared with the shear demand in Fig. 13.21. The minimum amount of the CFRP strips was evidently sufficient to improve the shear strength. This was proved by the experiment, were the shear failure of the column was not observed.

Since the shear failure, which was the weakest link of the as-built column, was prevented, the other failure mechanisms were activated. They were caused by other

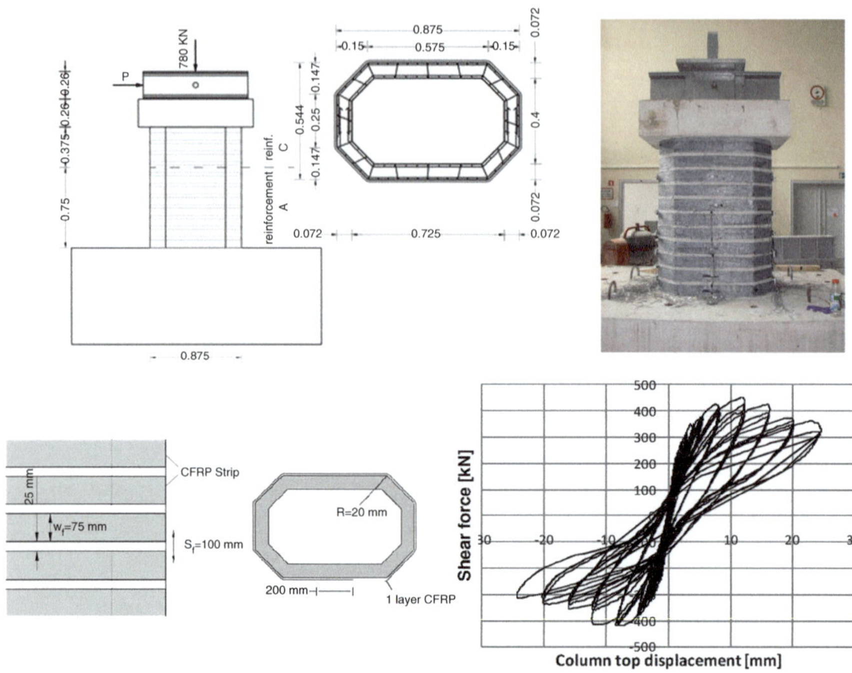

Fig. 13.20 The layout of the strengthened column and its cyclic response

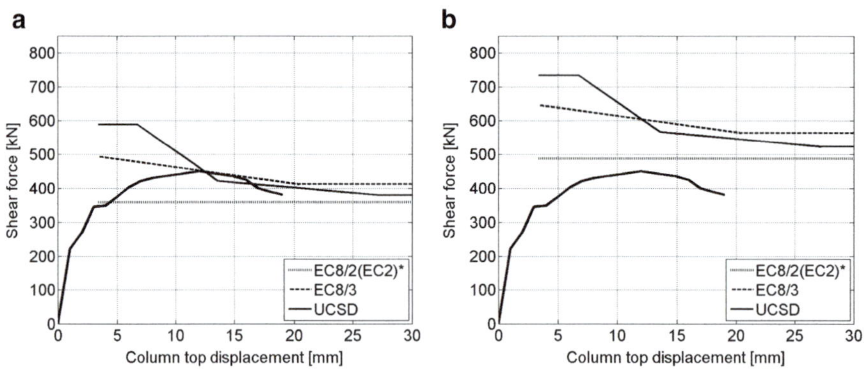

Fig. 13.21 The shear capacity of the column (a) without and (b) with CFRP jacket compared with the shear demand

construction deficiencies (lap splices in the critical regions and transverse reinforcement inside the longitudinal bars – for more details see Sect. 13.4.1 and (Isaković et al. 2008b)). The strength degradation started relatively early (at the displacement of 12 mm) due to the slip of the longitudinal bars at the inner side of the column. This was followed by buckling and fracturing of the longitudinal bars. Consequently, considerable pinching in the hysteresis loop can be observed. It can be concluded that the minimum amount of CFRP, which successfully increased the shear strength of the columns could not also prevent other unfavourable types of failure.

13.4.6 Concluding Remarks

The experimental and analytical studies of the cyclic response of short reinforced concrete hollow box columns, constructed in a typical viaduct in Central Europe in the 1970s, were performed. The column comprises several construction details, which are nowadays considered inappropriate for seismic regions. The shear strengthening of such columns using concrete and CFRP jacketing was investigated analytically and experimentally.

The experiment of the as built column demonstrated that it had a quite good displacement ductility capacity, which was acceptable for moderate seismic demand regions (e.g. Central Europe).

The shear strength of as built column was estimated by using standard models, which yielded quite different results. The procedures proposed in the European standards EC8/2 (EC2) and EC8/3, and UCSD were taken into account.

The shear strength estimated according to the standard EC8/2 (EC2) was too low, because the shear strength of the concrete V_c was neglected when the demand exceeds this value. When, V_c was taken into account, the shear strength of the column was comparable with the results obtained by using the other two methods, but only to those values which corresponded to larger displacement ductility demands. In the region of lower ductility demands the value of the shear strength was well below the values estimated by the other two methods.

A more suitable procedure for the estimation of the shear strength of investigated column is that proposed in the standard EC8/3. Using this procedure the shear strength of the bottom part of the column was estimated quite accurately. However, the standard underestimated the shear strength of the top of the column, where the ductility demand was low.

Good estimates of the type and location of failures were obtained by using the UCSD method.

Based on the large differences between the considered procedures, as well as other results presented elsewhere, it can be concluded that the problem of shear strength and deformability is, in general, not adequately solved and it demands further studies. It may be appropriate to reconsider this problem in the future developments of the Eurocode standards.

The as built column was strengthened by using concrete jacket and CFRP strips. Both ways of strengthening improved the shear strength of column and prevented its shear failure. The minimum amount of the CFRP strips could not prevent the other unfavourable types of failure caused by the other construction deficiencies of the column (lap splices in the critical regions and transverse reinforcement inside the longitudinal bars).

13.5 Seismic Isolation Using New Semi-Active Magnetically Controlled Elastomeric Bearings

Many standard isolation systems are based on rubber bearings which are typically designed to resist a large reference (design) earthquake. The rubber bearings change their stiffness under different seismic intensities. When they are subjected to earthquakes of intensities lower than the design earthquake, their stiffness typically increase. This can reduce the efficiency of the isolation.

To improve the response of the rubber isolators at the lower seismic intensities a new isolation device, adjustable to the intensity of the load has been developed. It is fabricated from a magnetically controlled elastomer (MCE), that is one for which the stiffness can be varied by applying a magnetic field. It was developed in the frame of the 5th frame European project VAST-IMAGE (principal Maurer and Sohne, Germany). It is briefly described in Sect. 13.5.1.

In bridges which are partially isolated, the isolation devices are typically located only on the short stiff piers. An increase of the stiffness of the isolation device under the weak earthquakes increases also the total stiffness of the pier. Consequently in some cases the shear forces in such columns can be larger at weak earthquakes than that corresponding to the design earthquake. In such cases the MCE isolation can be used (see Sect. 13.5.2).

13.5.1 Short Description of the MCE Bearing

The MCE bearing (see Fig. 13.22a) consists of an elastomeric bearing core which is sensitive to a magnetic field; an arrangement of permanent magnets that applies a permanent magnetic field to the bearing core and; an arrangement of coils that once

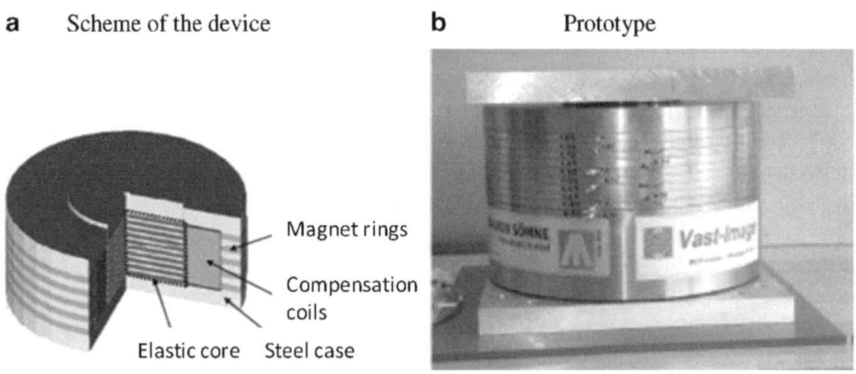

Fig. 13.22 MCE bearings; (**a**) the scheme, (**b**) the prototype (courtesy of Maurer and Shöne)

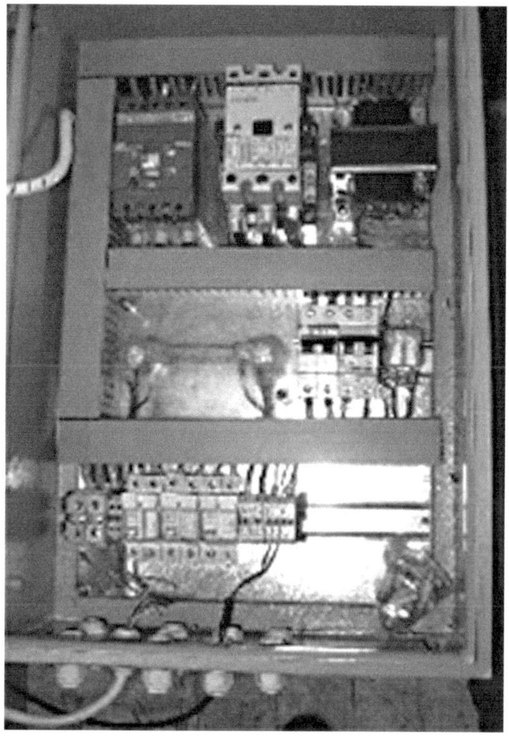

Fig. 13.23 The control unit (courtesy of ISMES)

activated counteract the magnetic field so that the bearing core is free from magnetic induction.

For fail-safe reasons the MCE isolators show their maximum stiffness under passive conditions, i.e. they show the design stiffness for the high design level earthquakes. In the case of low level earthquakes the coils are activated and the MCE isolators show reduced stiffness. A sensor indicates the level of the earthquake and an electronic control board with current driver regulates the stiffness of the device.

The elastic core of the device consists of alternating MCE and steel layers. The magnets are placed inside the outer shell of the device, and are not subjected to external mechanical loads. The fabricated prototype (produced by Maurer and Shöne, Germany in collaboration with TARRC, UK, and IFW, Germany) of the MCE bearings is presented in Fig. 13.22b.

A special control unit manages the changes in the stiffness of the MCE bearings, based on the specially designed control algorithm (the unit and the algorithm were designed at CESI/ISMES, Italy). The control unit is presented in Fig. 13.23. Full power activation of the MCE isolator needs an electrical supply of approximately 2,000 W. This was realized by an array of eight 12V lead batteries.

The control algorithm specially developed for the Magnetic Controlled Elastomer-based devices is able to "recognize" an earthquake and assign to it a

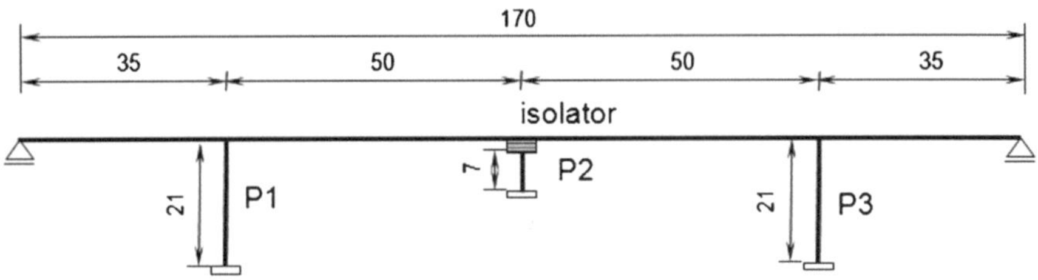

Fig. 13.24 An example of the bridge which is partially isolated

certain status in order to properly turn on the electrical power supply which in turn creates the magnetic field required to activate the iron-loaded elastomer.

The general approach is based on a three-status concept, which includes "status 0" for which the system is in stand-by configuration, "status 1", which indicates the detection of a ground acceleration above a given threshold, and "status 2", which corresponds to the detected strong part of the earthquake, which is recognized by the exceeding of a given threshold in the relative displacement of the isolation devices.

This three-status strategy corresponds to the fail-safe configuration adopted by the MCE isolator device, which is magnetically active (and therefore stiffer) when in its usual non-activated configuration (due to the presence of permanent magnets), to withstand static and thermal loads. It becomes softer when the electrical activation generates an internal magnetic field counteracting that generated by the permanent magnets.

13.5.2 Applicability in Bridges

Earthquakes weaker than the design earthquake can be critical in partially isolated bridges, similar to the structure presented in Fig. 13.24. The isolation devices are placed at the top of the short pier in order to reduce its stiffness and consequently, to reduce its shear forces. Typically the stiffness is reduced in such amount to get almost uniformly distributed shear demand in all piers.

The properties of isolation devices are typically determined based on the design seismic level. When the structure is subjected to weaker earthquakes the stiffness of the rubber and consequently the stiffness of the short central pier increase. Due to the increased stiffness of the central pier, the redistribution of the seismic demand between piers occurs. Thus the short pier can be subjected to considerably larger forces than the other piers supporting the bridge. At weak earthquakes the shear forces in this pier can be even larger than those at the design earthquakes in spite of the reduced seismic intensity.

The effects of the increased rubber stiffness to the response of different piers are illustrated in Fig. 13.25. The solid and dotted lines represent forces in central and

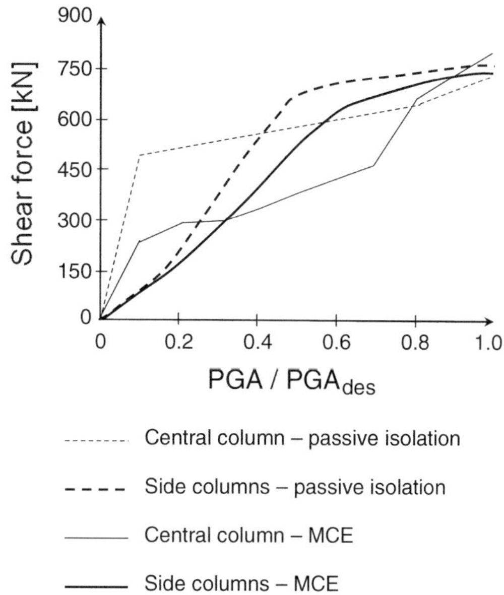

Fig. 13.25 Shear forces in columns of the bridge presented in Fig. 13.24 for different seismic intensity levels and for the case of isolation with conventional rubber bearings (*dotted lines*) and with MCE bearings (*solid lines*)

side piers at different seismic intensities, respectively. It can be observed that in the central short pier the shear force at 10% of the design seismic intensity is increased to the almost 70% of the value, which correspond to the design earthquakes. It is more than three times larger than the shear forces in the side piers.

When the standard rubber bearings are replaced with the MCE device the stiffness of the central pier can be regulated also at the weak earthquakes. The response of central and side columns is presented in Fig. 13.25 with dotted lines. It can be observed that shear forces can be substantially reduced at weak earthquakes, e.g. at 10% of the design earthquake the shear force is reduced to approximately 50% of that corresponding to the passive isolation and same seismic intensity

More details about the MCE bearings, about its applicability in other types of structures as well as about the adequate numerical models, which were developed at the University of Ljubljana can be found in Distl (2006) and Isaković et al. (2010).

13.5.3 Discussion

In order to improve the response of rubber isolators at lower seismic intensities, a new isolation device, which can adjust itself to the intensity of the load, has been developed. It is fabricated from an MCE elastomer whose stiffness can be varied by changing a magnetic field. It was illustrated that it can be used to improve the efficiency of the seismic isolation in partially isolated bridges when they are subjected to earthquakes weaker than the design earthquake.

13.6 Conclusions

An overview of the research related to the seismic analysis and design of RC bridges, recently performed at UL FGG is made. Four main topics are addressed: (1) Pushover based analysis of bridges, (2) Modelling of RC bridge columns, (3) Estimation of the shear strength and shear strengthening of typical RC hollow box bridge columns with substandard construction details, (4) Seismic isolation of RC bridges using new semi-active device – magnetically controlled elastomer.

1. It was found that single-mode pushover based methods can be used for the analysis of bridges which have the fundamental mode with the effective mass of at least 80% of the total mass and the value of index of regularity of less than 5%. In general they are accurate enough for the analysis of regular short and medium span bridges. For longer bridges (with total length in excess of 500 m) multi-mode pushover methods or nonlinear response history analysis should be employed.
2. All investigated macro models were found to be suitable for modelling the global flexural behaviour of viaduct columns. They differ regarding the model sophistication. The beam-column element with lumped plasticity is the simplest but very successful in the prediction of the global response. Thus, according to the authors' opinion, it is can be the preferred choice. However, when the strains or stress in some parts of the structure are of the interest, or coupled bi-directional response is investigated; fiber or MVL elements were found to be more efficient.
3. It was found that the problem of shear strength and deformability is, in general, not adequately solved and it demands further studies. The methods, used for estimation of the shear strength of hollow box column differed significantly. Based on the comparison with the experimental data, the method, proposed at UCSD was found the most appropriate in the investigated case. It was found that it may be appropriate to reconsider the problem of the estimation of the shear strength in the future developments of the Eurocode standards. The as built column was strengthened by using concrete jacket and CFRP strips. Both ways of strengthening improved the shear strength of column and successfully prevented its shear failure.
4. To improve the response of rubber isolators at lower seismic intensities, a new isolation device, fabricated from an MCE elastomer and which can adjust itself to the intensity of the load, has been developed. It could be used to improve the efficiency of the seismic isolation in partially isolated bridges subjected to earthquakes weaker than the design earthquake.

Acknowledgments The work presented in Sects. 13.2 and 13.3 was funded by Ministry of Higher Education, Science and Technology of the Republic of Slovenia.

The work presented in Sect. 13.4 was partly funded by the Company for Motorways in the Republic of Slovenia (DARS). The presented experiments were performed in the cooperation with Slovenian National Building and Civil Engineering Institute (ZAG). The CFRP jacketing was provided by SIKA d.o.o., Slovenia.

The work presented in Sect. 13.5 was funded by the European Union through the 5th Framework research project VAST-IMAGE (Variable Stiffness Seismic Isolators and Vibration Mitigation Dampers Based on Magnetically Controlled Elastomer) under the coordination of Maurer Söhne, Germany as the principal investigator (grant number EV G1-CT-2002-00063). The work of the authors was also supported by the Ministry of Higher Education, Science and Technology of the Republic of Slovenia.

Some results presented in the chapter are obtained in the frame of the Ph. D thesis of Jaka Zevnik, Zlatko Vidrih and Mauro Nino Popeyo Lazaro.

References

Aydinoğlu MN (2003) An incremental response spectrum analysis based on inelastic spectral displacement for multi-mode seismic performance evaluation. Bull Earthquake Eng 1:3–36

CEN (2004a) Eurocode 8: Design of structures for earthquake resistance. Part 1: General rules, seismic action and rules for buildings. EN 1998–1, Euro Commit for Stand, Brussels, Dec 2004

CEN (2004b) Eurocode 2: Design of concrete structures - Part 1–1: General rules and rules for buildings. EN 1992-1-1. Comité Européen de Normalisation, Brussels

CEN (2005a) Eurocode 8: Design of structures for earthquake resistance. Part 2: Design of structures for earthquake resistance – Part 2: Bridges, EN 1998–2:2005, European Committee for Standardization, Brussels

CEN (2005b) Eurocode 8: Design of structures for earthquake resistance. Part 3: Strengthening and repair of buildings. EN 1998–3, Euro Commit for Stand, Brussels

Chopra AK, Goel RK (2002) A modal pushover analysis for estimating seismic demands for buildings. Earthquake Eng Struct Dyn 31:561–582

Distl J (2006) Development of variable stiffness seismic isolators and vibration mitigation dampers based on magnetically controlled elastomer. Final Report, Maurer Sohne GmbH & Co. KG

Fajfar P (2000) A nonlinear analysis method for performance-based seismic design. Earthquake Spectra 16:573–592

Fischinger M, Vidic T, Fajfar P (1992) Non-linear seismic analysis of structural walls using the multiple-vertical-line-element model, Non-linear seismic analysis and design of reinforced concrete buildings. Elsevier, Bled, pp 191–202

Fischinger M, Isakovic T, Kante P (2004) Implementation of a macro model to predict seismic response of RC structural walls. Comput Concr 1(2):211–226

Isaković T, Fischinger M (2006) Higher modes in simplified inelastic seismic analysis of single column bent viaducts. Earthquake Eng Struct Dyn 35(1):95–114. doi:10.1002/eqe.535

Isaković T, Fischinger M (2011) Applicability of pushover methods to the seismic analyses of an RC bridge, experimentally tested on three shake tables. J Earthquake Eng 15(2):303–320. doi:10.1080/13632461003802009

Isaković T, Fischinger M, Kante P (2003) Bridges: when is single mode seismic analysis adequate? Proc Inst Civ Engrs Structs Bldgs 156(2):165–173

Isaković T, Popeyo Lazaro MN, Fischinger M (2008a) Applicability of pushover methods for the seismic analysis of single-column bent viaducts. Earthquake Eng Struct Dyn 37(8):1185–1202. doi:10.1002/eqe.813

Isaković T, Bevc L, Fischinger M (2008b) Modelling the cyclic flexural and shear response of the R. C. hollow box columns of an existing viaduct. J Earthquake Eng 12(7):1120–1138

Isaković T, Zevnik J, Fischinger M (2010) Floor response spectra in isolated structures, subjected to earthquakes weaker than the design earthquake – PART II: isolation with magnetically controlled elastomeric bearings. Struct Control Health Monit, Published online, DOI: 10.1002/stc.391

Kabeyasawa T, Shiohara H, Otani S, Aoyama H, (1983) Analysis of the full-scale seven-story reinforced concrete test structure. J Facul Eng, University of Tokyo, pp 431–478

Kanaan AE, Powell GH (1973) A general purpose computer program for dynamic analysis of planar structures, Report UBC/EERC-73/6, University of California, Berkeley

McKenna F, Fenves GL (2007) Open system for earthquake engineering simulation, Pacific Earthquake Engineering Research Center, Berkeley. http://opensees.berkeley.edu

Paraskeva TS, Kappos AJ, Sextos AG (2006) Extension of modal pushover analysis to seismic assessment of bridges. Earthquake Eng Struct Dyn 35(11):1269–1293

Pinto AV, Negro P (1995) Seismic testing of large-scale bridges. In: Proceedings of the US national seismic conference on bridges and highways – progress in research and practice, San Diego

Prakash V, Powell GH, Filippou FC (1993) DRAIN-3DX: Base program users guide, Department of Civil Engineering, University of California

Priestley MJN, Seible F, Calvi GM (1996) Seismic design and retrofit of bridges. Wiley, New York

RC elements under cycling loading – State of the art report (1996) CEB, Bulletin d'information no 230. Thomas Telford Service Ltd., London

Scott MH, Fenves GL (2006) Plastic hinge integration methods for force-based beam-column elements. ASCE J Struct Eng 132(2):244–252

Spacone E, Filippou FC, Taucer FF (1996a) Fibre beam-column model for non-linear analysis of R/C frames: Part I. Formulation. Earthquake Eng Struct Dyn 25(7):711–725

Spacone E, Filippou FC, Taucer FF (1996b) Fibre beam-column model for non-linear analysis of R/C frames: Part II. Application. Earthquake Eng Struct Dyn 25(7):727–742

Vulcano A, Bertero VV, Caloti V (1989) Analytical modelling of R/C structural walls. In: Proceedings of the 9th WCEE, vol 6, Tokyo-Kyoto, Maruzen, pp 41–46

Chapter 14
A Multi-Platform Simulation Alternative for the Performance-Based Design of Interactive Soil-Bridge Systems

Anastasios G. Sextos

Abstract An approach for combining the merits of different computational tools with respect to modeling the embankment-abutment-deck and the soil-foundation-pier-superstructure interaction of large, flexibly supported bride systems is presented herein. In particular, the Multi-Platform Simulation (MPS) concept is applied for the deterministic and probabilistic assessment of two real bridges supported on soft soil conditions and the limitations and advantages of MPS are comparatively outlined and discussed. The experience gained indicates that depending on the problem studied and the specific objectives of seismic assessment for a given bridge, multi-platform simulation can contribute towards the more accurate representation of the soil-foundation-bridge system as a whole and thus, reveal aspects of the coupled system response that are otherwise suppressed using conventional analysis methods.

Keywords Multi-platform analysis • Bridge engineering • Seismic design • Probabilistic assessment • Soil-structure interaction

14.1 Introduction

The importance of Soil-Structure-Interaction (SSI) for the assessment of the dynamic response of bridges has been widely recognized in numerous research studies. Despite the extensive research over the last 30 years though, common practices and codified approaches are still approximate while the problem is often treated as a conditionally beneficial phenomenon (Mylonakis and Gazetas 2000) on the basis of the anticipated period elongation of the structure (and the monotonic decrease of spectral accelerations of the design spectra), as well as on the energy

A.G. Sextos (✉)
Department of Civil Engineering, Division of Structural Engineering,
Aristotle University of Thessaloniki, Thessaloniki, Greece
e-mail: asextos@civil.auth.gr

M. Dolšek (ed.), *Protection of Built Environment Against Earthquakes*,
DOI 10.1007/978-94-007-1448-9_14,

dissipation at the foundation level caused by wave radiation and hysteretic damping, thus leading to a common assumption that any structure can be conservatively assumed to be fixed at its base. In fact, this perception has been long proven to be misleading since the foundation is flexible, dissipates energy and interacts with the surrounding soil and the superstructure in such a way, that it filters seismic motion (kinematic interaction) while it is subjected to inertial forces generated by the vibration of the superstructure (inertial interaction). This phenomenon is very complex and its beneficial or detrimental effect on the dynamic response of the bridge is dependent on a series of parameters such as Pender (1993), Wolf (1994), Gazetas and Mylonakis (1998), Finn (2005) the intensity of ground motion, the dominant wavelengths, the angle of incidence of the seismic waves, the stromatography, stiffness and damping of soil as well as the size, geometry, stiffness, slenderness and dynamic characteristics of the structure.

For the case of bridges in particular, the problem becomes even more challenging due to the longer dimensions of the superstructure and the subsequent spatial variability of soil properties along its length and also due to the significant contribution of bridge lateral boundary conditions in the overall seismic response of bridges (Pecker 2004; Beltrami et al. 2005; Combault et al. 2005). The significant role played by the embankment-foundation-abutment system has been illustrated by numerous researchers not only in terms of the resulting modified dynamic characteristics of the bridge (Goel and Chopra 1997; Dicleli 2005; Kotsoglou and Pantazopoulou 2007) but also with regard to the modification of the incoming seismic motion itself by the embankment presence (Zhang and Makris 2002). Earthquake damage reports and laboratory tests have also indicated that the consideration of soil-structure interaction both at the location of the pier and at the lateral supports of the bridge is required in order to obtain a reliable estimate of the bridge response under earthquake loading.

Ideally, this would have been feasible through the detailed modeling of the whole soil domain surrounding and supporting the bridge through the simulation of both the (non-linear) dynamic pier-foundation-subsoil and deck-abutment-embankment interaction, the shear deformation and flexural failure of RC members (i.e. piers and piles) as well of any potential geometric non-linearity that commonly arises under large seismic forces (i.e. closure of gaps or joints at the deck level). Nevertheless, the literature related to such a 'holistic' finite element modeling is indeed very limited primarily for two main reasons: (a) the coupled modeling of all these systems (i.e., embankments, abutments, pier foundation, subsoil and superstructure) still requires extensive computational effort due to the model size and the soil behavior complexity and (b) it is rather subjective whether a single software package indeed exists that could possibly combine all the features required for the advanced simulation of the non-linear response of all the aforementioned soil and structural sub-systems in order to predict their response with equal rigor.

Laboratory tests on the other hand, are a fundamental source of knowledge that could in principle resolve most of the aforementioned limitations. Nevertheless, due to the dimensions of bridges, experiments are usually conducted on the most vulnerable components only, i.e., piers (Biskinis and Fardis 2006) or bearing devices (Bousias et al. 2008). In certain cases, the pseudo-dynamic (PSD) testing method is used, according to which, a part of the structure can be physically tested while the rest is numerically modeled with finite elements, using an appropriate time-integration algorithm for the equation of motion. Again though, due to the scale of the problem and the required capacity of the laboratory in which the experiment is to be conducted, most components are tested at a reduced scale and as such, the number of full-scale complete structure tests is very limited (Negro et al. 1996; Molina et al. 1999; Pinho and Elnashai 2000; Chen et al. 2003; Jeong and Elnashai 2005). Given the above limitations, typically, the foundations and soil are also not considered at all.

A system by which a number of laboratories could combine their capabilities to undertake a set of integrated component tests of structural and geotechnical elements is an exceptionally attractive option that has been developed recently in the United States for the assessment of complex interacting systems. In this framework, which has been supported by the National Science Foundation (NSF) through the Network for Earthquake Engineering Simulation (NEES) scheme Watanabe et al. (1999), National Science and Network for earthquake engineering simulation (NEES) (2000), Tsai et al. (2003), Kwon et al. (2005), Pan et al. (2005), Takahashi and Fenves (2006), the limitations related to the size of the problem and the laboratory capacities are essentially raised. On the contrary, there is no need any more for using a single experimental facility neither for physical proximity of the multiple sub-components tested. Moreover, since communication is solely web-based, using the same information technology advances and protocols, some components of the system can be analyzed numerically. This multi-site "*Hybrid Simulation*" approach is therefore a challenging but equally difficult task that requires in-depth knowledge of specialized experimental and analytical tools, their detailed requirements, and necessitates considerable programming effort (Elnashai et al. 2007). According to this approach, the dynamic response of full scale specimens that are discretized into sub-structures is properly controlled with the use of purpose-specific coordination software. Two such purposely developed software platforms exist, i.e., the OpenFresco Takahashi and Fenves (2006) and UI-SimCor Spencer et al. (2006).

The latter, developed by the research group of the University of Illinois, is the first platform that has been used for multi-site testing of bridges including SSI phenomena: it consists of an enhanced Matlab based script which coordinates either software or hardware supporting NEESgrid Teleoperation Control Protocol (NTCP) as well as TCP-IP connections outside of the NEES system.

The basic concept of UI-SimCor is that analytical models of some parts of the structure or experimental specimens representing other parts of the same structure are considered as super-elements with many degrees of freedom (DOFs). The elements – analytical or experimental - are treated on different networked computers and, can thus be located practically anywhere in the world. Specially developed interface programs, controlled by UI-Simcor allow the interaction with

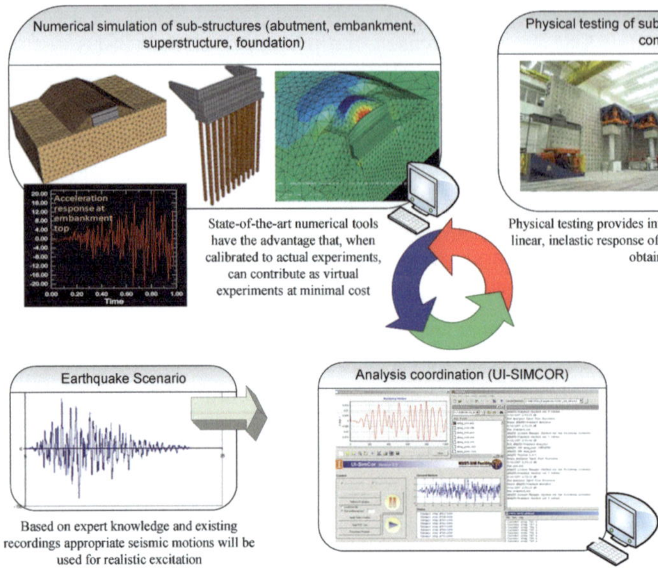

Fig. 14.1 Overview of the multi-platform analysis and/or hybrid experimentation scheme

different analysis software such as Zeus-NL (Elnashai et al. 2002), OpenSees (McKenna and Fenves 2001), FedeasLab (Filippou and Constantinides 2004) and ABAQUS (Hibbit and Sorenson 2006) through different communication protocols.

The concept of Hybrid simulation has been also applied in Korea (Watanabe et al. 1999) and Taiwan (Tsai et al. 2003) for earthquake engineering research purposes. In the European Research Area, it was first introduced at the ELSA laboratory (Pegon and Pinto 2000; Pinto et al. 2004). A similar to the NEES initiative is the UK Network for Earthquake Engineering Simulation (UK-NEES) (Neild et al. 2005) comprising the research laboratories at the Universities of Bristol, Oxford and Cambridge and aiming to provide the main UK earthquake engineering experimental laboratories with the necessary equipment to become nodes of the NEES network. Hybrid experiments have also been performed by University Patras, Greece, for a multi-span bridge structure.

Although this concept has been initially introduced to coordinate both experimental and analytical modules, it has also been successfully applied (Kwon and Elnashai 2008) for the coordination of purely numerical analysis modules (no physical testing is performed) in the framework of the assessment of real bridges in the U.S. for various soil conditions, as well as for the study of the potential impact of liquefaction susceptibility (Kwon et al. 2008, 2009). This so called, "*Multi-Platform Simulation*" (MPS) is also a promising alternative to the aforementioned Hybrid simulation approach, primarily because it permits the sub-structured analysis of a complex system using purely analytical tools, similarly physically distributed as was the previous case (Fig. 14.1). The advantage of this approach is that the appropriate selection and combination of different analysis packages, has the minimum assumptions and enables the concurrent use of the most suitable and sophisticated constitutive laws, element types and features of each package for each corresponding part of the system. In other words, different software can be used for different system components (i.e. abutments, superstructure and supporting pile groups) depending on the foreseen inelastic material behavior, level and nature of the seismic forces and the geometry of the particular problem. It is believed that this approach leads to combined capabilities that no finite element program currently provides, nor is probable to provide in the near future.

On the contrary, it has the minimum assumptions possible and permits the best available option to simulate each component using the most appropriate analytical model, while integrating the various contributions into a fully interacting system. Along these lines the objective of this chapter is to demonstrate the applicability of the Multi-platform analysis for the deterministic and probabilistic assessment of large soil-bridge systems and discuss their relative drawbacks and merits.

14.2 Coordination of the Multi-Platform Analysis

In the MPS framework described above, and by adopting UI-SIMCOR as the analysis coordinator, Pseudo-Dynamic (PSD) testing of the bridges to be studied consists of four different stages (Spencer et al. 2006):

(a) *initialization*: connection is made to each simulation module and the related variables are initialized,
(b) *initial stiffness formulation*: the global stiffness matrix is formed. In case that the structural dimension is not excessive, the analysis coordinator sends predefined displacement for each degree of freedom to each module and takes measured forces to establish the initial stiffness of the whole structure. In case that the structure is large, then the stiffness matrix can be loaded from separate files that define the stiffness matrix of each module.
(c) *static loading*: displacements due to gravity forces are imposed and the initial (pre-seismic) stress condition of the concrete members is established.
(d) *dynamic loading*: the analysis is made using an α-OS scheme (Combescure and Pegon 1997) which is a non-iterative implicit time integration scheme that provides the unconditional stability needed in case of large number of DOFs, while preserving the implementation simplicity of explicit schemes.

It is recalled that according to the principle of the Pseudo-Dynamic method, the dynamic response of a structure can be represented by a discrete parameter model with a finite number of degrees of freedom and as such, the corresponding equation of motion of the MDOF model can be written as a function of the (non-linear) restoring forces:

$$Ma(t) + Cv(t) + r(d(t)) = f(t) \tag{14.1}$$

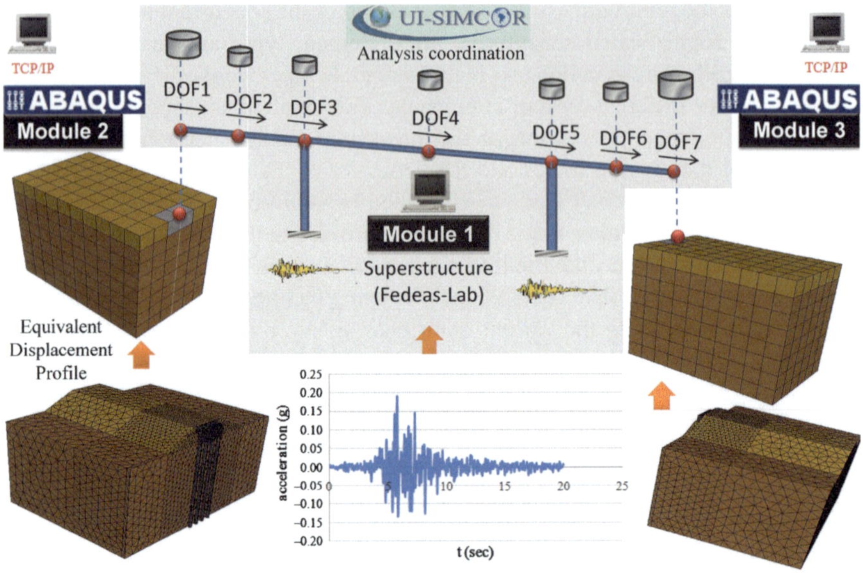

Fig. 14.2 Example of breakdown of various modules for the abutment-embankment system and the superstructure of a typical bridge analyzed using different analysis software

where M is the mass matrix, C the damping matrix (typically set equal to zero), $a(t)$ the acceleration vector, $v(t)$ the velocity vector, $r(d(t))$ the structural restoring force vector (typically obtained through the experimentally tested specimen) and $f(t)$ the external force vector.

In case of purely numerical sub-structuring, the predicted displacements are applied at the control points and the non-linear restoring forces are obtained. Apparently, force equilibrium and displacement compatibility has to be satisfied at all interfaces between the sub-structures (handled as system modules). The novel feature of the MPS approach is that, in contrast to conventional PSD simulation where time integration scheme is combined with a single analysis platform, it coordinates several restoring force modules using different communication protocols (Kwon et al. 2005). In the particular framework developed, six communication protocols are implemented, namely, NTCP, TCP/IP, LabVIEW1, LabVIEW2, OpenFresco1D and NHCP, the first (NEESgrid Teleoperation Control Protocol) being the standard communication protocol. Each of these modules is considered as a super element with many degrees of freedom and can be separately analysed on a single or multiple network computers.

An example of the analysis coordination made by UI-Simcor, of various numerical modules defined for the abutment-embankment systems and the superstructure of a typical bridge analyzed using different analysis software is illustrated in Fig. 14.2. In this example, the superstructure is modeled with the structural analysis software Fedeas-Lab (Filippou and Constantinides 2004) while the embankment-abutment system is modeled with the commercial finite element software ABAQUS

(Hibbit and Sorenson 2006). Seven dynamic degrees of freedom are considered along the corresponding seven control points of the bridge deck, while the edge DOFs (i.e., DOF1 and DOF7 act as the interface with the embankment-abutment sub-structures). A non-linear material law (Mohr-Coulomb) is assigned for soil while the structural inelasticity is captured trough the distributed plasticity model of Fedeas-Lab. In order to accelerate the analysis, an alternative where a simpler embankment-abutment module exhibiting the same non-linear force-displacement response with the refined 3D model is also used. Similar analysis frameworks are described in detail in the following two sections where the dynamic response of two real bridges was investigated.

14.3 MPS Application for Deterministic Assessment: A Typical Egnatia Highway Overcrossing

14.3.1 Overview of the Bridge Studied

A first application (Sextos and Taskari 2008) of the above Multi-platform analysis was performed for the deterministic assessment of an overpass (overcrossing) along the Egnatia highway, a large road network that has been constructed in northern Greece with more than 646 bridges built of a total 40 km length. The particular bridge adopted for study (Kappos et al. 2007) is a three-span, symmetric structure of 70 m length (span lengths are 19, 32 and 19 m respectively) curved in elevation (maximum camber of 8%), that intersects the highway axis at an angle of 75.3°. The deck is 11 m wide and 1.60 m high. The prestressed deck has a hollow T-beam-like section and is supported on two circular piers of 1.70 m diameter and 8.50 m height which are monolithically connected to the superstructure and the foundation. At the abutments (which have a 10.50 × 1.20 m wall section of 5.0 m height), the deck is connected through two pot bearings that permit sliding along the two principal bridge axes and a sliding joint separates the deck from the backwall. Seismic forces are also resisted by the activation of stoppers (in the transverse direction) which are constructed at the seating of the abutments. The foundation on the other hand is deep, due to the soft clay formations characterizing the overall area. The pier foundation consists of a 2 × 2 pile group of 28.0–32.0 m long piles, connected with a 1.60 × 5.0 × 5.0 m pile cap, while the abutments are supported on a 1 × 4 pile row 27–35.0 m long at 2.80 m axial spacing, all piles having equal diameter of 1.0 m (Fig. 14.3).

The bridge was designed for normal loads according to the German Norms (i.e. DIN 1055, 1045, 1072, 1075, 1054, 4227, 4085, 4014) while the seismic design was carried out according to the Greek Seismic Code EAK 2000 and the relevant Greek standards E39/99 for the seismic design of bridges. The bridge site is located in the Seismic Zone I which is equivalent to a peak ground acceleration of 0.16g. The behaviour factors of the system adopted for design according to the E39/99

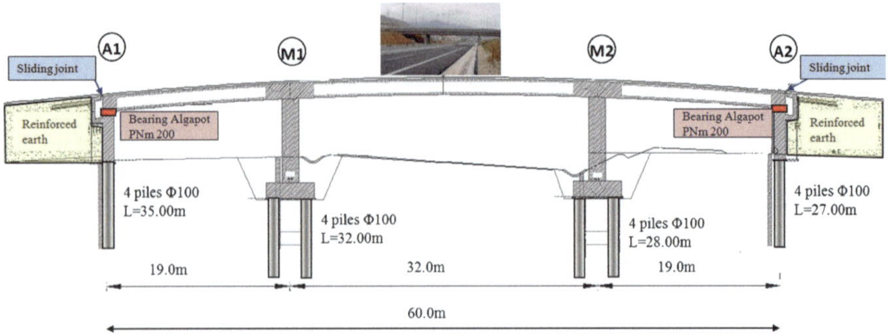

Fig. 14.3 Longitudinal cross-section of the bridge (*above*) and indicative overview of a typical overcrossing along Egnatia Highway (*bottom*) (Kappos et al. 2007)

document were $q_x = 2.50$, $q_y = 3.50$ and $q_z = 1.00$ for the response in the three principal directions, respectively. The target displacements of the bridge under study for the longitudinal direction are 7 cm for the case of stiff and 9.5 cm and soft soil conditions while they are equal to 5 cm and 6 cm respectively in the transverse direction (the complete calculation process can be found in Kappos et al. 2007). It is noted that the joint in the longitudinal direction is expected to close for twice the design earthquake.

14.3.2 MPS Application in the Linear and Non-linear Range

Given the detailed data available for the particular case-study, an effort was made to use common assumptions regarding earthquake excitations and solution algorithms. Along these lines, the Kozani, Greece earthquake (PGA = 0.19 g) was uniformly applied in all cases, while the Hilbert-Hughes-Taylor integration method was used, with time step $\Delta t = 0.01$ s and a total of 1000 steps (10 s of input). A uniform damping value of 5% was assumed for the first and second modes of vibration, defined through the Rayleigh alpha and beta corresponding factors. Gaps and stoppers that have been designed for the particular structure were ignored to ensure maximum possible activation of the embankment-abutment system. All analyses were conducted in the elastic range and the excitation was performed in the longitudinal direction. Parametric analysis was also performed to investigate the relative influence of various modelling assumptions. It was concluded that the parameter related to the maximum level of uncertainty was the critical embankment mass that was expected to be activated during the particular earthquake excitation and most importantly, the means to simulate its effect in the framework of the four different analysis strategies adopted. Four different models of increasing analysis complexity in considering the effect of embankment-abutment-superstructure interaction were developed (Sextos and Taskari 2008), starting from simple beam-type, dynamic spring and dashpot supported bridge models, to coupled multi-platform modules (the latter seen in Fig. 14.4).

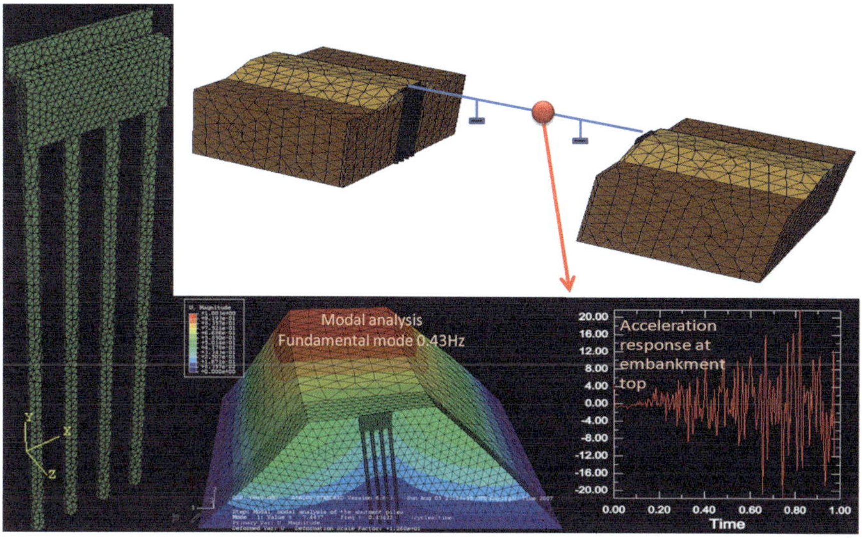

Fig. 14.4 System sub-structuring for multi-platform analysis

The results (Sextos and Taskari 2008) of the comparative analysis indicate that the maximum longitudinal displacement of the deck lies in the range of 0.6–1.0 cm (for all modelling approaches), whereas the fundamental period of the overall system may also differ by more than 100% despite the effort to use compatible properties where available. Further response measures (i.e. middle pier stresses) present dispersion is of the same order. It is also shown that Multi-Platform analysis predicted a bridge response which lays within the envelope of the response predicted by the three simpler (i.e., beam-type, dynamic spring and dashpot) approaches (Sextos and Taskari 2008).

Following the assessment of the four different approaches examined in the linear range, an effort was also made to compare the response of the bridge in the non-linear range. Along these lines, dynamic impedance matrices of the three simpler models which represent the abutment-embankment as well as the dynamic pile group stiffness were properly modified based to consider soil non-linear behaviour under strong ground shaking.

With respect to the MPS model, the Mohr-Coulomb constitutive model implemented in ABAQUS was utilised to simulate the non-linear soil behaviour. It is noted that in order to reduce the computational time required (which was approximately 7 h for a 10 s duration of uniform ground excitation at 0.01 s step executed on a Core 2 Duo processor with 2 GB RAM., the secant stiffness based on the detailed pushover analysis results along the longitudinal direction was applied and MPS was eventually repeated as an equivalent linear analysis. The later assumption is reduces the computational time to 2.5 h without introducing any additional uncertainty to the problem since it essentially imposes an 'exact' (i.e., derived through refined pushover analysis) non-linear, force-displacement

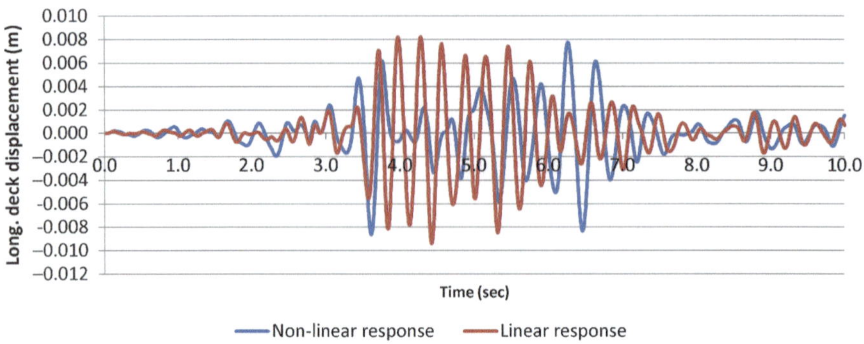

Fig. 14.5 Longitudinal displacement of the deck considering linear and non-linear soil response at the abutments, embankments and pier supports

relationship at each abutment control point that is used through a Multi-Platform process which is also inherently pseudo-static. The longitudinal displacement of the deck considering linear and non-linear soil response at the abutments, embankments and pier supports is presented in Fig. 14.5 where it is seen that Multi-Platform analysis is not only feasible in the non-linear range, but it leads to comparable results with those derived using linear elastic analysis with the expected increase in the response amplitude. Moreover, it is noted that the Multi-Platform simulation coordinated by UI-Simcor, may also have significant advantages for capturing the inelastic response of the R/C piers (not triggered in the case studied), because it can combine different software platforms and hence, implement different specialised constitutive models for the soil and structure. 4. MPS application for probabilistic assessment: Meloland Road Overcrossing.

14.4 MPS Application for Probabilistic Assessment: Meloland Road Overcrossing

14.4.1 Overview of the Bridge Studied

Having applied a pilot Multi-platform analysis for the deterministic assessment of a short overpass, an effort was made to perform the probabilistic vulnerability assessment of another, well-studied bridge, giving emphasis on the flexibility of the subsoil volume, the complaisance of the abutments and the liquefaction potential of sand deposits below the superstructure. Along these lines, the Meloland Road Overcrossing (Zhang and Makris 2002) was adopted for study. It is noted that Multi-Platform simulation was first performed for the particular bridge by Kwon and Elnashai (Kwon and Elnashai 2008) but without consideration of soil liquefaction. The MR overcrossing was built in 1971 and is located over Interstate 8 approximately 0.5 km from the fault rupture of the 1979 Imperial Valley earthquake as seen in Fig. 14.6.

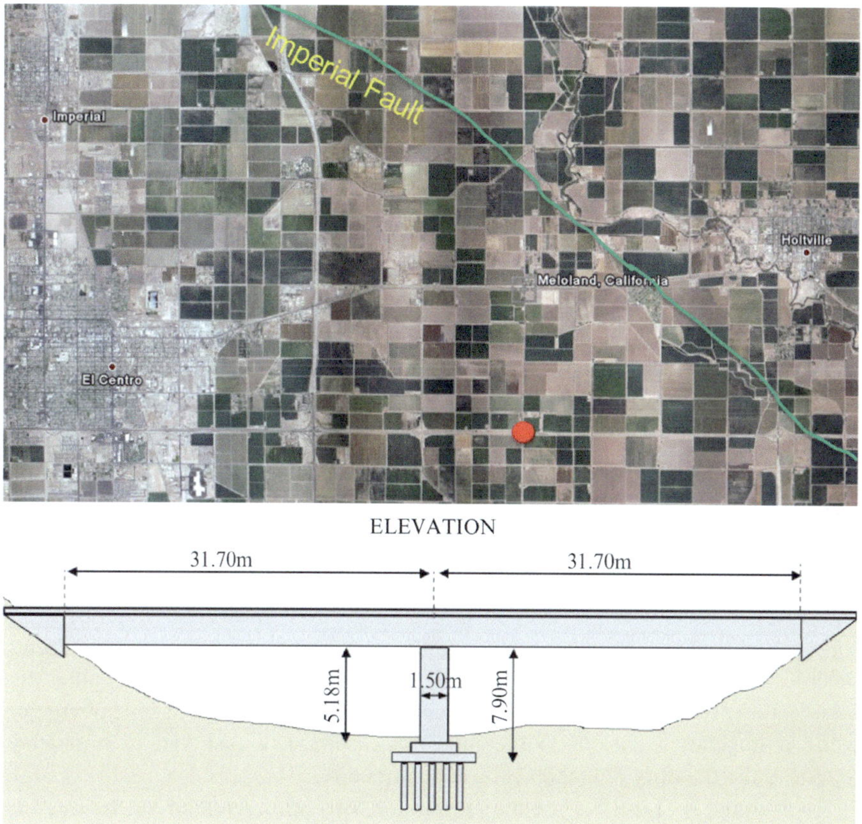

Fig. 14.6 Overview of the Meloland Road Overcrossing (MRO)

The bridge consists of two spans of pre-stressed box-girder decks monolithically connected to the center pier. The abutments are placed on fill. Seven piles support each abutment. Each side of abutment has 5.9 m of wing-wall. The pier at the center of the bridge has a diameter of 1.5 m and is 7.9 m high measuring from the top of piles. A total of 18 longitudinal reinforcement bars are used in the pier, the foundations of which are supported on 25 timber piles spaced at 0.91 m. The procedure of earthquake selection, site response and liquefaction consideration, analysis environment and vulnerability assessment is presented below.

14.4.2 Ground Motion Selection Process

Currently, various methods and procedures are applied in order to account for the inherent uncertainty of earthquake ground motion and select an adequately representative set of records to be used in the framework of the fragility analysis

Table 14.1 High and low pass filters rounded-off values for an event of magnitude M = 6.0

PSDF $S_0\,(cm^2\,/\,sec^3)$	KT-filter $f_g(Hz)$	KT-filter ξ_g	HP-filter ξ_g	HP-filter ξ_g	LP-filter ξ_g	LP-filter ξ_m
150.0	5.0	0.7	0.8	1.0	10.0	1.4

Table 14.2 High and low pass filters rounded-off values for an event of magnitude M = 6.5

PSDF $S_0\,(cm^2\,/\,sec^3)$	KT-filter $f_g(Hz)$	KT-filter ξ_g	HP-filter ξ_g	HP-filter ξ_g	LP-filter ξ_g	LP-filter ξ_m
300.0	6.0	3.0	0.3	1.0	5.0	0.7

Table 14.3 High and low pass filters rounded-off values for an event of magnitude M = 7.0

PSDF $S_0\,(cm^2\,/\,sec^3)$	KT-filter $f_g(Hz)$	KT-filter ξ_g	HP-filter ξ_g	HP-filter ξ_g	LP-filter ξ_g	LP-filter ξ_m
800.0	4.0	3.5	0.2	1.0	3.5	0.7

(a review of which can be found in Katsanos et al. (2010). In the particular case studied though and unlike the common practice, earthquake records were generated and selected directly at the bedrock level in order to consider the effect of strain softening of soil layers under cyclic loading and the potential liquefaction during seismic wave propagation on the overall fragility of the bridge. It is recalled that liquefaction affects both the frequency content and the amplitude of seismic motion while it imposes a *cut-off* on the input *acceleration*, a fact that in turn tends to reduce the ductility demand of the superstructure.

To this end, six levels of ground motion intensity were adopted for the fragility analysis, i.e., 0.05g, 0.1g, 0.2g, 0.3g, 0.4g, 0.5g, all assumed at the bedrock level. In total, 18 artificial records were generated at the bedrock level, corresponding to three surface wave Magnitudes (m_s) namely 6.0, 6.5 and 7.0. Records at the bedrock level were generated as white noise emanating from the earthquake source using Monte-Carlo simulations for stationary processes (Manolis et al. 2007) and were rendered non-stationary through use of a Eurocode 8 prescribed time envelope function. Records were filtered through a modified Kanai-Tajimi (KT) filter:

$$S(f) = \left[\frac{1+4\xi_g^2(f/f_g)^2}{\left[1-(f/f_g)^2\right]^2+4\xi_g^2(f/f_g)^2}\right]\left[\frac{(f/f_f)^4}{\left[1-(f/f_f)^2\right]^2+4\xi_f^2(f/f_f)^2}\right] \tag{14.2}$$

where ξ_g, f_g, ξ_f and f_f control the shape of the spectrum (ξ_g, f_g referring to the soil damping and frequency respectively) and S_0 is the white noise intensity factor typically related to the peak ground excitation and distance R by various expressions (i.e., (Shinozuka and Sato 1967; Shinozuka 1987)). Records were further filtered using a Magnitude-dependent high-pass (HP) filter, and a low-pass (LP) filter which were derived according to the values proposed by Papageorgiou and Aki (1983) as summarized in Tables 14.1–14.3.

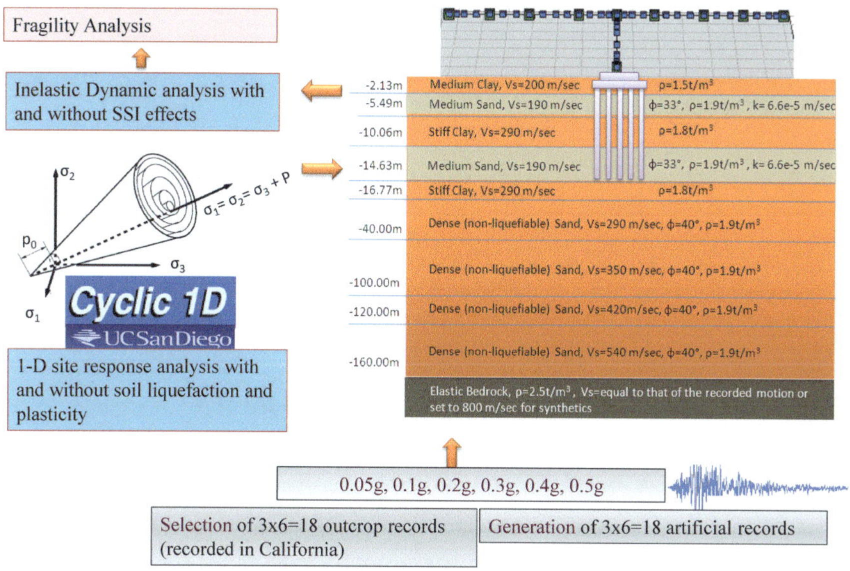

Fig. 14.7 Application of MPS for the Fragility analysis of MRO considering soil liquefaction and soil-structure interaction (Kwon et al. 2009)

No interaction phenomena between the incoming wave and the structure of the upper soil deposits (e.g., cavities, cracks, free surface and surface topography) were taken into account. The duration was also considered as a variable, according to the corresponding magnitude M and epicentral distance R. From the numerous definitions of strong ground motion duration, the 'significant duration' was adopted (Trifunac and Brady 1975) appropriately modified by a factor of 1.3 in order to derive the total signal duration required for the modulating function.

It is noted that despite of the convenience of the Kanai-Tajimi power spectrum, numerous researchers have stressed its inability to represent the high frequency content of near field motions. Nevertheless, the reason that it was adopted in the present study was that, it provides an incremental level of earthquake intensity which is well controlled by the user, a fact crucial for the foreseen fragility analysis. The target level of progressively increasing bedrock PGA (i.e., 0.05g, 0.1g, 0.2g, 0.3g, 0.4 g and 0.5g) was therefore established (Fig. 14.6). In order to account for the shortcoming in the representation of near-field motions, the frequency and damping of the HP filter were appropriately modified for all records assumed at distances $R \leq 20$ km. A complete set of the Magnitude M and distance R values used can be found in Kwon et al. (2008).

As a means to complement the artificially-generated ground motions, it was deemed necessary to select an additional set of 18 records from a large database of available recordings (http://peer.berkeley.edu). Ground motions recorded at rock sites with PGA of 0.05g, 0.1g, 0.2g, 0.3g, 0.4g and 0.5g were again selected (Fig. 14.7). The selected ground motions are used as bedrock reference motion

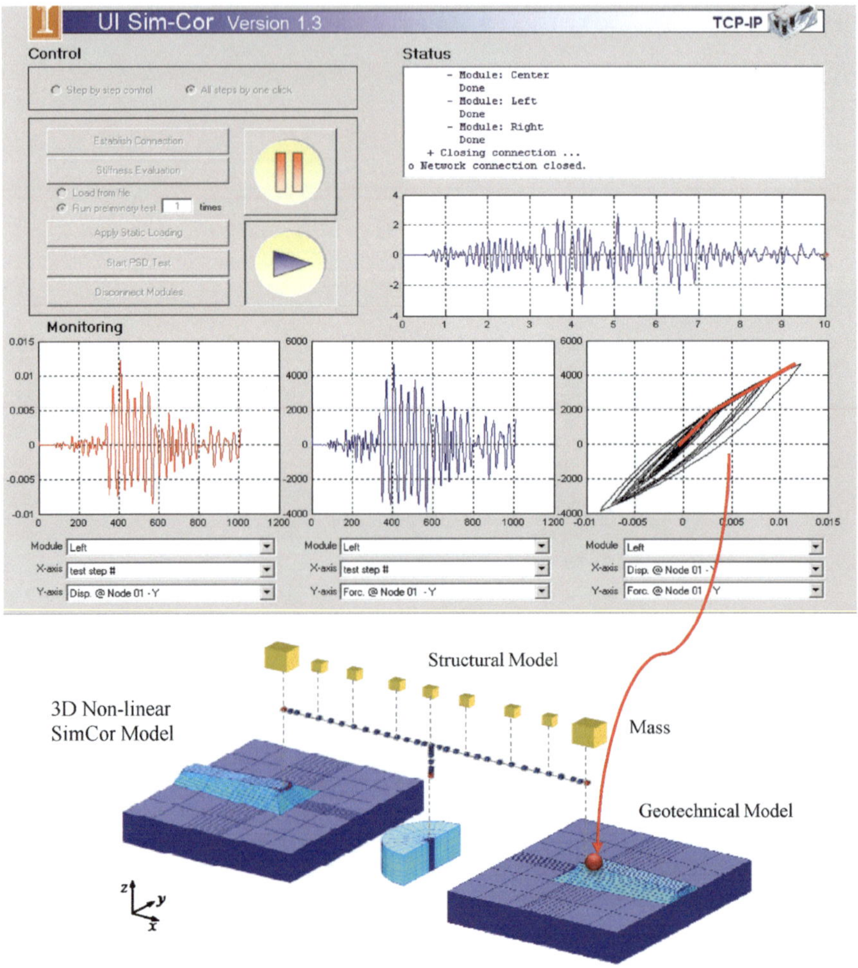

Fig. 14.8 System sub-structuring for multi-platform simulation (Kwon et al. 2008)

with minor scaling. Both near- and far-field records were sought in order to enrich the earthquake ground motion sample and further mitigate the inherent limitation of the artificial ground motions for near-field conditions due to the limitations of the Kanai-Tajimi spectrum. The catalogue of the ground motions used is presented in Kwon et al. (2008).

14.4.3 Probabilistic Assessment Framework and Sample Results

With the complete set of ground motions and calibrated inelastic finite element models, a total of 144 MPS analyses (Fig. 14.8) were carried out. These comprise

4 different representations of the bridge system corresponding to four different levels of analysis complexity (Kwon et al. 2009) subjected to 36 ground motions. Further modelling uncertainties we not considered, due to the fact that all analyses were computationally rather expensive. It was initially assumed that damage to the bridge occurs only at the bottom of the pier. Hence damage of an element corresponds to damage of the system. In Kwon et al. (2009) new damage indices are further introduced to take account of excessive foundation lateral displacement, rotation as well as potential pile damage at the interface between liquefiable and non-liquefiable layers. The Structural Damage Index, herein denoted as SDI to be distinguished from ground-related Damage Indices (DI) for a pier, was computed based on the Moment-Curvature relationship from response history analysis of the bridge system and the modified Park-Ang relationship (Park and Ang 1985):

$$SDI = \frac{\varphi_m - \varphi_y}{\varphi_u - \varphi_y} + \beta_e \frac{E}{M_y \varphi_u} \tag{14.3}$$

where φ_m is the maximum curvature achieved during cyclic loading, φ_u is the ultimate curvature, φ_y and M_y is the yield curvature and bending moment of the concrete section respectively, E is the cumulative energy absorbed in the hysteresis loops and β_e is a strength loss parameter set equal to 0.1.

After calculating the median DI values for each level of input intensity parameter (i.e. bedrock PGA), the Damage Index and Damage Limit States were correlated according to Stone and Taylor (1993), i.e. damage limit state I ('repairable damage') was considered for values of $0.11 < \mathrm{DI} \leq 0.40$, damage limit state II ('irreparable damage') for values $0.40 < \mathrm{DI} \leq 0.77$ and damage limit state III ('collapse') for $0.77 < \mathrm{DI} \leq 1.00$. Three different fragility curves were then derived for each Damage State, after plotting the corresponding histogram of its occurrence as a function of seismic intensity. The probability that damage will eventually exceed a certain Damage State, given a level of intensity was then computed though the following expression (Nielson 2005):

$$P[(D>C)|IM] = \Phi\left[\frac{in\left(\frac{\widehat{D}}{\widehat{C}}\right)}{\sqrt{\beta_{D|IM}^2 + \beta_C^2}}\right] \tag{14.4}$$

Where $\widehat{D}$, $\beta_{D/IM}$ and $\widehat{C}$, β_C is the median and the standard deviation of the demand and the capacity respectively and IM the appropriate intensity measure (i.e., bedrock PGA, in this case). It is noted herein that a factor:

$$\beta = \sqrt{\beta_{D|IM}^2 + \beta_C^2} = 0.6 \tag{14.5}$$

was adopted to account for all sources of demand and supply uncertainty. Detailed inelastic site response analyses considering liquefaction effects were carried out.

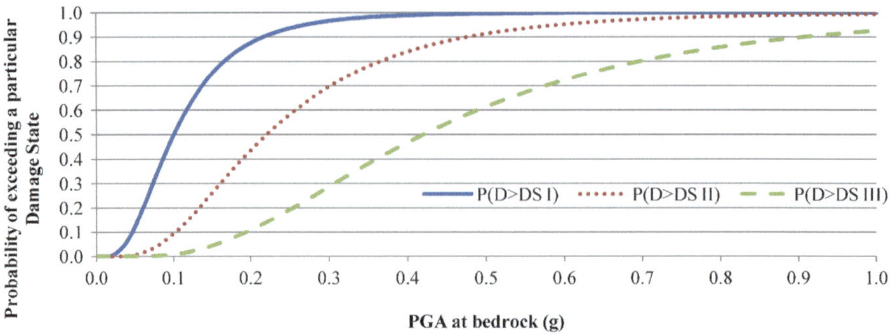

Fig. 14.9 Fragility curves of the Meloland Road Overcrossing

The results indicate that the inelastic dynamic response of the investigated bridge-foundation-soil system is significantly affected by the liquefaction of upper sand layers. In addition to the significance of accounting for liquefaction, the spatial extent of liquefaction is also shown to be important because it can be considered as an additional source of spatial variation of earthquake input (Sextos et al. 2003). It was also shown that both liquefaction and its spatial distribution are influential on determining the characteristics of the input motion and the demand imposed on foundations and superstructure. Sample results of the bridge fragility for various levels of earthquake loading are illustrated in Fig. 14.9. Apparently, the assumptions made to facilitate the application of the proposed Multi-platform analysis framework for liquefaction-sensitive fragility assessment (i.e., constitutive laws and parameters adopted for considering liquefaction, procedure for site response analysis etc.) are not integral to the procedure and may be improved by researchers seeking to quantify the probabilistic response of soil-foundation-structure systems.

14.5 Concluding Remarks on the Applicability of Multi-Platform Simulation for the Deterministic and Probabilistic Seismic Response Assessment of Soil-Bridge Systems

The analyses presented herein demonstrate the applicability of Multi-Platform Simulation for the deterministic and probabilistic assessment of bridges considering soil-structure interaction and soil liquefaction. The experience gained by these two pilot studies is that MPS is an effective and feasible analysis alternative that presents the advantage of permitting the use of different software for different system components. This is particularly valuable when the non-linear inelastic response of concrete members and/or soil is of importance because specialized tools in structural and geotechnical engineering can be coupled effectively in a way that is not currently feasible when using a single computer program. On the other hand, in contrast to purely dynamic analysis of systems using a single software, MPS is by definition inherently limited by the assumptions of the Pseudo-Dynamic method and hence, particular soil components inevitably respond and contribute in a static manner. It is believed that MPS is a promising approach that provides rigorous results at acceptable computational cost. As a result it is currently deemed preferable compared to single platform analysis, at least for large and complex soil-bridge systems that exhibit highly non-linear behaviour. The future advances in numerical analysis, material modelling, integration schemes and finite element software as a whole, together with the anticipated rapid increase of computational power will dictate whether single or multi-platform analysis will become eventually the standard tool for assessing the seismic response of complex systems.

Acknowledgements The author would like to thank Prof. Amr Elnashai and Assist. Prof. Oh-Sung Kwon for their precious assistance regarding the application of the analysis coordinator UI-Simcor, developed at the University of Illinois. Moreover, it is noted that most of the work presented herein relates to joint publications. Acknowledgments are also ought to Prof. Andreas Kappos and Petros Potikas at Aristotle University Thessaloniki for kindly providing the data and results derived during the non-linear static assessment of the first bridge studies as well as to Olympia Taskari for her significant contribution in the MPS of the particular overpass.

References

Beltrami C, Lai CG, Pecker A (2005) A kinematic interaction model for a large-diameter shaft foundation. An application to seismic demand assessment of a bridge subject to coupled swaying rocking excitation. J Earthquake Eng 9(Special issue 2):355–393

Biskinis D, Fardis M (2006) Assessment and upgrading of resistance and deformation capacity of concrete Piers. In: 1st European conference on earthquake engineering and seismology (a joint event of the 13th ECEE and the 30th General Assembly of the ESC), Geneva, paper 315

Bousias SN, Strepelias E, Palios X, Fardis MN (2008) Experimental study of bridge seismic isolation systems with or without supplemental energy. In: 14th world conference on earthquake engineering, Beijing

Chen C-H, Lai W-C, Cordova P, Deierlein GG, and Tsai K-C (2003) Pseudo-dynamic test of full-scale RCs frame: Part 1 – Design, construction and testing. In: International workshop on steel and concrete composite constructions, Taipei, pp 107–118

Combault J, Pecker A, Teyssandier J-P, Tourtois J-M (2005) Rion-antirion bridge, Greece – Concept, design, and construction. Struct Eng Int J Int Assoc Bridge Struct Eng (IABSE) 15 (1):22–27

Combescure D, Pegon P (1997) α-Operator splitting time integration technique for Pseudo-dynamic testing error propagation analysis. Soil Dyn Earthquake Eng 16:427–443

Dicleli M (2005) Integral abutment-backfill behaviour on sand soil – pushover analysis approach. J Bridge Eng 10(3):354–364

Elnashai AS, Papanikolaou V, Lee D (2002) Zeus NL – a system for inelastic analysis of structures, Technical Report. Mid-America Earthquake Center, University of Illinois at Urbana-Champaign

Elnashai AS et al (2007) Distributed earthquake response history analysis of interacting and multi-resolution structural and geotechnical systems. ECCOMAS Thematic conference on computational methods in structural dynamics and earthquake engineering, COMPDYN2007, Rhodes

Filippou FC, Constantinides M (2004) FEDEASLab Getting Started Guide and Simulation Examples Technical Report NEESgrid-2004-22. www.nees-grid org
Finn WD (2005) A study of piles during earthquakes: issues of design and analysis. Bull Earthquake Eng 3(2):141–234
Gazetas G, Mylonakis G (1998) Seismic soil-structure interaction: new evidence and emerging issues, Geotechnical special publication 75, Geotechnical earthquake engineering and soil dynamics III. Am Soc Civil Eng 2:1119–1174
Goel RK, Chopra A (1997) Evaluation of bridge abutment capacity and stiffness during earthquakes. Earthquake Spectra 13(1):1–23
Hibbit K, Sorenson E (2006) ABAQUS ver 6.6. User's manual, Pawtucket
Jeong S, Elnashai AS (2005) Analytical assessment of an irregular RC frame for full-scale 3D Pseudo-dynamic testing part I: analytical model verification. J Earthquake Eng 9(1):95–128
Kappos A, Potikas P, Sextos A (2007) Seismic assessment of an overpass bridge accounting for non-linear material and soil response and varying boundary conditions', Computational methods in structural dynamics and earthquake engineering, COMPDYN 2007. paper 1580
Katsanos E, Sextos A, Manolis G (2010) Selection of earthquake ground motion records: a state-of-the-art review from a structural engineering perspective. Soil Dyn Earthquake Eng 30(4):157–169
Kotsoglou A, Pantazopoulou S (2007) Bridge–embankment interaction under transverse ground excitation. Earthquake Eng Struct Dyn 36:1719–1740
Kwon O-S, Elnashai AS (2008) Seismic analysis of Meloland Road Overcrossing using multiplatform simulation software including SSI. J Struct Eng 134(4):651–660
Kwon O-S, Nakata N, Elnashai AS, Spencer B (2005) A framework for multi-site distributed simulation and application to complex structural systems. J Earthquake Eng 9(5):741–753
Kwon O-S, Sextos A, Elnashai AS (2008) Liquefaction-dependent fragility relationships of complex bridge-foundation-soil systems. In: International conference on earthquake engineering and disaster mitigation, Jakarta
Kwon O-S, Sextos A, Elnashai AS (2009) Seismic Fragility of a bridge on liquefaction susceptible soil. In: 10th international conference on seismic safety and reliability, Osaka, 13–17 Sept 2009
Manolis GD, Rangelov TV, Dineva PS (2007) Free-field wave solutions in a half-plane exhibiting a special-type of continuous inhomogeneity. Wave Motion 44(4):304–321
McKenna F, Fenves GL (2001) The open sees command language manual, version 1.2. Pacific Earthquake Engineering Research Center, University of California at Berkeley
Molina FJ, Verzeletti G, Magonette G, Buchet P, Geradin M (1999) Bi-directional pseudodynamic test of a full-Size three-storey building. Earthquake Eng Struct Dyn 28:1541–1566
Mylonakis G, Gazetas G (2000) Seismic soil-structure interaction: beneficial or detrimental? J Earthquake Eng 4(3):277–301
National Science Foundation, Network for Earthquake Engineering Simulation (NEES) (2000) System integration, program solicitation Report NSC00-7, USA
Negro P, Pinto AV, Verzeletti G, Magonette GE (1996) PsD test on four-story R/C building designed according to Eurocodes. J Struct Eng 122(12):1409–1471
Neild SA, Stoten DP, Drury D, Wagg DJ (2005) Control issues relating to real-time substructuring experiments using a shaking table. Earthquake Eng Struct Dyn 34(9):1171–1192
Nielson BG (2005) Analytical fragility curves for highway bridges in moderate seismic zones. Ph.D. thesis, Georgia Institute of Technology, Atlanta
Pan P, Tada M, Nakashima M (2005) Online hybrid test by internet linkage of distributed test-analysis domains. Earthquake Eng Struct Dyn 34:1407–1425
Papageorgiou AS, Aki K (1983) A specific barrier model for the quantitative description of inhomogeneous faulting and the prediction of strong ground motion: i description of the model. Bul Seism Soc Am 73(3):693–722
Park Y-J, Ang AH-S (1985) Mechanistic seismic damage model for reinforced concrete. J Struct Eng 111(4):722–739

Pecker A (2004) Design and construction of the Rion-Antirion bridge. Geotech Spec Pub 126 (I):216–240
Pegon P, Pinto AV (2000) Pseudo-dynamic testing with substructuring at the ELSA laboratory. Earthquake Eng Struct Dyn 29(7):905–925
Pender MJ (1993) Aseismic pile foundation design analysis. Bull N Z Natl Soc Earthquake Eng 26 (1):49–161
Pinho R, Elnashai AS (2000) Dynamic collapse testing of a full-scale four storey RC frame. ISET J Earthquake Technol 37(4):143–163
Pinto AV, Pegon P, Magonette G, Tsionis G (2004) Pseudo-dynamic testing of bridges using non-linear substructuring. Earthquake Eng Struct Dyn 33(11):1125–1146
Sextos A, Taskari O (2008) Comparative assessment of advanced computational tools for embankment-abutment-bridge superstructure interaction. In: 14th world conference on earthquake engineering, Beijing
Sextos A, Kappos A, Pitilakis K (2003) Inelastic dynamic analysis of RC bridges accounting for spatial variability of ground motion, site effects and soil-structure interaction phenomena, Part 2: parametric analysis. Earthquake Eng Struct Dyn 32(4):629–652
Shinozuka M (1987) Stochastic fields and their digital simulation, in stochastic methods in structural dynamics. Martinus Nijhoff, Dordrecht
Shinozuka M, Sato Y (1967) Simulation of non-stationary random processes. J Eng Mech Div ASCE 93(EM1):11–40
Spencer BF, Elnashai AS, Park K, Kwon O-S (2006) Hybrid test using UI-simCor, three-site experiment . Final report to NEESit for Phase I project of hybrid simulation framework development, University of Illinois at Urbana-Champaign
Stone WC, Taylor AW (1993) Seismic performance of circular bridge columns designed in accordance with AASHTO/CALTRANS standards, vol 170, NIST building science series. National Institute of Standards and Technology, Gaithersburg
Takahashi Y, Fenves GL (2006) Software framework for distributed experimental computational simulation of structural systems. Earthquake Eng Struct Dyn 35:267–291
Trifunac MD, Brady AG (1975) A study on the duration of strong earthquake ground motion. Bull Seismol Soc Am 65(3):581–626
Tsai K, Hsieh S, Yang Y, Wang K, Wang S, Yeh C, Cheng W, Hsu C, Huang S (2003) Network platform for structural experiment and analysis (I), NCREE-03–021. National Center for Research on Earthquake Engineering, Taiwan
Watanabe E, Sugiura K, Nagata K, Yamaguchi T, Niwa, K (1999) Multi-phase Interaction testing system by means of the internet. In: 1st international conference on advances in structural engineering and mechanics, Seoul, pp 43–54
Wolf JP (1994) Foundation vibration analysis using simple physical models. PTR Prentice Hall in Englewood Cliffs, NJ
Zhang J, Makris N (2002) Kinematic response functions and dynamic stiffnesses of bridge embankments. Earthquake Eng Struct Dyn 31:1933–1966

Chapter 15
The Role of Probabilistic Methods in Evaluating the Seismic Risk of Concrete Dams

Alessio Lupoi and Carlo Callari

Abstract A recent research on seismic assessment of concrete dams is illustrated in this chapter, including a brief comparison with other available approaches to the subject. The work of the authors has been focused on development and validation of a probabilistic methodology taking into account the uncertainties affecting structural data and external actions as well as the physical complexity of the dam-foundation-reservoir system. The seismic response of such a system is estimated from a reduced number of dynamic finite element analyses and the corresponding fragility curves are obtained via a Monte Carlo simulation procedure. The main results of the application of the proposed methodology to the case of an existing concrete gravity dam are finally summarized.

Keywords Seismic assessment • concrete gravity dams • fragility curves • fault tree • risk analysis • finite elements • dynamics

15.1 Introduction

The performance of concrete gravity dams in past earthquakes has been, overall, satisfactory. Among the few structures where substantial damages were observed there is the Koyna Dam (India), a 103 m-high gravity dam, built in 1954–1963, which was in 1967 exposed to a nearby earthquake of magnitude M = 6.5 leading to peak accelerations of 0.49 g in dam body (up-downstream direction). Significant longitudinal cracks appeared at the neck, close to the abrupt change of the

A. Lupoi (✉)
Department of Structural Engineering & Geotechnics,
University of Rome "La Sapienza", Via Gramisci, 53-00197 Rome Italy
e-mail: alessio.lupoi@uniroma1.it

C. Callari
Facoltà di Ingegneria, University of Molise, Termoli (CB), Italy
e-mail: carlo.callari@unimol.it

M. Dolšek (ed.), *Protection of Built Environment Against Earthquakes*,
DOI 10.1007/978-94-007-1448-9_15,

Fig. 15.1 Cracks at downstream and upstream faces of Sefid Rud Dam (From Wieland 2009)

downstream face slope (Chopra and Chakrabarti 1972). Cracks were also observed in the inspection gallery located at dam base, where an almost 100% increase of drain discharges was registered after the earthquake.

More recently, the 106 m-high Sefid Rud buttress dam (Iran) built in 1958–1962 experienced the 1990 Manjil earthquake, whose epicentre was at less than 16 km from dam site (M = 7.7, estimated PGA = 0.7 g). Severe cracking appeared at horizontal lift joints in the upper part of the dam body, with a 20 mm shear displacement oriented in downstream direction (Fig. 15.1).

In Taiwan, during the 1999 Chi-Chi earthquake (M = 7.3), two of the 18 spillways composing the 25 m-high Shih-Kang Weir collapsed as a consequence of fault movements at one abutment. This is the unique reported example of a dam exposed to substantial fault movements. In particular, over two thirds of the dam body were uplifted up to 9 m and displaced 2 m horizontally. The dam experienced horizontal accelerations up to 0.50 g.

In summary, the Shih-Kang Weir is the unique case of earthquake-induced collapse of a large concrete gravity dam. It is also generally acknowledged that the seismic performance of the other two aforementioned dams was mainly motivated by poor design and construction (Koyna Dam) or by significant underestimation of seismic actions (Sefir Rud Dam). In both cases, as a consequence of earthquake damage, there was no significant release of reservoir water and the structures were rehabilitated and upgraded. These data suggest a minor seismic vulnerability of concrete gravity dams. However, it has to be noted that less than 20 concrete dams (including arch dams) have been exposed to events characterized by a PGA greater than 0.2 g (Wieland et al. 2003; Bureau 2003).

As it was for Koyna and Sefir Rud Dams, many other existing concrete gravity dams were designed employing out-of-date analysis methods and/or assuming seismic actions that are nowadays considered as unacceptably underestimated. As a consequence, the assessment of the current safety level of these structures is a main concern for owners and reclamation agencies.

Motivated by these considerations, a tool for the seismic assessment of existing dams has been recently presented by Lupoi and Callari (2011).

Table 15.1 Classification in terms of the *hazard* (ICOLD 1989)

Condition	Hazard class and (Hazard rating)
$PGA<0.10g$	I (Low)
$0.10g \le PGA \le 0.25g$	II (Moderate)
$PGA>0.25g$ (but no active fault within 10 km of site)	III (High)
$PGA>0.25g$ (active fault closer than 10 km from site)	IV (Extreme)

Table 15.2 Dam vulnerability classification: weighting points in brackets (ICOLD 1989)

Risk factor	Extreme	High	Moderate	Low
Capacity [100 m^3]	>120 (6)	120-1 (4)	1-0.1 (2)	<0.1 (0)
Height [m]	>45 (6)	45-30 (4)	30-15 (2)	<15 (0)
Evacuation requirements (No. of persons)	>1,000 (12)	1,000-100 (8)	100-1 (4)	None (0)
Potential downstream damage	High (12)	Moderate (8)	Low (4)	None (0)

The proposed methodology can be employed to support cost-effective decisions on rehabilitation and retrofitting strategies. Using concepts similar to those proposed for buildings and bridges by Lupoi et al. (2006), the method is able to face both the complexity of the dam-foundation-reservoir system and the several uncertainties affecting the problem. These uncertainties are a consequence of the lack of knowledge of "physical" data, such as dam geometry, rock mass profile, material properties as well as of inaccuracies of the models employed to evaluate the system failure modes.

Some conventional seismic classifications of dams and of other available probabilistic approaches to the evaluation of seismic risk of concrete dams are briefly reviewed in the next two sections. The recently proposed procedure for the probabilistic evaluation of seismic risk is summarized in Sect. 15.4 and its application to a real case is reported in Sect. 15.5. The chapter ends with some concluding remarks.

15.2 Traditional Seismic Classifications of Dams

The evaluation of the seismic safety of existing dams is still carried out largely by empirical methods. The use of a reliability approach is still limited to the scientific community. Empirical methods are typically based on statistical analyses and provide a "class of risk" as function of the site and of the type and geometry of the dam. A representative example is provided by the traditional method by ICOLD (1989): a "Hazard Class" is assigned as function of the PGA and of the fault distance (Table 15.1).

Similarly, a "Risk class" is evaluated on the basis of: reservoir capacity; dam height; number of persons to be evacuated; potential damage in downstream valley (Table 15.2). The "Risk Factor" is given by the sum of the ratings corresponding to these technical and social parameters (Table 15.3).

Table 15.3 Dam risk classification (ICOLD 1989)

Total risk factor	Risk class and (risk rating)
$0 \div 6$	I (Low)
$7 \div 18$	II (Moderate)
$19 \div 30$	III (High)
$31 \div 36$	IV (Extreme)

This simple method is applicable to "ordinary" dams, while it is recommended to carry out a detailed analysis for more complex structures (dam height greater than 90 m, reservoir capacity greater than 120,000 m^3, poor ground conditions, etc.).

It can be noted that the ICOLD approach keeps separate the (seismic) hazard from the (structural) vulnerability, which represents an important limitation for the method. The "total risk factor" (*TRF*), introduced by Bureau (2003), overcomes this shortcoming. The following expression is proposed:

$$TRF = [(CRF + HRF + ARF) + DHF] \times PDF$$

where:

- *CRF* (*capacity risk factor*) is related to reservoir volume;
- *HRF* (*height risk factor*) is related to the dam height;
- *ARF* (*age rating factor*) is related to the dam age.
- *DHF* (*downstream hazard factor*) is related to the potential damages at downstream;
- *PDF*(*predicted damage factor*) accounts for the "damageability" of the dams.

The sum of $(CRF + HRF + ARF)$ accounts for the dam structural behaviour. The parameter *DHF* account for the social consequences and it is given by:

$$DHF = ERF + DRI$$

where *ERF* (*evacuation requirements factor*) depends on the exposure of the population and *DRI* (*downstream damage index*) is based on the value of private, commercial, industrial, or government property in the potential flood path.

The parameter *PDF* is obtained from the "*dam vulnerability curves*" developed by Bureau and Ballentine (2002) from the observed performance of dams during earthquakes. The curves depend on the dam type and on the site seismic hazard and tectonic environment.

The dams are classified in four "Risk Classes" as function of the *TRF* (Table 15.4).

This method has been employed by Tosun et al. (2007) to evaluate the seismic risk of 32 dams located in the Euphrates basin (Turkey). All the examined dams belong to Risk Class II and III.

Table 15.4 Risk class as function of the total risk factor (Bureau 2003)

Total risk factor	Risk class (risk rating)
2 ÷ 25	I (Low)
25 ÷ 125	II (Moderate)
125 ÷ 250	III (High)
>250	IV (Extreme)

15.3 Probabilistic Approaches to Seismic Assessment of Concrete Dams

A brief review of three relevant publications, among the very few available on this subject, is given below focusing on the following features: system model and seismic analysis; reliability method and random variables; failure mechanisms. A general review of reliability analysis for dam safety is reported by Westberg (2010).

15.3.1 Araújo and Awruch (1998)

A fully probabilistic methodology and its application to the case of the Tucuruí dam in Brasil are presented by Araújo and Awruch (1998). Dam body, foundation rock and reservoir water are discretized in finite elements. An Eulerian formulation is employed for the reservoir water. A dynamic analysis is carried out for the evaluation of system response. The seismic action is considered as a non-stationary stochastic process.

A standard Monte Carlo approach is employed to solve the reliability problem. In dam body, the concrete compressive strength is described by means of a spatially distributed random model. Such an accurate solution is motivated by the large volume of the dam body. The mean values of other material properties (concrete tensile strength, Young's modulus and the adhesion at the dam-foundation interface) are derived from the compressive strength through CEB-FIP (1993) deterministic expressions.

Cracking, concrete crushing and sliding at the dam-foundation interface are the three investigated mechanisms of failure. Results are presented in terms of safety factors, i.e. the ratios between demand and capacity. A set of 50 simulations is carried out. The adequacy of such a simulation set is not explicitly proven and the point-wise assessment of the stress state in dam body is not representative of the global state of the dam.

15.3.2 Yanmaz and Beser (2005)

Yanmaz and Beser (2005) describe a risk-based safety evaluation of concrete gravity dams. As a case study, the 50 m-high Porsuk dam (Turkey) is analysed using the commercial software CADAM (Leclerc et al. 2004).

This software performs the analysis of gravity dam-foundation systems modelling lift joints as well as drain location and effectiveness. An added mass-mass approach (Westergaard 1933) is employed to account for the effects of the reservoir water. Pseudo-Static and Pseudo-Dynamic analyses are carried out for the earthquake load condition. In the latter, the maximum response due to the fundamental mode of vibration is represented by equivalent lateral forces and is computed directly from the earthquake design spectrum without a response history analysis.

The probability of failure of the dam-foundation system is carried out using the Monte-Carlo simulation. The tensile strength, the peak friction angle and the peak cohesion in lift joints and dam base are treated as random variables. Drain efficiency, ice load and the PGA are the other random variables considered in the assessment.

Tensile cracking at the upstream face, sliding (both peak and residual) at lift joints and dam base as well as overturning are the investigated failure mechanisms. Results are presented in terms of safety factors.

15.3.3 Tekie and Ellingwood (2003)

Tekie and Ellingwood (2003) presented a methodology for developing seismic fragility curves of concrete gravity dams. The method is applied to the 53 m high Bluestone concrete gravity dam built in the'40 s in USA.

Dam body and foundation are modelled by finite elements; the reservoir is modelled using the Darbe's approach (nodal masses in series with dampers). A non-linear response at soil-structure interface is modelled by a Mohr-Coulomb friction law. The system response is evaluated by dynamic analysis using recorded accelerograms as seismic input.

A Monte Carlo method is used with Latin Hypercube sampling to reduce the computational cost of simulation. The material random variables are compressive strength of concrete; friction angle, cohesion and dilation angle of foundation rock. In addition, the efficiency of vertical drains, the efficiency of grout curtains and the effective uplift area are also considered as random variables. For the hazard side, the spectral acceleration at reference period is the hazard random variable.

Several failure mechanisms are investigated: drift deformation of the dam body, cracking at the dam neck, material failure for compression either in foundation or in concrete at the toe; compression stress at the dam-soil interface; sliding at the dam-soil interface (for the seismic assessment only), pool elevation and resultant outside of the dam's base (for the flood assessment only). Results consist of fragility curves for the investigated mechanisms.

15.3.4 Comments

The stochastic approach by Araújo and Awruch (1998), although very accurate, is not suitable for practical applications due to its high computational costs and also to the poor representation of the seismic input achieved by the artificial records employed.

On the contrary, the numerical model of the system employed by Yanmaz and Beser (2005) as well as the type of analysis for the earthquake load conditions are too simplified for an accurate assessment of the seismic risk. In addition, results are not suitable to be presented in the form of fragility curves.

In summary, among the examined papers, the methodology by Tekie and Ellingwood (2003), based on the combination of dynamic analyses and recorded accelerograms, is certainly the most suitable for carrying out a seismic assessment of a concrete dam. However, improvements in the derivation of the fragility curves and in the modelling of the reservoir-dam-foundation system may be introduced, as illustrated in the next section.

15.4 The Probabilistic Methodology by Lupoi and Callari (2011)

Motivated by the above considerations and following concepts similar to those proposed for buildings and bridges by Lupoi et al. (2006), the authors developed a probabilistically-based methodology for the seismic assessment of existing dams. A brief summary of the method is provided in the following section. A detailed description can be found in (Lupoi and Callari 2011).

The reservoir-dam-foundation is described as a system consisting of as many components as its *failure* mechanisms. The term "failure" denotes herein the exceedance of a pre-defined limit state. The state of the system is defined in terms of the state of its components through a function expressing the logical arrangement of failure mechanisms.

The state of a generic i-th component is described by a limit-state function of the form:

$$g_i = C_i(\mathbf{x}) - \underset{t}{max}\left[D_i(\mathbf{x}, \mathbf{y}, t)\right] \tag{15.1}$$

where D_i is the structural demand induced by the external actions and C_i is the corresponding capacity. Demand and capacity are function of the "structural" random variables and of the "external action" random variables, separately collected in the two vectors named **x** and **y**, respectively. In Eq. 15.1 the capacity is assumed to be constant over the time duration of the seismic event and to be independent on the external action. Thus, the failure of the component occurs for $g_i < 0$.

The separation of the "structural" (e.g. geometry, strength of materials, etc) and the "external action" random variables (e.g. frequency content, duration, intensity, reservoir water level) is particularly useful for the derivation of the so-called fragility curves, i.e. the relations between the dam failure probability and the intensity of the external actions. In fact, following a well-established trend (Cornell et al. 2002; Pinto et al. 2004), the variability in D_i induced by **y** is established by means of numerical analyses carried out for a limited number of pre-selected recorded ground motions and reservoir water levels. These analyses are performed for the mean values of random structural properties **x**. The uncertainty in the demand related to structural parameters **x** may be efficiently accounted, though in approximate fashion, by a linear expansion of the demand vector with respect to the mean of **x** (Lupoi et al. 2006).

Regarding the terms $C_i(\mathbf{x})$, the available capacity formulas are generally based on relatively weak mechanical basis, integrated with empirical knowledge. In the proposed method the capacity terms are expressed in the multiplicative format:

$$C_i(\mathbf{x}, \varepsilon_C) = \overline{C}_i(\mathbf{x})\, \varepsilon_{C_i} \tag{15.2}$$

where $\overline{C}_i(\mathbf{x})$ is the value obtained by semi-empirical formulas available in the literature and $\varepsilon_{\mathrm{Ci}}$ is a model-error term accounting for scatter and, when necessary, for bias as well. The type of distribution of $\varepsilon_{\mathrm{Ci}}$ is based on expert judgment.

By evaluating demands and capacities as described above, the system-reliability problem is reduced to a time-invariant one that can be written, in a general cut-set formulation, as:

$$P_f(\mathbf{y}) = Pr\left\{ \bigcup_{J=1}^{n_C} \bigcap_{i \in I_{C_j}} C_i(\mathbf{x}, \varepsilon_{C_i}) \leq D_i(\mathbf{x}|\mathbf{y}) \right\} \tag{15.3}$$

where n_c is the number of cut-sets and I_{Ci} is the index set for the modes belonging to the *j*-th cut-set. The cut-set formulation is a conventional way to describe the logical arrangement of components for the evaluation of the state of the system (Der Kiureghian 2005).

The probability of failure in Eq. 15.3 is evaluated by a standard Monte Carlo simulation, which is simple and comparatively inexpensive, since it does not require any structural analysis. In practice, for each combination of the seismic intensity, *IM*, and of the reservoir water level, H_w, the Monte Carlo simulation consists of the following steps:

- sampling from the basic random variables **x**;
- evaluating the capacities from the corresponding models;
- sampling randomly from the *N* demand vectors obtained from dynamic analyses (where *N* is the number of recorded ground motions);
- estimating the state of the elements Eq. 15.1;
- estimating the state of the system Eq. 15.3.

Monte Carlo simulation with a sufficient number of samples yields the conditional exceedance probability $P_f(\mathbf{y})$. The complete fragility curve is obtained by repeating this step for a convenient number of **y**. The seismic risk of the system can be obtained by convolution of the fragility curves with the hazard curves of the seismic action and of the reservoir water level.

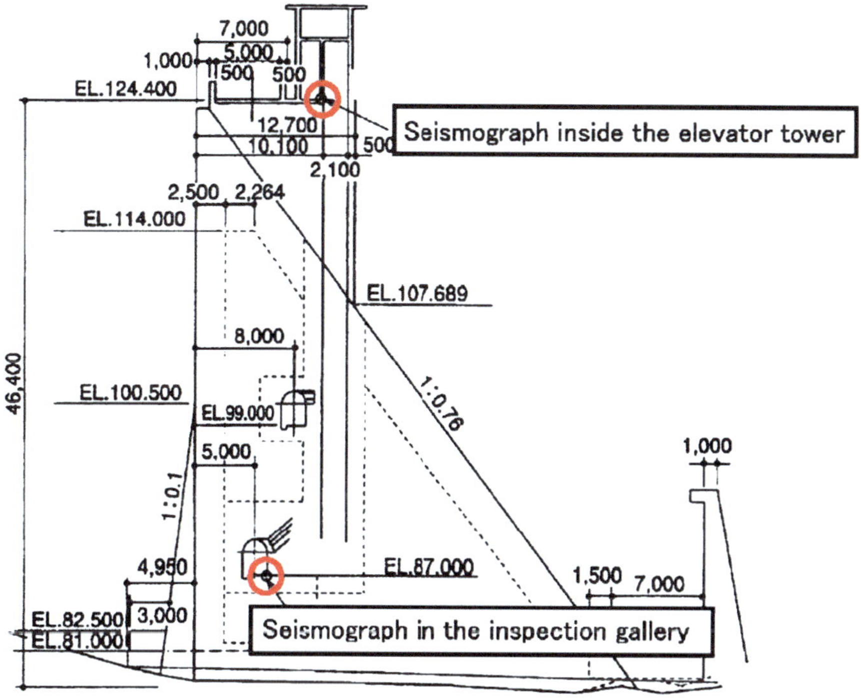

Fig. 15.2 Transversal section of Kasho Dam with accelerometer locations (Takasu et al. 2002)

The employed procedure is able to evaluate the system failure probability accounting for all relevant failure mechanisms and the uncertainties as described by **x** and **y**. It is noted that the correlation between structural demands is taken implicitly into account by the employed methodology through Monte Carlo simulations.

15.5 Application to a Case Study: The Kasho Dam

The procedure described in the previous section has been applied to the assessment of the Kasho Dam, a 46.4 m high concrete gravity dam located in the western part of the Japanese island of Honshu (Fig. 15.2). The dam construction was completed in 1989. The seismic design was based on a pseudo-static analysis method for a peak ground acceleration (*PGA*) of 0.12 g (Ohmachi et al. 2003). On October 6th, 2000, the Kasho Dam was subjected to the Western Tottori Prefecture earthquake of Magnitude 7.3, whose epicentre was located at about 3 km from the Kasho site. The seismic event was recorded by two accelerometers located at the crest and in proximity of the dam base. Horizontal accelerations up to 0.54 g and 2.09 g were recorded at the base and at the crest of the dam, respectively. In spite of this, no damage was observed in dam body and machineries, but only in appurtenant structures (i.e. concrete cracking in the gate chamber located above the crest centre).

15.5.1 The Numerical Model

In this section, only basic information on the numerical model is provided. A detailed presentation and discussion of modelling issues is reported by Lupoi and Callari (2011). Available monitoring data allowed a calibration of the several parameters involved in the analysis, especially those related to the definition of "failure" mechanisms. More specifically, these mechanisms are considered to represent the operational limit state (sliding at the dam-rock interface, material failure in the dam body, excessive deflection/drift deformation, material failure at the neck, etc.), for which the *functionality* of the dam should be preserved.

Regarding this limit state, it has to be noted that even the localized appearance of cracks in dams can be reason for serious concern about the safety of downstream valley, usually leading to cost-effective decisions such as the limitation of reservoir operation. As observed in Sect. 15.1, in concrete dams these operational limit states are often reported as a consequence of high-intensity earthquakes, in contrast with the extremely rare attainment of ultimate limit states.

According to the above consideration, and consistently with the observed response of Kasho Dam to the high-intensity Tottori earthquake, the operational limit state is identified with the *incipient* appearance of non linear phenomena. In other words, the system components (dam body, dam/foundation interface) exhibit a linear behaviour up to failure for the selected mechanisms, where the term "failure" denotes the attainment of the limit state. Therefore, in the numerical analysis, it has been consistently assumed a linear elastic response for involved materials and a perfect adhesion at concrete-rock interface, since the numerical results obtained after the attainment of the operational limit state are not employed in the proposed assessment method.

All the components of the dam-foundation-reservoir system at Kasho site have been modelled by means of four-noded quadrilateral finite elements (Figs. 15.3 and 15.4) available in FEAP code (Taylor 2002). A Lagrangian formulation has been considered for reservoir water, assuming displacements as nodal unknowns. As illustrated in Fig. 15.3, at interfaces of water with dam body and rock mass, zero normal components of relative displacements have been imposed, while no restraint has been applied to tangential displacements (Zienkiewicz et al. 1986).

A Rayleigh damping is employed in numerical simulations and the corresponding coefficients are obtained from the viscous damping ratios using a standard procedure (Clough and Penzien 1993). It is noted that, in view of the simplifications involved in the numerical model, these coefficients have not to be understood as merely representative of material damping but they also account for other

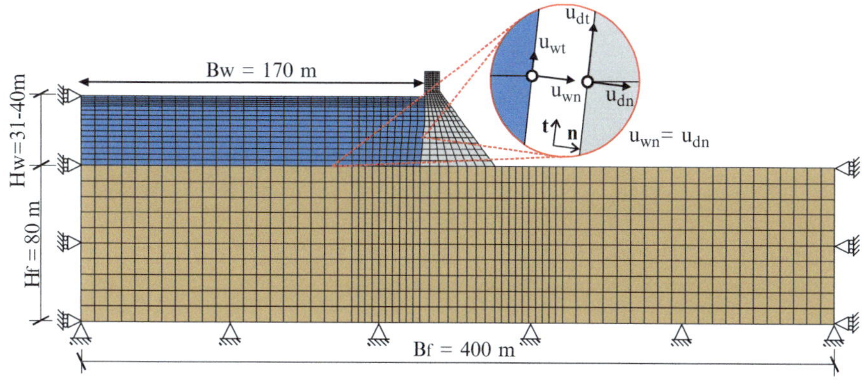

Fig. 15.3 Finite element discretization of the dam-foundation-reservoir domain with illustration of the conditions imposed on displacements at boundaries and at interfaces of water with dam body and rock mass ($H_f = 80$ m, $B_f/H_f = 5$ and $B_w/H_w \geq 4$)

important mechanisms of energy dissipation characterizing the problem at hand (radiation damping in both foundation and reservoir). In fact, these coefficients are strongly related to the size of the numerical domain and to the conditions applied at lateral boundaries (Clough and Zienkiewicz 1978).

By means of preliminary simulations, we have investigated several settings for geometry, boundary conditions and viscous parameters, concluding that the effects on dam response due to spurious wave reflections at bottom and lateral mesh boundaries are practically negligible for the configuration shown in Fig. 15.3.

Another set of preliminary dynamic analyses has been performed to investigate the potential influence of water cavitation phenomena. It has been concluded that reasonably accurate solutions are obtained for the earthquake intensity range of interest by means of the considered linear model, which is unable to reproduce cavitation.

With the problem setting described above, a very good agreement with available monitoring data has been assessed for the results of a time-history analysis of the Kasho Dam system excited by an adequately deconvolved Tottori natural record. This comparison has shown the effectiveness of the considered finite-element modelling of the reservoir water, in contrast with the observed underestimation of measured crest accelerations obtained by means of the added mass approach based on Westergaard theory (Westergaard 1933). The fundamental period calculated by modal analysis for the dam-foundation system is equal to 0.12 s, a result consistent with the value (0.10 s) obtained by Omachi *et al.* from data recorded during the Tottori earthquake (Ohmachi et al. 2003).

Also the effect of uplift pressures has been considered in finite element analyses. In view of the drainage holes located at about 5.5 m from the dam upstream face, a typical bilinear uplift distribution has been employed. Results of several preliminary static simulations have shown a negligible influence of drain effectiveness variation (0÷65%) on stresses at dam base. Hence, a constant effectiveness of 25%

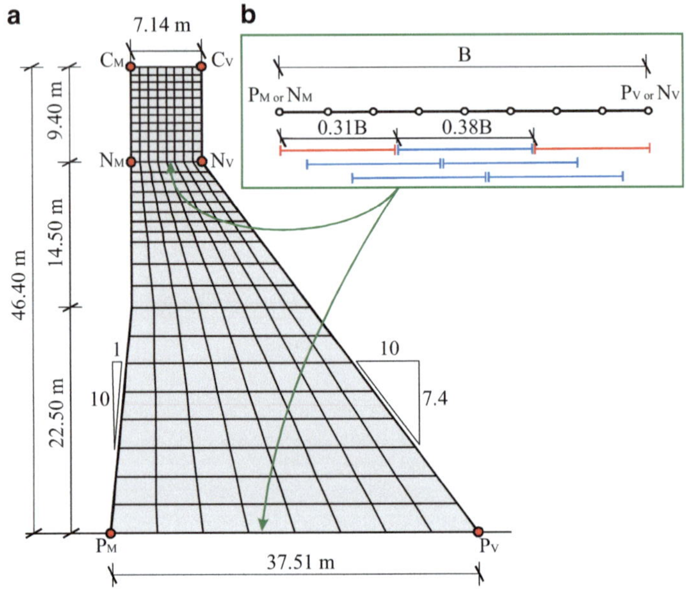

Fig. 15.4 (**a**) Detail of dam body geometry and of its finite element discretization. (**b**) Illustration of horizontal surfaces at dam base (P_MP_V) and neck (N_MN_V) considered in "parallel" option

(USACE 2000) has been assumed in dynamic analyses. Furthermore, as suggested by USACE (1995a) and FERC (2002), the dynamic effects on the static distribution of uplift pressures have been neglected.

15.5.2 Definition of Failure Mechanisms

For the operational limit state of interest, the following failure mechanisms have been identified:

- Excessive deformation of the dam body inducing service limitation for equipment and installations;
- Cracking or sliding at dam base (P_MP_V in Fig. 15.4);
- Cracking at the dam neck (N_MN_V in Fig. 15.4);
- Cracking at the upstream face (P_MC_M in Fig. 15.4).

The occurrence of any of the mechanisms outlined above causes the failure of the whole system, i.e. the system's components are arranged as a series.

The excessive deformation of the dam body is checked in terms of the drift deformation between the crest and the base, that is, in terms of the difference between horizontal displacements of points C_M and P_M in Fig. 15.4. The mean value of the drift capacity has been taken equal to 0.02%, which is considered as appropriate for very stiff structures such as concrete dams.

The assessment of the appearance of a crack mechanism in the numerical response is a delicate issue for the multi-dimensional problem at hand. In this study, the crack formation is checked in terms of maximum tensile normal stress $\sigma_{t,D}$ evaluated at mesh nodes by means of lumped projection. The formation of a crack is conventionally defined as the contemporaneous occurrence of condition $\sigma_{t,D} > \sigma_{t,C}$ at a given number n of nodes. Two different options have been investigated:

(a) one single node (series arrangement of nodes);
(b) $n > 1$ nodes aligned along a surface of given length (parallel arrangement of nodes).

In the present study, this critical length has been assumed approximately equal to 1/3 of the dam width B at a specific elevation. More precisely, as depicted in Fig. 15.4b, this surface extension is 0.31B in the proximity of dam faces and 0.38B in the inner part of the dam body. This setting corresponds to 3 contiguous nodes for the mesh employed in the simulations. A similar treatment has been used for the sliding mechanism, represented by condition $\tau_D > \tau_C$, where τ is the tangential stress acting on the interface between concrete and rock mass.

For further details on the failure mechanisms and on the random variables briefly described in the next section the interested reader is referred again to Lupoi and Callari (2011).

15.5.3 Random Variables

The basic "material/structural" random variables for the probabilistic assessment are the following:

- characteristic strength R_{ck} of concrete;
- Geological Strength Index (*GSI*) of the rock mass (Hoek et al. 2002);
- error term ε_δ in the drift capacity model;
- error term ε_f in the capacity model for tensile strength at concrete-rock interface.

For the numerical analyses, ten recorded accelerograms have been selected among a large database according to the following criterion: the difference between the *mean* spectrum and a *reference* spectrum is less or equal to 10%. The *mean* spectrum is evaluated from the spectra of the candidate natural records (scaled at the same *PGA*). The Eurocode 8 elastic spectrum (soil type A) (CEN 2004) has been employed as *reference* spectrum for the purposes of this case study. The mean spectrum and the reference spectrum are shown in Fig. 15.5, together with the ten spectra and the mean ± standard deviation spectra of the selected records. A uniform random variable (*l* in Table 15.5) is used in Monte Carlo simulation for sampling the demand vectors.

The basic random variables, and corresponding distribution's parameters, employed in the Monte Carlo simulation are listed in Table 15.5.

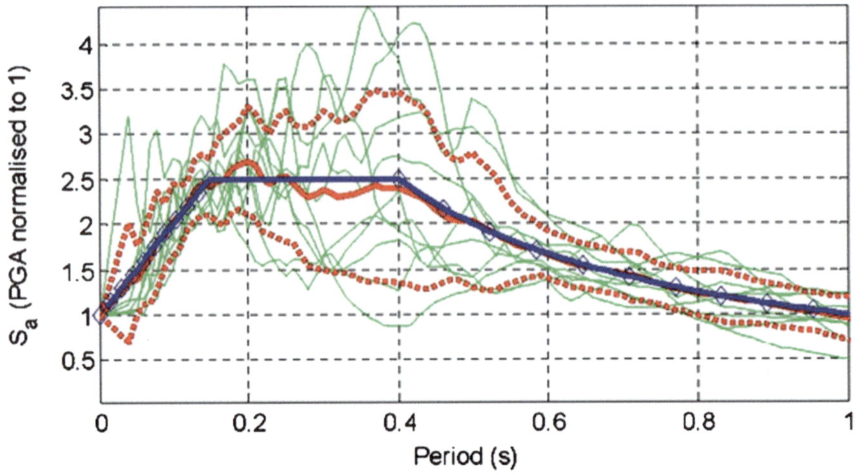

Fig. 15.5 Reference spectrum (*blue bold line*), mean spectrum (*red bold line*), mean ± 1 st.dev. spectra (*red dotted lines*) and natural records spectra (*thin lines*)

Table 15.5 Basic random variables

Variable	Type	Range	Mean	CoV
Rck	Lognormal	–	25 MPa	0.2
GSI	Uniform	60 ÷ 80	–	–
ε_δ	Lognormal	–	1	0.2
ε_f	Lognormal	–	1	0.2
l	Uniform	1 ÷ 10	–	–

The uncertainties related to the *PGA* and to the reservoir water level H_w are represented by the corresponding hazard curves. An experimental cumulative distribution function (CDF) of the reservoir level H_w has been derived from available data relative to 2005–2007 period (Tottori Prefecture 2008). The corresponding mean value of H_w is 36 m, the 5% lower fractile and the 95% upper fractile values are 31 m and 40 m, respectively. These three reservoir levels have been used to evaluate the conditional probability of failure $P_f(\mathbf{y})$ by means of Eq. 15.3.

15.5.4 Material Properties and Capacity Models

The Young modulus of concrete (E_C) and of rock mass in static ($E_{r,st}$) and dynamic ($E_{r,dyn}$) conditions are given as function of the basic random variables by standard relations (Hoek et al. 2002; NTC2008 2008; JNC 2000).

The mean value of the drift capacity has been taken equal to 0.02%, which is considered as appropriate for very stiff structures such as concrete dams.

The concrete tensile strength f_{ct} in the dam body is obtained as a function of the concrete compressive strength (USACE 1995b).

The tensile strength at concrete-rock interface it is assumed equal to $\bar{f}_{ct,b} =$ 1.0 MPa, consistently with the experimental data collected by Ruggeri (2004).

The sliding capacity at dam base $\tau_{lim}(t)$ has been evaluated employing the Mohr-Coulomb model. The cohesion c_b and the friction angle φ_b at the interface between concrete and rock mass have been in turn calculated through linear interpolation of the model by Barton et al. (1985) in the range of stresses $\sigma_b(t)$ characterizing the numerical results. The Barton model parameters have been related to the rock joint conditions corresponding to the aforementioned *GSI* classification of the rock mass. The so-obtained values of cohesion and friction angle are fully consistent with the experimental data reviewed by Ruggeri (2004).

15.5.5 Dynamic Analyses

The dynamic time-history analyses have been performed for the combination of the ten recorded accelerograms and the three different reservoir levels, thus leading to a total of 30 simulations. The analyses have been carried out for $PGA = 0.1$ g. In view of the assumed linear behaviour, the dam response at higher values of PGA has been obtained by scaling up the results calculated for $PGA = 0.1$ g. The analyses have been carried out employing the mean values of basic random variables and the corresponding values of material properties.

15.5.6 Fragility Curves

The probability of failure $P_f(\mathbf{y})$ relative to the operational limit state for the Kasho Dam has been evaluated conditional to 9 values of *PGA* (ranging from 0.1 to 0.9 g) and to 3 values of H_w (31, 36 and 40 m). The corresponding fragility curves are shown in Figs. 15.6–15.9. The plots include the fragility of the whole dam system and also those of the investigated mechanism of failure.

The curves shown in Fig. 15.6, which refer to the case of $H_w = 36$ m, have been evaluated without accounting for the uncertainties due to structural parameters and assuming a "localized" option (*a*) for crack/sliding formation (see Fig. 15.4). Under this assumption, the tensile cracking at foundation is by far the critical failure mechanism for the dam system, yielding a P_f close to unity for $PGA = 0.45$ g. This estimate of the dam vulnerability based on option (*a*) seems to be rather conservative, especially in view of the absence of damage induced in Kasho Dam by the Tottori earthquake, whose intensity was about 0.5 g.

The curves in Fig. 15.7, which refer again to the case of $H_w = 36$ m, have been evaluated assuming the "diffuse" option (*b*) for crack/sliding formation (see Fig. 15.4). A significant difference in the results is noted, both in absolute

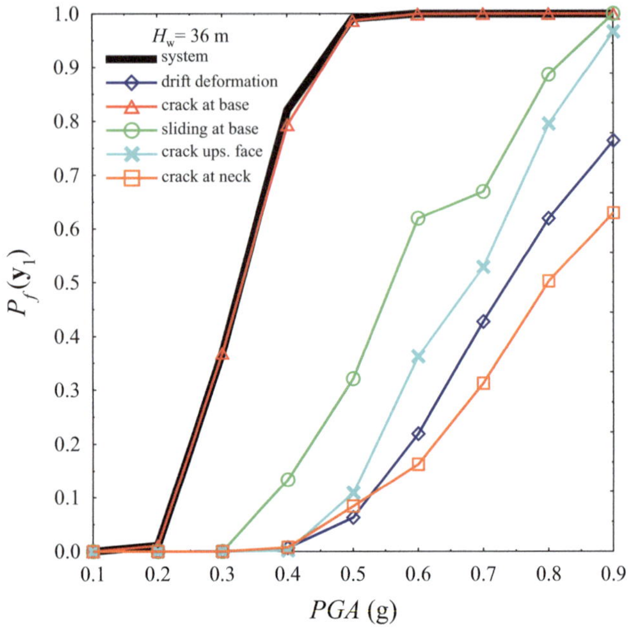

Fig. 15.6 Fragility curves for $H_w = 36$ m, series arrangement for components

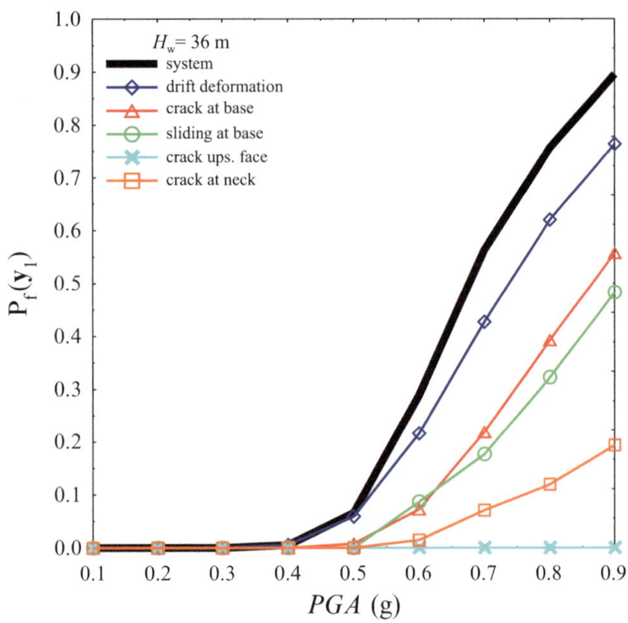

Fig. 15.7 Fragility curves for $H_w = 36$ m, parallel arrangement for components

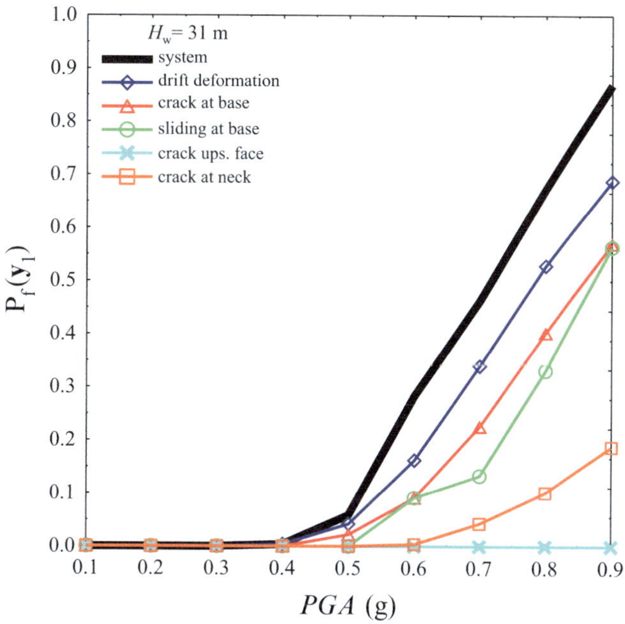

Fig. 15.8 Fragility curves for $H_w = 31$ m, parallel arrangement for components

terms and in relative importance of the failure mechanisms. The drift deformation results to be the most probable critical mechanism; the cracking and sliding failure mechanisms at dam base come next, with comparable P_f's. It is also noted that the event of cracking at the dam upstream face has a negligible probability of occurrence. The estimated system P_f at $PGA = 0.45$ g is now consistent with experimental evidence. These results provide strong evidence in support of the *diffuse* mechanism (*b*) and also indications about the extension of the cracking surface. With respect to the case of option (*a*), the reduction in P_f was obviously expected and the probabilistic approach applied herein is also capable of providing realistic quantitative estimates of such reduction.

The fragility curves for the cases of $H_w = 31$ m and of $H_w = 40$ m are shown in Figs. 15.8 and 15.9, respectively. All curves have been evaluated for the "diffuse" crack-formation option and neglecting the structural uncertainties. The effect of reservoir level on the system P_f is significant: for instance, at $PGA = 0.5$ g, the P_f for the higher water level is about 2.5 times the P_f for the lower one. By comparing the components' fragility curves, it can be noted that the increment of dam vulnerability is almost entirely due to the increment in the drift deformation mechanism. Nevertheless, the reservoir water level has no influence on the rest of the failure mechanisms.

The effect of uncertainties in material parameters has also been investigated. Two additional sets of 30 analyses have been performed to evaluate the response gradients of the demands with respect to the basic random variables: one for a

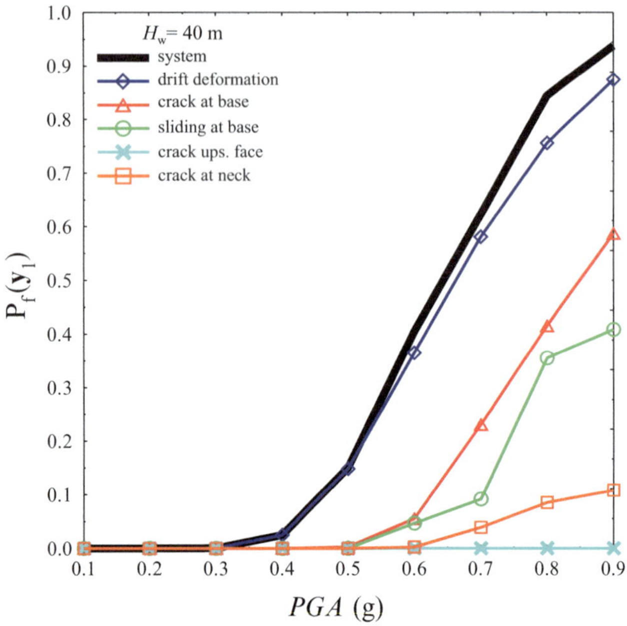

Fig. 15.9 Fragility curves for $H_w = 40$ m, parallel arrangement for components

perturbed value of the basic random variable R_{ck} and one for a perturbation of the basic random variable *GSI*. Results are not shown here since, for this particular application, the overall effects on P_f due to structural uncertainties result to be negligible. For the problem at hand, this result can be attributed to the uniform randomness of material properties all over the dam body and to the assumed linear behaviour of involved materials.

15.6 Concluding Remarks

Recent advances in seismic assessment of concrete dams by means of probabilistic methods have been illustrated in this chapter after a brief literature review on the subject.

The empirical methods traditionally used in practice are adequate to provide a general indication on the seismic risk of dams. Very few studies based on a probabilistic approach are available in literature for the seismic assessment of a concrete dam and many of they lack either of the necessary accuracy or of the practical applicability.

The probabilistic methodology proposed by the authors tackles the main issues in dam assessment, i.e. large uncertainties and system complexity, employing advanced techniques. In particular, the procedure is able to account for uncertainties both in material properties and in external actions (ground motion and reservoir level) and also for multiple failure mechanisms. The numerical model employed to simulate the dynamic interaction between the dam structure, the foundation rock mass and the reservoir water has been extensively tested, leading to a very good agreement with available monitoring data.

Attention has been focused on the operational limit state, which is of primary importance for dam structures. Appropriate failure mechanisms have been identified. Fragility curves have been computed for three reservoir levels. The increment of the reservoir level has the effect of a noticeable increment of the system P_f. By comparing single mechanism fragilities, it is noted that the relative contribution of deformation mechanism to system fragility remarkably increases with reservoir level.

Results provide evidence in support of a *diffuse* formation of mechanisms, as opposed to a "point-wise failure" approach. As expected, the probabilistic approach applied herein is able to accurately estimate the corresponding reduction in the probability of failure. Under this perspective, the crucial advantage over traditional "deterministic" assessment methods is the quantitative information about seismic risk that can be provided to decision makers, obtained by convolution of fragility curves with hazard curves.

Acknowledgements C. Callari was supported by MIUR in the context of research projects on existing concrete dams coordinated by G. Maier (PRIN 2004) and G. Novati (PRIN 2007). The authors are grateful to G. Lupoi for fundamental suggestions and to former students F. Ceglie, R. Conforti, S. Tomasi and P. Di Filippo involved in the preliminary phases of the research.

References

Araújo JM, Awruch AM (1998) Probabilistic finite element analysis of concrete gravity dams. Adv Eng Softw 29(2):97–104

Barton NR, Bandis SC, Bakhtar K (1985) Strength deformation and conductivity coupling of rock joints. Int J Rock Mech Min Sci 22(3):121–140

Bureau GJ (2003) Dams and appurtenant facilities. In: Chen WF, Scawthorn C (eds) Earthquake engineering handbook. CRC Press, Boca Raton, Chapter 26

Bureau GJ, Ballentine GD (2002) A comprehensive seismic vulnerability and loss assessment of the state of south Carolina using HAZUS. Part IV. Dam inventory and vulnerability assessment methodology. In: 7th national conference on earthquake engineering, Boston, 21–25 July 2002. Earthquake Engineering Research Institute, Oakland

CEB-FIP (1993) Model code 1990. Comite Euro-International du Beton, Switzerland

CEN (2004) EN 1998–1:2004, Eurocode 8: Design of structures for earthquake resistance – Part 1: General rules, seismic actions and rules for buildings, Comité Européen de Normalisation Brussels

Chopra AK, Chakrabarti P (1972) The earthquake experience at Koyna Dam and stresses in concrete gravity dams. Earthquake Eng Struct Dyn 1(2):151–164

Clough RW, Penzien J (1993) Dynamics of structures, 2nd edn. McGraw-Hill, New York

Clough RW, Zienkiewicz OC (1978) Finite element methods in analysis and design of dams. International Commission of Large Dams, Paris, Bulletin 30

Cornell AC, Jalayer F, Hamburger R, Foutch D (2002) Probabilistic basis for the 2000 SAC/FEMA steel moment frame guidelines. J Eng Mech 128(4):526–533

Der Kiureghian A (2005) First- and second-order reliability methods. In: Nikolaidis E et al (eds) Engineering design reliability handbook. CRC Press, Boca Raton

FERC (2002) Engineering guidelines for the evaluation of hydropower projects, chapter III, gravity dams. Federal Energy Regulatory Commission/Office of Energy Projects/Division of Dam Safety and Inspections, Washington, DC

Hoek E, Carranza-Torres C, Corkum B (2002) Hoek-Brown failure criterion – 2002 Edition. In: Hammah R et al (eds) Proceedings of 5th NARMS and the 17th TAC conference (NARMS-TAC 2002): mining and tunnelling innovation and opportunity, 07–10 July 2002, University of Toronto Press, Toronto, pp 267–273

ICOLD (1989) Selecting seismic parameters for large dams, guidelines. International Commission of Large Dams, Paris, Bulletin 72

Japan Nuclear Cycle Development Institute (JNC) (2000) H12: Project to establish the scientific and technical basis for HLW disposal in Japan. http://www.jaea.go.jp/04/tisou/english/report/H12_report.html. Accessed 25 oct 2010

Leclerc M et al (2004) CADAM software, version 1.4.13. École Polytechinque de Montréal, Quebec

Lupoi A, Callari C (2011) A probabilistic method for the seismic assessment of existing concrete gravity dams. Structural and Infrastructural Engineering, DOI: 10.1080/15732479.2011.574819.

Lupoi G, Franchin P, Lupoi A, Pinto PE (2006) Seismic fragility analysis of structural systems. J Eng Mech ASCE 132(4):385–395

NTC2008 (2008) Nuove Norme Tecniche per le Costruzioni. Ministero delle Infrastrutture, D.M. 14/01/2008, Rome: Gazzetta Ufficiale n. 29 del 4 febbraio 2008 – Suppl. Ordinario n. 30 (in Italian)

Ohmachi T, Kojima N, Murakami A, Komaba N (2003) Near field effects of hidden seismic faulting on a concrete dam. J Nat Disaster Sci 25(1):7–15

Pinto PE, Giannini R, Franchin P (2004) Seismic reliability analysis of structures. IUSS Press, Pavia

Ruggeri G (2004) Sliding safety of existing gravity dams. Final report of ICOLD working group. http://www.britishdams.org/2004conf/reports/sliding.pdf. Accessed 25 Oct 2010

Takasu S, Yoshida H, Yamaguchi Y, Sasaki T, Iwashita T (2002) Behaviour of dams due to the western Tottori-Prefecture earthquake in 2000. Tech Note Natl Inst Land Infrastructure Manage 41:157–170

Taylor RL (2002) FEAP – a finite element analysis program – version 7.4 user manual. University of California at Berkeley, Department of Civil and Environmental Engineering, Berkeley

Tekie PB, Ellingwood B (2003) Seismic fragility assessment of concrete gravity dams. Earthquake Eng Struct Dyn 32:2221–2240

Tosun H et al (2007) Seismic hazard and total risk analyses for large dams in Euphrates basin. Turk Eng Geol 89:155–170

Tottori Prefecture (2008) Available from: http://www.pref.tottori.lg.jp/dd.aspx?menuid=73986. Accessed 25 Oct 2010 (in Japanese), http://www.pref.tottori.lg.jp/dd.aspx?menuid=74025. Accessed 25 oct 2010 (in Japanese), http://www7.apionet.or.jp/dam/kasho/index.php. Accessed 25 oct 2010 (in Japanese)

USACE (1995a) Gravity dam design. US Army Corps of Engineers, Washington, DC, Engineering Manual no. EM 1110-2-2200

USACE (1995b) Seismic design provisions for roller compacted concrete dams. US Army Corps of Engineers, Washington, DC, Engineering Provision no. EP 1110-2-12

USACE (2000) Evaluation and comparison of stability analysis and uplift criteria for concrete gravity dams by three federal agency. US Army Corps of Engineers, Washington, DC. Technical Report no. ERDC/ITL TR-00-1

Westberg M (2010) Reliability-based assessment of concrete dam stability. Ph.D. thesis, Report TVBK-1039. Lund University, Lund

Westergaard HM (1933) Water pressures on dams during earthquakes. Trans ASCE 98:418–472

Wieland M (2009) Seismic aspects of large dams. ANCOLD 2009 conference. Keynote Lecture, Adelaide, Australia, 11-15/11/2009. In: Australian national committee on large dams conference proceedings

Wieland M, Brenner RP, Sommer P (2003) Earthquake resiliency of large concrete dams: damage, repair, and strengthening concepts. In: Proceedings of the 21st international congress on large dams, ICOLD, Montreal

Yanmaz AM, Beser MR (2005) On the reliability-based safety analysis of the Porsuk dam. Turk J Eng Environ Sci 29:309–320

Zienkiewicz OC, Clough RW, Seed HB (1986) Earthquake analysis procedures for dams – state of the art. International Commission of Large Dams, Paris, Bulletin 52

Index

M. Dolšek (ed.), *Protection of Built Environment Against Earthquakes*,
DOI 10.1007/978-94-007-1448-9,

Q

R

S

T

MIX
Papier aus verantwortungsvollen Quellen
Paper from responsible sources
FSC® C105338

If you have any concerns about our products,
you can contact us on
ProductSafety@springernature.com

In case Publisher is established outside the EU,
the EU authorized representative is:
Springer Nature Customer Service Center GmbH
Europaplatz 3, 69115 Heidelberg, Germany

Printed by Libri Plureos GmbH
in Hamburg, Germany